# Connections Between Diabetes Mellitus, Other Metabolic and Endocrine Dysfunctions and Cardiovascular Pathologies—Second Edition

# Connections Between Diabetes Mellitus, Other Metabolic and Endocrine Dysfunctions and Cardiovascular Pathologies—Second Edition

Guest Editors

**Dragos Cozma**
**Cristina Tudoran**

MDPI

Basel • Beijing • Wuhan • Barcelona • Belgrade • Novi Sad • Cluj • Manchester

*Guest Editors*
Dragos Cozma
Department of Cardiology
"Victor Babes" University of
Medicine and Pharmacy
Timisoara
Romania

Cristina Tudoran
Discipline of Cardiology
University of Medicine and
Pharmacy "Victor Babes"
Timisoara
Romania

*Editorial Office*
MDPI AG
Grosspeteranlage 5
4052 Basel, Switzerland

This is a reprint of the Special Issue, published open access by the journal *Biomedicines* (ISSN 2227-9059), freely accessible at: https://www.mdpi.com/journal/biomedicines/special_issues/1IB45123O2.

For citation purposes, cite each article independently as indicated on the article page online and as indicated below:

Lastname, A.A.; Lastname, B.B. Article Title. *Journal Name* **Year**, *Volume Number*, Page Range.

**ISBN 978-3-7258-6774-5 (Hbk)**
**ISBN 978-3-7258-6775-2 (PDF)**
**https://doi.org/10.3390/books978-3-7258-6775-2**

# Contents

# About the Editors

**Dragos Cozma**

Dragos Cozma works at the University of Medicine and Pharmacy "Victor Babes", Timisoara, and the Institute of Cardiovascular Diseases, Timisoara. His area of expertise is heart failure. He published over 70 manuscripts in ISI-indexed medical journals and has extensive editorial experience.

**Cristina Tudoran**

Cristina Tudoran is a Lecturer at the University of Medicine and Pharmacy "Victor Babes", Timisoara, and a senior physician at the County Emergency Hospital "Pius Brînzeu." Her research focuses on cardiovascular alterations in patients diagnosed with other comorbidities, with extensive publications and editorial experience.

biomedicines

MDPI

*Editorial*

# Connections Between Diabetes Mellitus, Other Metabolic and Endocrine Dysfunctions and Cardiovascular Pathologies—Second Edition

**Cristina Tudoran [1,2,3,*] and Dragos Cozma [4,5,6]**

1 Department VII, Internal Medicine II, Discipline of Cardiology, Victor Babeș University of Medicine and Pharmacy, E. Murgu Square, Nr. 2, 300041 Timisoara, Romania
2 County Emergency Hospital "Pius Brinzeu", L. Rebreanu, Nr. 156, 300723 Timisoara, Romania
3 Center of Molecular Research in Nephrology and Vascular Disease, Faculty of the Victor Babeș University of Medicine and Pharmacy, E. Murgu Square, Nr. 2, 300041 Timisoara, Romania
4 Institute of Cardiovascular Diseases Timisoara, 300310 Timisoara, Romania; dragos.cozma@umft.ro
5 Cardiology Department, Victor Babeș University of Medicine and Pharmacy, 300041 Timisoara, Romania
6 Research Center of the Institute of Cardiovascular Diseases Timisoara, 300310 Timisoara, Romania
* Correspondence: tudoran.cristina@umft.ro

The association between cardiovascular (CV) pathologies and metabolic/endocrine dysfunctions was first observed a long time ago, and great emphasis has been given to this relationship. Several guidelines of the European Society of Cardiology on coronary artery disease, arterial hypertension, heart failure (HF), and dyslipidemia include chapters on diabetes mellitus (DM) and chronic kidney disease (CKD) [1–6]. However, there are still unanswered questions regarding common pathophysiologic pathways, risk profile, and management of these comorbid associations of cardio-endocrine/metabolic pathologies. Another subject of interest is the benefits of newly developed drugs, and also their safety profiles.

In recent decades, the connections between endocrine dysfunctions and the CV system (CVS) have captivated the attention of several researchers [7,8]. Recent studies highlight that this morbid relationship begins even sooner than initially assumed. Besides genetic factors [9], multiple hormones (thyroid and sexual hormones, cortisol, insulin-like growth factor, etc.) influence the evolution of the CVS starting from the intrauterine evolution and play a pivotal role in its structural and functional development throughout childhood. On one side, hormones are involved in the maintenance of blood pressure, fluid balance, and myocardial remodelling, and, on the other side, endocrine imbalance may lead to CV alterations. Calcaterra et al. explored the heart's intrinsic endocrine activity and how hormonal signals interact with the development of the CVS starting from the fetal state and continuing through early childhood [7]. Metabolic, thyroid, and other endocrine dysfunctions could lead to the earlier onset of CV pathologies in adult life [10]. The early interplay between the endocrine/metabolic system and CVS could account for the clustering of several metabolic and CV pathologies in the same patient, during adult life.

Another topic is the impact of sex-related differences on CV and metabolic pathologies. Stoiculescu et al. analyzed the influence of sex and DM on the rehospitalization rate in 1018 patients with HF with mildly reduced or/and preserved ejection fraction [11]. They determined that women without DM represent a low-risk phenotype, whereas diabetic men exhibit the highest risk of recurrent hospitalizations due to HF decompensation [11]. Metabolically-dysfunction-associated steatotic liver disease (MASLD) is common in patients with DM [12], and was for a long time considered to be more frequently diagnosed in men. Up-to-date research has demonstrated that, in women, MASLD often develops later, typically after menopause, being associated with a greater burden of comorbidities, higher mortality, increased cirrhosis prevalence, and a higher likelihood of requiring liver transplantation. New data, supported by detailed hepatic evaluations, suggest that postmenopausal women with visceral obesity showed similar or worse liver and cardiometabolic profiles than men, despite the failure of routine clinical and laboratory data (transaminases and the Fatty Liver Index) to diagnose MASLD. In women over 50 years old, anthropometric height-adjusted measurements (body mass index, waist circumference, waist-to-height ratio) and non-invasive markers of fibrosis and steatosis, together with the application of sex-specific cut-offs, may help reveal a comparable or slightly higher burden of disease associated with visceral fat accumulation. These data indicate that a more tailored approach, including sex-specific clinical and diagnostic criteria, for the detection of MASLD is needed in postmenopausal women, particularly in those with central obesity, to ensure early identification and management of this liver disease [13].

With the availability of modern tools for the diagnosis and management of HF, researchers realized that there is a bidirectional relationship between HF and metabolic/endocrine dysfunctions. As HF worsens, more and more organs are affected. Dilated cardiomyopathy (DCM) is often diagnosed in patients with severe signs and symptoms of HF. As a connection between metabolic/endocrine and CVD has been established, researchers are focusing on a possible correlation between type 2 DM (T2DM), obesity, thyroid dysfunction, and the prevalence of DCM [14]. In their retrospective cohort study on 1079 patients, Tica et al. [15] demonstrated that in non-insulin-treated T2DM patients, hypothyroidism and male sex were independently associated with an increased risk of developing DCM. These findings support the hypothesis that the coexistence of an increased metabolic burden (DM, obesity, metabolic syndrome), hypothyroidism, and even sex-related risk factors in the same patient group could favour a higher incidence of DCM or other CV alterations [16,17].

In recent years, researchers have observed the co-occurrence of multiple diseases in the same patient, which at first glance appear to have little in common. Another example of multiple cardio-metabolic dysfunctions and other peculiar diseases is the co-existence of DM, CV pathologies, and periodontal disease (PD) in the same individual [18–20]. The increased incidence of all three pathologies observed in recent years could partially explain this phenomenon. Another possible explanation could result from common pathophysiological mechanisms, such as chronic inflammation (elevated levels of Interleukin (IL)-6, tumour Necrosis Factor (TNF)-$\alpha$, and reactive protein C (CRP)), immune dysregulation, oxidative stress, and microbial dysbiosis. DM exacerbates PD and accelerates tissue destruction via hyperglycaemia, while periodontitis worsens glycemic control and insulin resistance (IR). Both conditions independently elevate CVR. Periodontal therapy has led to lower glycemic levels and improved endothelial dysfunction, while cardiometabolic therapies (statins, Glucagon-Like Peptide (GLP-1) receptor agonists, Sodium-Glucose Cotransporter-2 inhibitors (SGLT2i)) ameliorated PD [20].

Peculiar adverse hormonal effects such as tertiary hyperparathyroidism (THPT), often encountered in CKD [21] due to prolonged secondary hyperparathyroidism, have been diagnosed in patients with diabetic nephropathy. THPT in DM favours vascular calcifications and increases the risk of developing major CV events. These patients raise diagnostic and therapeutic challenges, often requiring a multidisciplinary approach to address CV, renal, and metabolic pathologies [22]. The absence of standardized diagnostic criteria delays the diagnosis of THPT in patients with diabetic nephropathy, leading to inconsistent treatment strategies. Management must therefore integrate targeted metabolic and CV strategies, alongside specific therapies for THPT, including vitamin D analogues, calcimimetics, and

https://doi.org/10.3390/biomedicines14010250

phosphate-lowering drugs [22]. Patients with diabetic nephropathy and THPT undergoing haemodialysis are particularly difficult to treat as current guidelines offer wide reference ranges of optimal parathormone targets.

Referring to ESC guidelines, highly recommended HF therapies such as first-generation SGLT2i (empagliflozin and dapagliflozin) have substantially reshaped the management of HF and also of T2DM, owing to their glucose-lowering properties and to their consistent CV and renal protective effects [2,23,24]. The purpose of developing next-generation SGLT2-based therapies is to refine pharmacological selectivity, optimize pharmacokinetic profiles, and expand therapeutic applicability in patients without DM. Next-generation SGLT2is are characterized by structural and pharmacokinetic innovations, refined or deliberately balanced SGLT2/SGLT1 selectivity profiles, and enhanced engagement of extra-renal and glucose-independent pathways, including myocardial energy consumption, inflammatory modulation, vascular function, and gut–heart signalling. Dual SGLT1/SGLT2 inhibition further broadens therapeutic use by incorporating incretin-mediated mechanisms and intestinal glucose modulation, which may be particularly relevant in patients with high CV risk or with acute/decompensated CV pathologies. Recent findings indicate that the renal benefits of SGLT2i are disease-specific, with differential responses due to the underlying disease. Similarly, in HF, SGLT2-based therapies target both acute hemodynamic decompensation and chronic myocardial remodelling, supporting their use in HF [25].

Concerning risk reduction strategies, statins are widely used lipid-lowering agents that significantly reduce CV morbidity and mortality by lowering low-density lipoprotein (LDL) levels. Although employed for decades in patients with CVD, researchers suggested that they are prone to drug-to-drug interactions, and even some vitamins could influence their effects [26]. In their study, Cozma et al. explore the complex relationship between statins, vitamin D, and their impact on CV outcomes [27]. Both molecules intersect metabolically at 7-dehydrocholesterol. While theoretical concerns exist about the capacity of statins to impair vitamin D synthesis, clinical studies suggest a neutral or modestly positive effect on circulating 25(OH)D levels. Statin therapy does not appear to induce vitamin D deficiency and may modestly increase serum 25(OH)D concentrations, particularly lipophilic statins such as rosuvastatin and atorvastatin. Although routine vitamin D supplementation is not indicated in all statin-treated individuals, targeted repletion in those with confirmed deficiency—especially in the context of statin-associated muscle symptoms—may enhance tolerability and therapeutic adherence. While low vitamin D levels are associated with higher CV risk, supplementation has demonstrated consistent benefit in reducing major CV events, underlying the need for targeted research in high-risk, vitamin D-deficient populations [27].

Several studies demonstrated that hepatic overproduction of very-low-density lipoprotein (VLDL) worsens dyslipidemia and increases CV risk [28,29]. As the levels of cyclic adenosine monophosphate (cAMP) are indirectly correlated to the activation of the microsomal triglyceride transfer protein (MTP), which is essential in the hepatic production of VLDL, theophylline, a methylxanthine found in tea, may alter the hepatic production of VLDL by increasing intracellular cAMP. Park et al. demonstrated in their ex vivo rat model that tea-derived theophylline could directly inhibit MTP activity in rat hepatocytes after diet-induced increased MTP levels [30]. This inhibition decreases the secretion of triglycerides, total cholesterol, and VLDL while simultaneously doubling the high-density lipoprotein (HDL) output, thereby improving the TC/HDL-C and TG/HDL-C ratios and the atherogenic index. These results highlight the potential of theophylline to improve diet-induced dyslipidemia and reduce CV risk. Considering that theophylline is naturally abundant in common dietary sources, such as green tea, cacao, and kola nuts, its regular

https://doi.org/10.3390/biomedicines14010250

consumption may improve lipid profile and reduce the associated CV risk, allowing the reduction in lipid-lowering drug doses [30].

Future studies should focus on early identification and targeted management of CV risk factors including sex-specific hormonal dysregulation and refinement of CV risk-profiles strategies that address combined genetic, demographic, sexual, and metabolic characteristics [7,15]. Large-scale double-blind randomized studies, combined with multidisciplinary approaches and targeted health strategies, are required to address coexisting endocrine/metabolic, renal, and CV pathologies and to improve long-term outcomes for this category of very high-risk patients [20,22,25,27,30].

Regarding peculiar therapeutic options discussed in this reprint, such as the effects of theophylline on lipid metabolism [30], of vitamin D on statins [27], and the CV, renal, and metabolic effects of next-generation SGLT2i [25], future research should rigorously evaluate their effectiveness, optimal dosing regimens, possible adverse effects, and long-term safety in specific populations, such as patients with cardiomyopathies, liver diseases, advanced CKD, or acute hemodynamic instability, and the long-term implications of early initiation strategies in acute HF. Additional challenges relate to implementation, including cost-effectiveness, access, and integration into guideline-directed therapy without unintended adverse interactions [22,25,27,30].

**Author Contributions:** Conceptualization, C.T. and D.C.; methodology, C.T. and D.C.; software, C.T. and D.C.; validation, C.T. and D.C.; formal analysis, C.T. and D.C.; investigation, C.T. and D.C.; resources, C.T. and D.C.; data curation, C.T. and D.C.; writing—original draft preparation, C.T. and D.C.; writing—review and editing, C.T. and D.C.; visualization, C.T. and D.C.; supervision, C.T. and D.C.; project administration, C.T. and D.C.; funding acquisition, C.T. and D.C. All authors have read and agreed to the published version of the manuscript.

**Funding:** This research received no external funding. Internal funding: We would like to acknowledge Victor Babes University of Medicine and Pharmacy Timisoara for their support in covering the publication costs for this research paper.

**Institutional Review Board Statement:** Not applicable.

**Informed Consent Statement:** Not applicable.

**Data Availability Statement:** All of the data are presented in the manuscript.

**Acknowledgments:** We express our gratitude to the entire *Biomedicines* editorial team. Without their expertise and professionalism, "Connections Between Diabetes Mellitus, Other Metabolic and Endocrine Dysfunctions, and Cardiovascular Pathologies—Second Edition" would never have been so successful. We would also like to express our appreciation to all of the contributors who chose to submit their valuable work to this Special Issue, making it possible for us to publish this book.

**Conflicts of Interest:** The authors declare no conflicts of interest.

## References

1. Cosentino, F.; Grant, P.J.; Aboyans, V.; Bailey, C.J.; Ceriello, A.; Delgado, V.; Federici, M.; Filippatos, G.; Grobbee, D.E.; Hansen, T.B.; et al. 2019 ESC Guidelines on Diabetes, Pre-Diabetes, and Cardiovascular Diseases Developed in Collaboration with the EASD. *Eur. Heart J.* **2020**, *41*, 255–323. [CrossRef]
2. 2023 Focused Update of the 2021 ESC Guidelines for the Diagnosis and Treatment of Acute and Chronic Heart Failure. Available online: https://www.escardio.org/Guidelines/Clinical-Practice-Guidelines/Focused-Update-on-Heart-Failure-Guidelines (accessed on 28 April 2024).
3. Mach, F.; Koskinas, K.C.; Roeters van Lennep, J.E.; Tokgözoğlu, L.; Badimon, L.; Baigent, C.; Benn, M.; Binder, C.J.; Catapano, A.L.; De Backer, G.G.; et al. 2025 Focused Update of the 2019 ESC/EAS Guidelines for the Management of Dyslipidaemias: Developed by the Task Force for the Management of Dyslipidaemias of the European Society of Cardiology (ESC) and the European Atherosclerosis Society (EAS). *Eur. Heart J.* **2025**, *46*, 4359–4378. [CrossRef] [PubMed]

https://doi.org/10.3390/biomedicines14010250

4. Williams, B.; Mancia, G.; Spiering, W.; Agabiti Rosei, E.; Azizi, M.; Burnier, M.; Clement, D.L.; Coca, A.; de Simone, G.; Dominiczak, A.; et al. 2018 ESC/ESH Guidelines for the Management of Arterial Hypertension: The Task Force for the Management of Arterial Hypertension of the European Society of Cardiology (ESC) and the European Society of Hypertension (ESH). *Eur. Heart J.* **2018**, *39*, 3021–3104. [CrossRef] [PubMed]
5. ESC Guidelines for the Management of Elevated Blood Pressure and Hypertension. Available online: https://www.escardio.org/Guidelines/Clinical-Practice-Guidelines/Elevated-Blood-Pressure-and-Hypertension (accessed on 11 January 2026).
6. 2023 ESC Guidelines for the Management of Acute Coronary Syndromes. Available online: https://www.escardio.org/Guidelines/Clinical-Practice-Guidelines/Acute-Coronary-Syndromes-ACS-Guidelines (accessed on 3 January 2026).
7. Calcaterra, V.; Mannarino, S.; Puricelli, F.; Fini, G.; Cecconi, G.; Evangelista, M.; Baj, B.; Gazzola, C.; Zuccotti, G. Cardiac Endocrine Function and Hormonal Interplay in Pediatrics: From Development to Clinical Implications. *Biomedicines* **2025**, *13*, 2225. [CrossRef]
8. Tudoran, M.; Tudoran, C. Particularities of Endothelial Dysfunction in Hypothyroid Patients. *Kardiol. Pol.* **2015**, *73*, 337–343. [CrossRef]
9. Sánchez-Soriano, C.; Pearson, E.R.; Reynolds, R.M. The Role of Genetics in Fetal Programming of Adult Cardiometabolic Disease. *J. Dev. Orig. Health Dis.* **2022**, *13*, 292–299. [CrossRef]
10. Ghaemmaghami, Z. Endocrine Disorders and the Cardiovascular System. In *Practical Cardiology*; Elsevier: Amsterdam, The Netherlands, 2022; pp. 685–691.
11. Stoiculescu, F.-M.; Hădăreanu, D.-R.; Hădăreanu, C.-D.; Iovănescu, M.-L.; Târtea, G.-C.; Donoiu, I.; Cojocaru, P.-A.; Militaru, S.; Istrătoaie, O.; Florescu, C. A Four-Phenotype Model for Risk Stratification in Heart Failure with Preserved and Mildly Reduced Ejection Fraction: The Role of Sex and Diabetes. *Biomedicines* **2026**, *14*, 173. [CrossRef]
12. Leith, D.; Lin, Y.Y.; Brennan, P. Metabolic Dysfunction-Associated Steatotic Liver Disease and Type 2 Diabetes: A Deadly Synergy. *touchREV. Endocrinol.* **2024**, *20*, 5–9. [CrossRef]
13. Milani, I.; Parrotta, M.E.; Colangeli, L.; Chinucci, M.; Palleschi, S.; Rossi, B.; Sbraccia, P.; Mantovani, A.; Leonetti, F.; Guglielmi, V.; et al. Sex Differences in MASLD After Age 50: Presentation, Diagnosis, and Clinical Implications. *Biomedicines* **2025**, *13*, 2292. [CrossRef]
14. Tudoran, C.; Tudoran, M.; Vlad, M.; Balas, M.; Pop, G.N.; Parv, F. Echocardiographic Evolution of Pulmonary Hypertension in Female Patients with Hyperthyroidism. *Anatol. J. Cardiol.* **2018**, *20*, 174–181. [CrossRef]
15. Țica, O.; Șandor, M.I.; Huniadi, A.; Daina, C.; Filip, S.M.; Brihan, I.; Sabău, M.; Zaha, I.; Țica, O. Intersecting Endocrine Pathways in Cardiomyopathy: The Role of Metabolic Burden in Structural Heart Disease. *Biomedicines* **2025**, *13*, 2364. [CrossRef] [PubMed]
16. Zhao, X.; Liu, S.; Wang, X.; Chen, Y.; Pang, P.; Yang, Q.; Lin, J.; Deng, S.; Wu, S.; Fan, G.; et al. Diabetic Cardiomyopathy: Clinical Phenotype and Practice. *Front. Endocrinol.* **2022**, *13*, 1032268. [CrossRef]
17. Bordejevic, D.A.; Pârvănescu, T.; Petrescu, L.; Mornoș, C.; Olariu, I.; Crișan, S.; Văcărescu, C.; Lazăr, M.; Morariu, V.I.; Citu, I.M.; et al. Left Ventricular Remodeling Risk Predicted by Two-Dimensional Speckle Tracking Echocardiography in Acute Myocardial Infarction Patients with Midrange or Preserved Ejection Fraction in Western Romania. *Ther. Clin. Risk Manag.* **2021**, *17*, 249–258. [CrossRef]
18. Leng, Y.; Hu, Q.; Ling, Q.; Yao, X.; Liu, M.; Chen, J.; Yan, Z.; Dai, Q. Periodontal Disease Is Associated with the Risk of Cardiovascular Disease Independent of Sex: A Meta-Analysis. *Front. Cardiovasc. Med.* **2023**, *10*, 1114927. [CrossRef]
19. Sanghvi, M.M.; Ramírez, J.; Chadalavada, S.; Aung, N.; Munroe, P.B.; Donos, N.; Petersen, S.E. The Association Between Periodontal Disease and Cardiovascular Disease. *JACC Adv.* **2024**, *3*, 101241. [CrossRef] [PubMed]
20. Țica, O.; Romanul, I.; Ciavoi, G.; Pantea, V.A.; Scrobota, I.; Șipoș, L.; Daina, C.M.; Țica, O. A Clinical Review of the Connections Between Diabetes Mellitus, Periodontal Disease, and Cardiovascular Pathologies. *Biomedicines* **2025**, *13*, 2309. [CrossRef] [PubMed]
21. Mejía Raudales, B.; Portillo Muñoz, C.A.; Chavez, G.S.; Gough, S. Severe Consequences of Uncontrolled Tertiary Hyperparathyroidism in a Limited Resources Setting: A Case Report. *Front. Nephrol.* **2025**, *5*, 1645622. [CrossRef]
22. Costea, M.; Tilici, D.-M.; Paun, D.L.; Nimigean, V.R.; Paun, S.C.; Danciulescu-Miulescu, R.E. Tertiary Hyperparathyroidism in Diabetic Nephropathy: An Underrecognized Complication—A Narrative Review. *Biomedicines* **2025**, *13*, 2604. [CrossRef]
23. McDonagh, T.A.; Metra, M.; Adamo, M.; Gardner, R.S.; Baumbach, A.; Böhm, M.; Burri, H.; Butler, J.; Čelutkienė, J.; Chioncel, O.; et al. 2021 ESC Guidelines for the Diagnosis and Treatment of Acute and Chronic Heart Failure. *Eur. Heart J.* **2021**, *42*, 3599–3726. [CrossRef]
24. Tudoran, C.; Tudoran, M.; Giurgi-Oncu, C.; Abu-Awwad, A.; Abu-Awwad, S.-A.; Voiţă-Mekereş, F. Associations between Oral Glucose-Lowering Agents and Increased Risk for Life-Threatening Arrhythmias in Patients with Type 2 Diabetes Mellitus—A Literature Review. *Medicina* **2023**, *59*, 1760. [CrossRef]
25. Movila, D.; Seiman, D.D.; Dragan, S.R. Next-Generation SGLT2 Inhibitors: Innovations and Clinical Perspectives. *Biomedicines* **2026**, *14*, 81. [CrossRef]
26. Corsini, A.; Bellosta, S. Drug-Drug Interaction with Statins. *Expert Rev. Clin. Pharmacol.* **2008**, *1*, 105–113. [CrossRef]


https://doi.org/10.3390/biomedicines14010250

27. Cozma, D.; Tudoran, C.; Văcărescu, C. Statins, Vitamin D, and Cardiovascular Health: A Comprehensive Review. *Biomedicines* **2025**, *13*, 2515. [CrossRef]
28. Heidemann, B.E.; Koopal, C.; Bots, M.L.; Asselbergs, F.W.; Westerink, J.; Visseren, F.L.J. The Relation between VLDL-Cholesterol and Risk of Cardiovascular Events in Patients with Manifest Cardiovascular Disease. *Int. J. Cardiol.* **2021**, *322*, 251–257. [CrossRef]
29. Lee, H.-C.; Akhmedov, A.; Chen, C.-H. Spotlight on Very-Low-Density Lipoprotein as a Driver of Cardiometabolic Disorders: Implications for Disease Progression and Mechanistic Insights. *Front. Cardiovasc. Med.* **2022**, *9*, 993633. [CrossRef] [PubMed]
30. Park, M.-K.; Jang, H.; Bae, J.-S.; Shin, J.-H.; Park, H.-J. Theophylline Attenuates the Release of Cardiovascular Disease-Related Triglyceride and Cholesterol by Inhibiting the Activity of Microsomal Triglyceride Transfer Protein in Rat Hepatocytes. *Biomedicines* **2025**, *13*, 2579. [CrossRef] [PubMed]



https://doi.org/10.3390/biomedicines14010250

biomedicines

MDPI

*Review*

# Cardiac Endocrine Function and Hormonal Interplay in Pediatrics: From Development to Clinical Implications

**Valeria Calcaterra [1,2,*,†], Savina Mannarino [3,†], Filippo Puricelli [3], Giulia Fini [3], Giulia Cecconi [3], Martina Evangelista [3], Beatrice Baj [3], Cassandra Gazzola [2] and Gianvincenzo Zuccotti [2,4,†]**

1 Department of Internal Medicine and Therapeutics, University of Pavia, 27100 Pavia, Italy
2 Pediatric Department, Pediatric Unit, Buzzi Children's Hospital, 20154 Milano, Italy; cassandra.gazzola@unimi.it (C.G.); gianvincenzo.zuccotti@unimi.it (G.Z.)
3 Pediatric Department, Pediatric Cardiology Unit, Buzzi Children's Hospital, 20154 Milan, Italy; savina.mannarino@asst-fbf-sacco.it (S.M.); filippo.puricelli@asst-fbf-sacco.it (F.P.); giulia.fini@asst-fbf-sacco.it (G.F.); giulia.cecconi@asst-fbf-sacco.it (G.C.); martina.evangelista@asst-fbf-sacco.it (M.E.); beatrice.baj@unimi.it (B.B.)
4 Department of Biomedical and Clinical Science, University of Milano, 20157 Milano, Italy
* Correspondence: valeria.calcaterra@unipv.it
† These authors contributed equally to this work.

**Abstract**

The endocrine system plays a pivotal role in all stages of cardiac development and in maintaining the structural and functional integrity of the heart. Notably, the heart itself functions as an endocrine organ, producing hormones that regulate blood pressure, fluid balance, and myocardial remodeling. This narrative review explores the endocrine mechanisms underlying cardiac development and function, with a focus on fetal and pediatric life. Special attention is given to the heart's intrinsic endocrine activity and how hormonal signals interact with the cardiovascular system during early development. Hormonal signaling is essential for maintaining physiological homeostasis and supporting proper heart development during growth. Disruptions in these signals may serve as silent precursors to structural or functional heart disease, potentially manifesting later in life. Understanding these interactions is clinically relevant, as endocrine imbalances can contribute to the onset, progression, and prognosis of pediatric cardiac disorders. Early identification of hormonal dysregulation can help prevent or mitigate adverse cardiovascular outcomes. Furthermore, recognizing age-specific patterns in hormone–heart interactions may enable the development of targeted diagnostic and therapeutic strategies.

**Keywords:** heart; cardiac; endocrine system; hormones; pediatrics; children

## 1. Introduction

The heart is the first organ to form and function during embryonic development, and its early maturation is essential for the viability of the growing embryo [1]. Cardiac development begins as early as the third week of gestation, guided by a complex interplay of molecular pathways, transcription factors, and morphogenetic cues that drive the differentiation and organization of cardiogenic progenitor cells into a structurally and functionally competent heart [2,3]. This process culminates in the formation of a four-chambered heart capable of supporting the fetal circulatory needs and preparing for the dramatic hemodynamic changes that occur at birth.

However, beyond the intrinsic genetic cardiogenesis programs, the endocrine system plays a fundamental and often underappreciated role in modulating every phase of cardiac


https://doi.org/0.3390/biomedicines13092225

development. Hormones such as thyroid hormones (THs) cortisol, insulin-like growth factors (IGFs), insulin, and sex steroids are key modulators of cardiac cell proliferation, metabolic reprogramming, sarcomere maturation, and mitochondrial biogenesis [4–7]. These hormones not only act systemically but also exert local paracrine effects within the fetal heart, creating a tightly regulated hormonal microenvironment that adapts dynamically throughout gestation.

For instance, THs are critical in promoting terminal differentiation of cardiomyocytes and the postnatal metabolic switch from glycolysis to oxidative phosphorylation [8]. Glucocorticoids facilitate structural remodeling and upregulate genes involved in calcium handling and contractile function [9]. Meanwhile, IGF-II and insulin drive early myocardial growth through proliferation, while IGF-I becomes more prominent in late gestation, supporting hypertrophy and maturation [10,11]. These processes are not isolated; rather, they are temporally coordinated through hormonal surges and feedback mechanisms involving the hypothalamic–pituitary–thyroid (HPT), hypothalamic–pituitary–adrenal (HPA), and pancreatic axes.

Disruption of this finely tuned endocrine environment, whether due to maternal disease, intrauterine stress, genetic syndromes, or iatrogenic factors, can significantly impair fetal cardiac development [12]. Emerging evidence links prenatal hormonal imbalances to congenital heart defects (CHDs), altered cardiomyocyte function, and long-term cardiovascular dysfunctions such as hypertension, metabolic syndrome, and arrhythmias [13,14].

Importantly, the heart itself acts as an endocrine organ, particularly through the production of atrial and brain natriuretic peptides (ANP and BNP), which regulate blood pressure, fluid homeostasis, and myocardial remodeling. These cardiac hormones have gained clinical significance in pediatric cardiology, not only as markers of heart failure but also as tools to assess disease severity, guide therapy, and predict outcomes in children with CHDs, cardiomyopathies, pulmonary hypertension, and inflammatory syndromes such as Kawasaki disease and multisystem inflammatory syndrome in children (MIS-C) [15–18].

Given the complex, bidirectional relationship between the endocrine and cardiovascular systems, understanding the hormonal regulation of cardiac physiology is crucial for both developmental biology and clinical practice. Hormonal signaling pathways not only shape cardiac architecture during development but also influence the heart's capacity to adapt, repair, and respond to physiological or pathological stress across the lifespan [6].

The aim of this narrative review is to explore the endocrine mechanisms involved in cardiac development and function, including the heart's intrinsic endocrine activity, with particular focus on how hormones interact with the cardiovascular system during fetal life and childhood. We also investigate the clinical relevance of these interactions in pediatric pathologies, highlighting how endocrine imbalances may contribute to disease onset, progression, and prognosis. This review integrates developmental endocrinology and pediatric cardiology to provide a comprehensive perspective on the hormone–heart relationship and its relevance to diagnosis, treatment, and preventive care.

## 2. Methods

A narrative review was conducted to explore the endocrine regulation of cardiac development and function, including the heart's own endocrine activity, with particular emphasis on hormonal interplay during fetal life and childhood, and its clinical implications in pediatrics.

The literature search was performed using PubMed and Scopus databases, covering publications from the past 10 years (2015–2025). The keywords used, either individually or in combination, included: hormones, endocrine system, endocrine function, heart, development, atrial natriuretic peptides, brain natriuretic peptides, heart function, children, pediatrics, congenital heart diseases.

Inclusion criteria comprised original research articles, clinical trials, meta-analyses, reviews, and cases series published in English within the last decade. Studies primarily focusing on pediatric populations were prioritized; however, research involving adult populations or animal models was also included if it contributed to understanding the relevant mechanisms or offered translational insights into pediatric conditions.

Exclusion criteria included editorials, commentaries, letters, case reports, and all articles not published in English.

The initial search identified 990 articles. Of these, 185 were excluded prior to screening due to duplication or other reasons, leaving 805 articles for screening. After reviewing titles and abstracts, 185 full-text articles were assessed for eligibility, and 130 met the inclusion criteria for the review. The reference lists of the included studies were also examined to identify any additional relevant publications.

The article selection process is summarized in Figure 1.

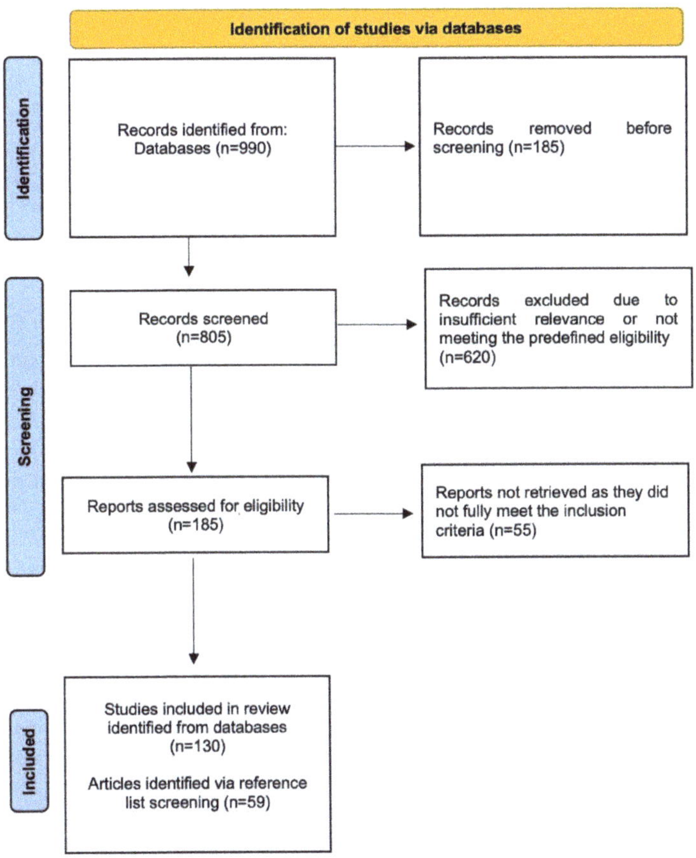


**Figure 1.** Flowchart of article selection.

## 3. Fetal Heart Development

The heart is the first functional organ to complete its formation developing from a group of specialized progenitor cells through the activation of specific cellular and molecular signals.

Heart formation begins during gastrulation (day 15–17 postconception), when mesodermal progenitors migrate from the primitive streak to the antero-lateral plate mesoderm [19,20]. These progenitors diversify into the first heart field (FHF) and second heart field (SHF). The FHF primarily generates the linear heart tube, from which the left ventricle and parts of the atria will develop, whereas the SHF is a source of cells that add to the poles of the heart tube contributing in the development of the right ventricle and outflow tract at the arterial pole of the heart and to part of the atria at the venous pole. Expression of specific transcription factors (such as NKX2-5, GATA4, ISL1, TBX5, and MEF2C) drives commitment and differentiation of cardiac lineage cells [21]. At around 19 days of gestation FHF-derived bilateral endocardial tubes fuse at the midline to form the primitive heart tube. This tube, comprising an inner endocardium, cardiac jelly, and outer myocardium, begins contracting around day 21 of gestational age [19,20]. By approximately day 22–28, rightward looping (also known as D-looping) occurs, transforming the linear structure into a curved configuration allowing the alignment of the atrio-ventricular canal and of the outflow tract for subsequent septation. At about 7 weeks of gestation the heart achieves a four-chambered structure with distinct systemic and pulmonary outlets. The intracardiac septations of the atria, the ventricles, and the outflow tracts are fundamental to complete heart formation and they all are interdependent [19,20]. The cardiac conduction system arises from specialized cardiomyocyte populations: the sinoatrial node originates from the posterior SHF and the atrioventricular node with the His–Purkinje system originates from the atrioventricular canal [22].

After birth, environmental conditions and demands on the heart change dramatically. Postnatally, cardiomyocytes are exposed to higher oxygen levels and must cope with increased blood pressure and mechanical load. Consequently, extensive changes in the myocardium occur during late gestation and the postnatal period in order to adapt to these new demands. During cardiac maturation, the vast majority of cardiomyocytes exit the cell cycle, losing the ability to reproduce and regenerate, and initiate essential metabolic, structural, and electrophysiological changes [23]. The major metabolic change is the shift from a metabolism based on glycolysis during fetal life to the more efficient oxidative phosphorylation. At the same time, there is also an increase in the number, size, and complexity of mitochondria [23]. Structural changes include binucleation and polyploidization, reorganization and hypertrophy of the sarcomeres, development of T-tubules, accumulation of gap junctions at the intercalated discs, and switching to mature isoforms of structural proteins and membrane channels that improve the electromechanical performance of the cardiomyocyte [23]. All these processes appear to be driven by endocrine signals related to normal fetal development and postnatal adaptation.

## 4. Hormonal Regulation of Heart Development and Function

### 4.1. Hormonal Control of Heart Development in Fetal Life

Hormonal signals have a crucial role in cardiomyocyte proliferation, maturation, and metabolic programming during fetal development and postnatal life. TH, IGFs, GH, insulin, and adrenal hormones each play critical and distinct roles in shaping the embryonic heart. This complex hormonal interplay ensures the heart develops in a way that supports both feto-maternal circulation and the transition to an independent circulation after birth. In the next subsections we will display how each hormone impacts on fetal heart development.

#### 4.1.1. Thyroid Hormones

During pregnancy and embryogenesis, thyroxine (T4) and triiodothyronine (T3) supply is entirely provided by maternal sources. These hormones are essential for proper organogenesis to occur, including cardiac embryogenesis [20,23,24]. From the 28th gestational week until term, there is a transition from complete maternal dependence to autonomous fetal synthesis of these hormones. This transition is granted and triggered by a physiological increase in fetal cortisol levels, which upregulates deiodinase enzymes, key regulators of thyroid hormone metabolism. The deiodinase family, consisting of three isoenzymes: D1, D2 and D3, regulate local activation or inactivation of thyroid hormones in a tissue-dependent manner. Specifically, D1 is active in the liver, kidney, thyroid, and pituitary gland and converts T4 to T3 and reverse T3 (rT3) whereas D3, expressed in the placenta and fetal tissues, inactivates T3 and converts T4 to rT3, thus protecting the fetus from hormone excess. During pregnancy, D3 predominates; near term the increase in cortisol reduces D3 activity in favor of D1; this switch is necessary to allow the fetus to meet postnatal metabolic demands through the autonomous production of thyroid hormones [24].

The hypothalamus–pituitary–thyroid gland axis governs TH production. Because of endogenous and environmental cues, the hypothalamus secretes thyrotropin-releasing hormone (TRH), which will activate the pituitary gland and it will release into the circulation thyroid-stimulating hormone (TSH). Finally, the thyroid-stimulating hormone stimulates follicular cells of the thyroid gland to synthesize the prohormone T4 and the activated form T3. Upon binding of T3, nuclear thyroid receptors (TR$\alpha$ and TR$\beta$) form heterodimers with the retinoic acid receptor and interact with the T3 response elements (TREs) to activate gene expression. Moreover, THs can bind cell surface receptors and mediate non-genomic pathways, such as the phosphatidylinositol 3-kinase (PI3K)/protein kinase B (AKT)/mTOR pathway and mitogen-activated protein kinase (MAPK)/extracellular signal-regulated kinase (ERK). Both signaling pathways are involved in controlling cell cycle progression and proliferation [5].

Thyroid hormones surge near term, triggering cardiomyocyte maturation, indeed $T_3$ engaging MAPK/PI3K pathways promotes structural and metabolic transitions within cardiomyocytes [5]. Specifically, $T_3$ downregulates cell cycle genes, reducing proliferation, and contributes to sarcomere maturation, improved calcium handling and enhanced mitochondrial biogenesis. In collaboration with cortisol, $T_3$ also helps the transition from IGF-II to IGF-I expression patterns, serving as a maturational switch [25].

There is increasing interest in studying how thyroid hormones interact with the cardiovascular system, especially in light of the hormones' possible ability to regenerate the myocardium. It has been demonstrated that T3 promotes the proliferation of cardiomyocytes, particularly during the early phases of development. Exogenous T3 administration seems to activate the ERK1/2 signaling pathway via IGF-1R, promoting cell proliferation in response to injury or trophic stimuli in neonatal mouse models, where cardiomyocytes retain immature characteristics and proliferative capacity [26].

This regenerative capacity, however, appeared to decrease after the sixth postnatal day; beyond this threshold, cardiomyocytes appeared to be refractory to mitogenic stimuli. Further investigations in the area of regional myocardial proliferation, conducted by Bogush et al. [10], showed that administration of T3 (3.5 ng/g/day for three consecutive days, from postnatal day 7 to day 9) induces a significant increase (of about 18%) in the number of cardiomyocytes in the left ventricle, with proliferation predominantly localized to the ventricular apex. The phosphatase DUSP5, which prevents ERK1/2 activation, controls this phenomenon. From the ventricular base to the apex, DUSP5 expression progressively rises during development, progressively restricting the proliferation of cardiomyocytes. The regional specificity of the T3-induced proliferative response at the apex was confirmed by clonal lineage tracing. Subsequent research revealed that T3 was able to restore cardiomyocyte proliferation through genetic inhibition of DUSP5, supporting the previously stated thesis and suggesting possible therapeutic uses for myocardial regeneration following infarction [27]. The idea that thyroid hormones may influence cardiac regeneration during neonatal life and possibly beyond is supported by these findings, which are consistent with larger studies investigating the activation of the ERBB2 pathway, which is linked to cardiomyocyte dedifferentiation and proliferation [28,29].

In addition to its proliferative and regenerative effects, thyroid hormones contribute significantly to postnatal cardiac maturation by promoting mitochondrial biogenesis and facilitating the metabolic switch to oxidative phosphorylation. T3 increases the number and density of mitochondrial cristae within cardiomyocytes. This increase is necessary to meet the increased energy requirements for cardiac growth and functional efficiency [30].

#### 4.1.2. Corticosteroid Hormones

Corticosteroid hormones (CHs) include mineralocorticoids and glucocorticoids. The pituitary gland releases adrenocorticotropic hormone (ACTH) which induces the secretion of mineralocorticoids and glucocorticoids in the cortex of the adrenal glands.

Cortisol is the predominant glucocorticoid in humans. Upon stimulation by ACTH or angiotensin II, cortisol is synthesized in the mitochondria via conversion of cortisone by type 1 11β-hydroxysteroid dehydrogenase.

Cortisol can bind to the glucocorticoid receptor (GR) and the closely related mineralocorticoid receptor (MR). The GR is represented in most tissues by two main isoforms (GRα and GRβ) [23,31]. Ligand binding triggers homodimerization and translocation to the nucleus, where the GR interacts with DNA response elements (GREs) to activate the genomic pathway [32]. GRs can also bind to the circular DNA of mitochondria, thereby regulating mitochondrial gene transcription. Cortisol can also impact gene expression via the interaction of the GR with other transcription factors, such as nuclear factor-κB (NF-κB) and activator protein-1 (AP-1) [4]. Membrane-associated GRs can also mediate non-genomic pathways such as PI3K/MAPK signaling, PLC signaling, and the $Ca^{2+}$/calmodulin-dependent protein kinase II pathway, promoting cardiomyocyte maturation [33].

The prepartum cortisol surge contributes to initiating structural, functional, and metabolic changes in the fetal myocardium, synergically with THs and IFG-I. In fact, cortisol stimulates local T3 production in the cardiac tissue and collaborates in the IGF-II to IGF-I switch at late gestation by downregulation of IGF-II [25].

The principal mineralocorticoids are aldosterone and deoxycorticosterone, secreted by the zona glomerulosa of the adrenal cortex under the regulation of ACTH, plasma potassium, and the renin–angiotensin–aldosterone system (RAAS) [34]. Aldosterone binds to the MR, classically regulating sodium reabsorption, potassium excretion, and extracellular fluid volume [35]. In the fetus, these actions are critical for maintaining blood pressure stability and optimizing placental perfusion. By influencing vascular resistance and intravascular volume, mineralocorticoids indirectly affect cardiac preload and systemic hemodynamics [36]. Importantly, MRs are expressed in the developing myocardium and vasculature. Their activation contributes to vascular remodeling and may influence cardiomyocyte growth [35].

Deoxycorticosterone, though less potent, appears to provide redundancy during fetal life, ensuring vascular tone and electrolyte balance [37].

#### 4.1.3. Catecholamines

The adrenal medulla secretes epinephrine and norepinephrine, whose concentrations increase during gestation and peak in the perinatal period [38]. This surge is essential for the dramatic cardiovascular adjustments at birth.

Catecholamines act primarily through β-adrenergic receptors, enhancing myocardial contractility, heart rate, and stroke volume, thereby ensuring adequate cardiac output under intrauterine stress conditions such as hypoxia or labor [39]. Through α-adrenergic receptors, they modulate vascular tone and systemic blood pressure, contributing to redistribution of blood flow towards vital organs such as the brain, heart, and adrenal glands during hypoxic episodes [40].

Beyond these immediate effects, catecholamines influence cardiac gene expression and structural remodeling, including hypertrophic growth and metabolic flexibility [41]. Their role in stress responsiveness during fetal life is crucial for survival at birth, where the sudden transition from placental to pulmonary circulation requires rapid cardiovascular adaptation [42].

Inadequate catecholamine signaling can impair neonatal blood pressure regulation, contractility, and thermogenesis, highlighting their essential role in perinatal cardiovascular resilience [43].

#### 4.1.4. Sex Hormones

SHs are mainly secreted by the testes and the ovaries, where they play a primary role in sexual differentiation and reproductive development. However, a fraction of SHs, especially adrenal androgens, is also synthesized in the adrenal cortex (zona reticularis), representing a minor but functionally relevant contribution during fetal life.

Among other hormones, the anterior pituitary secretes gonadotropins, which regulate the production of estrogens from the ovaries in the female and androgens from the testes in the male. In postmenopausal women and men, estrogens are converted predominantly from adrenal and ovarian or testicular androgens. Sex-hormone-binding globulin (SHBG) regulates the levels of free sex hormones that can readily enter the cell by passive diffusion, but it has also been shown that cellular uptake of SHBG-bound androgens and estrogens could be realized by endocytosis and that sex steroid signaling, at least in part, depends on this route [44].

Estrogens can bind to three forms of estrogen receptors (ERα, ERβ, and G-protein-coupled estrogen receptor (GPER)). In the nucleus, estrogen binding to ERs triggers the recruitment of coactivators that facilitate binding to DNA via estrogen response elements (EREs), mediating the effects of estrogens via the genomic pathway [44]. ERs can also bind to mitochondrial DNA and induce transcription of mitochondrially encoded genes directly or indirectly via nuclear respiratory factor-1 (NRF-1) [18]. ERs located in the plasma membrane or cytosol can mediate more rapid effects via non-genomic pathways, such as PI3K/AKT/mTOR and MAPK/ERK [44].

In the same way as estrogens, androgens cross the cell membrane and primarily bind to nuclear androgen receptors (ARα and ARβ), thereby inducing the genomic pathway via DNA response elements in target gene promoters. Testosterone can be converted intracellularly to dihydrotestosterone, which forms much more stable complexes with ARs and thus can amplify the androgen signaling in individual cells [23]. Membrane-bound receptors mediate non-genomic effects most prominently via the activation of protein kinase A (PKA) and protein kinase C (PKC) but also via activation of MAPK/ERK [23].

Besides well-known cardioprotective effects of estrogens during adult life, some studies also hint at the impact of sex hormones on the electrophysiological maturation of cardiomyocytes [23,45]. Estrogens also induce the transcription of mitochondria-encoded genes relevant for oxidative phosphorylation, enforcing the metabolic switch during cardiomyocyte maturation [33]. While the major hallmarks of cardiomyocyte maturation apply to both sexes, mitochondria demonstrate a clear sexual dimorphism, which is thought to originate from the exclusive maternal inheritance of their genome, resulting in an optimized function in female compared to male individuals [33]. This seems to contribute to gender differences in cardiac function, as well-summarized by Ventura-Clapier et al. [33].

#### 4.1.5. Growth Hormone/Insulin-like Growth Factors and Insulin

GH exerts significant effects on cardiac growth, function, and contractility. Many of these actions are mediated through IGF-I, which is primarily produced in the liver in response to GH stimulation. Therefore, the GH/IGF axis represents a central regulatory pathway in cardiac development and adaptation.

IGFs and insulin have an important role in the regulation of intrauterine growth [25,46]. IGF-I is primarily produced in the liver in response to GH, but the bioavailability of IGF-I/-II is regulated by their receptors (IGF-R), by the IGF-binding proteins, and by themselves. IGFs can bind either to their receptors or to the insulin receptor (IR) due to some structural homology. IGF-IRs belong to the receptor tyrosine kinase family and triggers phosphorylation of intracellular lipids, second messengers, and serine/threonine kinases upon ligand binding. Thereby, they activate non-genomic pathways such as PI3K/AKT/mTOR and MAPK/ERK [23,25]. IGF-I can also activate phospholipase C (PLC) via a heterotrimeric G protein, initiating an inositol 1,4,5-trisphosphate (IP3)-dependent signaling pathway that eventually affects intracellular calcium levels [23]. IGF-II can bind to IGF-IR but also to the mannose 6-phosphate receptor with high affinity, which is, therefore, known as the IGF-IIR. The IGF-IIR lacks the tyrosine kinase domain, therefore, binding this receptor results in internalization and degradation of the ligand instead of inducing phosphorylation cascade [23]. Accordingly, the bind of IGF-II to the IGF-IIR inhibits the signaling of IGF-II via the IGF-IR. Therefore, the main role of the IGF-IIR is to regulate IGF signaling.

IGF-II is the primary in utero growth factor, essential for cardiac mass expansion via cardiomyocyte proliferation. Near term, GH receptor expression increases, inducing IGF-I synthesis as IGF-II declines [25]. IGF-I promotes fetal cardiomyocyte hypertrophy and maturation, synergistically with THs and cortisol [5]. Insulin supports myocardial growth alongside and enhancing glucose transport and metabolic efficiency. Unlike IGF-I, insulin primarily stimulates mass growth rather than differentiation [25].

Beyond these indirect mechanisms, GH also exerts direct effects on the heart. GH receptors are expressed in cardiomyocytes, and their activation influences cardiac mass, geometry, and performance. Experimental and clinical studies have shown that GH can enhance myocardial contractility, improve diastolic function, and modulate vascular resistance [47,48]. Thus, the GH/IGF axis integrates both endocrine and paracrine/autocrine mechanisms, representing a central pathway in cardiac development and adaptation.

The endocrine regulation of fetal heart development is schematically represented in Figure 2.

Table 1 provides an overview of the major hormones implicated in cardiovascular development and function during fetal life.

### *4.2. Hormonal Effects on Heart Function in Childhood*

#### 4.2.1. Cardiac Effects of Thyroid Hormones

The thyroid mainly produces T3 and T4. Most circulating hormone is T4, an inactive prohormone, which is peripherally converted into the active T3. T3 acts by binding to nuclear receptors, stimulating the transcription of genes encoding contractile proteins (e.g., myosin isoforms) and proteins involved in intracellular calcium regulation (e.g., SERCA2 and phospholamban) [49]. In addition to genomic effects, thyroid hormones exert rapid non-genomic actions: they modulate membrane ion channels ($Na^+$, $K^+$, $Ca^{2+}$), influencing cardiac excitability and contractility. Moreover, thyroid hormones activate intracellular signaling pathways such as those mediated by phosphatidylinositol 3-kinase (PI3K) and protein kinase B (AKT); these signaling pathways enhance cell survival and physiological myocardial hypertrophy. Through the previously described mechanisms, which are activated with distinct temporal patterns, thyroid hormones exert an overall stimulatory effect on the cardiovascular system. This effect is manifested by positive chronotropic and inotropic actions, an increase in cardiac output, peripheral vasodilation, enhanced cellular metabolic activity, and elevated oxygen consumption [49]. In addition, THs improve diastolic relaxation (positive lusitropic effect) and increase cardiac preload through activation of the renin–angiotensin–aldosterone system. They also reduce systemic vascular resistance via direct vasodilatory actions on vascular smooth muscle. At the cellular level, thyroid hormones upregulate calcium-handling proteins such as SERCA2a and modulate β-adrenergic receptor density and sensitivity, thereby amplifying adrenergic responsiveness [49]. These combined mechanisms lead to higher heart rate, improved myocardial performance, and increased tissue oxygen delivery in order to sustain enhanced metabolic demands [50,51].

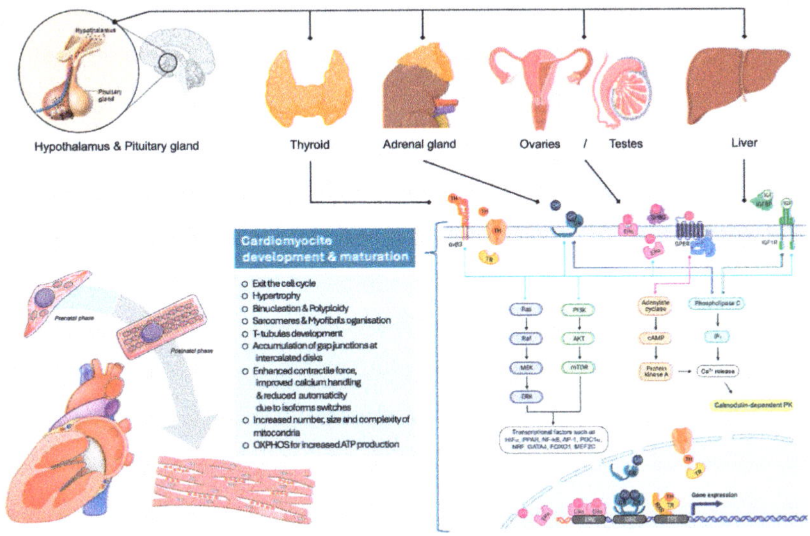


**Figure 2.** Endocrine regulation of fetal heart development. Various hormones contribute to the growth, maturation, and differentiation of cardiomyocytes during fetal development. Key hormonal players include thyroid hormones, adrenal and sex hormones, insulin-like growth factors, catecholamines, and natriuretic peptides. These hormones act through specific molecular signals and transduction pathways, modulating processes such as proliferation, hypertrophy, mitochondrial maturation, and cardiac energy metabolism. The coordinated interaction of these signals is essential to ensure the formation of a functionally competent heart at birth.

**Table 1.** Major hormones involved in cardiovascular development and function, with their roles and effects of dysfunction during fetal life.

| Hormone | Main Cardiovascular Functions | Effects of Dysfunction During Fetal Development |
| --- | --- | --- |
| Thyroid hormones | Promote cardiomyocyte terminal differentiation, sarcomere maturation, mitochondrial biogenesis, and metabolic switch from glycolysis to oxidative phosphorylation. Regulate calcium handling and contractile proteins. | Hypothyroidism → impaired cardiomyocyte proliferation, abnormal chamber geometry, reduced systolic/diastolic function, increased perinatal morbidity. Hyperthyroidism → tachycardia, arrhythmias, hypertrophy. |

**Table 1.** *Cont.*

| Hormone | Main Cardiovascular Functions | Effects of Dysfunction During Fetal Development |
|---|---|---|
| Glucocorticoids (cortisol) | Regulate cardiomyocyte maturation, stimulate local T3 production, support IGF-II → IGF-I switch, enhance calcium handling and contractility. | Excess → hypertrophy, disrupted architecture, arrhythmias, mitochondrial dysfunction. Deficit → delayed maturation, poor adaptation to birth hemodynamic stress. |
| Mineralocorticoids (aldosterone) | Modulate vascular tone and fluid balance via mineralocorticoid receptors. | Hyperaldosteronism → atrial enlargement, fibrosis, oxidative stress, ↑ increased atrial fibrillation risk. |
| Catecholamines | Essential for fetal cardiac contractility, heart rate regulation, and stress adaptation; support vascular tone and placental circulation. | Deficiency (e.g., impaired adrenal development) → reduced cardiac output, hypotension, poor stress response. Excess → tachycardia, arrhythmias, myocardial oxygen imbalance. |
| Growth hormone | Direct myocardial effects (↑ mass, geometry, contractility). Indirect effects via IGF-I production. | Growth hormone deficiency → reduced left ventricular mass, impaired contractility, endothelial dysfunction. |
| Insulin-like growth factors | IGF-II drives early cardiomyocyte proliferation; IGF-I supports hypertrophy and maturation near term; regulate PI3K/AKT/mTOR and MAPK/ERK signaling. | Low IGF signaling (e.g., intrauterine growth retardation) → reduced cardiomyocyte number, myocardial thinning, impaired growth and function. |
| Insulin | Supports myocardial growth, enhances glucose uptake and metabolic efficiency, promotes adaptive hypertrophy. | Insulin deficiency/resistance → metabolic inefficiency, risk of cardiomyopathy, association with maternal diabetes-related congenital heart diseases (CHDs). |
| Sex hormones (estrogens, androgens) | Estrogens: mitochondrial gene expression, oxidative metabolism, cardioprotection. Androgens: improve contractility, influence electrophysiological maturation. | Estrogen deficiency → impaired maturation, metabolic dysfunction. Androgen deficiency (e.g., delayed puberty) → impaired myocardial performance. |
| Natriuretic peptides | Regulate blood pressure, natriuresis, anti-hypertrophic and anti-fibrotic effects; biomarkers of stress and heart failure. | Deficiency → hypertrophy, fibrosis, increased arrhythmias; high neonatal brain natriuretic peptide predicts CHD severity or perinatal stress. |

Studies analyzing the interactions between THs and cardiac physiology in cohorts of healthy patients are limited. Among these, a prospective study examined a number of parameters of cardiological and endocrinological interest to determine how thyroid hormones, specifically FT4 and TSH, affected cardiovascular function in preschool-aged children [8]. TSH and FT4 levels, left ventricular (LV) mass by echocardiography, arterial stiffness by carotid–femoral pulse wave velocity (CFPWV), and systolic and diastolic blood pressure were measured in a group of 4251 children, all of whom had an average age of 6 years. Additionally, dual-energy X-ray absorptiometry (DEXA) was used to analyze body composition. The results showed an inverse association between FT4 and LV mass, as well as between FT4 and lean mass. About 55% of the effect of FT4 on LV mass was mediated by lean mass, indicating that FT4 indirectly influences cardiac structure through modulation of body composition. TSH was inversely associated with LV mass, particularly in males, and positively correlated with systolic and diastolic blood pressure [52]. These data suggest that elevated FT4 levels correlate with a reduction in LV mass and that this effect is mediated in large part by the action of hormones on lean mass composition. The differing associations of TSH and FT4 with cardiovascular parameters show different mechanisms by which THs rule cardiovascular function in childhood, highlighting the importance of the thyroid axis as an early regulator of cardiac and vascular physiology with possible implications for adult cardiovascular disease prevention [53].

The role of TH in cardiac physiology is also confirmed in "subclinical" conditions. Banu Rupani et al. [54] examined the effects of six-month treatment with levothyroxine on specific echocardiographic parameters in a cohort of 30 pediatric patients diagnosed with subclinical hypothyroidism. The parameters evaluated were: ejection fraction (EF), fractional shortening (FS), myocardial performance index (MPI), left ventricular end-diastolic diameter (LVEDD), and left ventricular end-systolic diameter (LVESD). The group consisted of 19 males (63.3%) and 11 females (36.7%), with a mean age of 6.60 ± 2.13 years. After treatment, some statistically significant changes were observed, including: the mean EF increased significantly from 63.13 ± 3.01% to 69.07 ± 4.50% ($p < 0.001$), and the FS increased from 31.83 ± 1.62% to 35.10 ± 1.13% ($p < 0.001$). MPI also showed a significant increase, from 0.28 ± 0.02 to 0.33 ± 0.03 ($p < 0.001$). LVEDD showed a significant reduction from 3.47 ± 0.46 cm to 3.05 ± 0.40 cm ($p < 0.001$). $E/E'$ and LVESD did not show statistically significant changes [54]. Overall, the results seem to suggest that levothyroxine therapy may lead to improvements in cardiac function in children with subclinical hypothyroidism.

Finally, the interaction between THs and cardiac function was also supported by data on the influence of THs on prognostic impacts of thyroid hormones on outcomes of cardiac pediatric patients [55,56] and severe illness [57,58] and the beneficial effects of thyroid hormone supplementation [57].

#### 4.2.2. Cardiovascular Control via the Hypothalamic–Pituitary–Adrenal (HPA) Axis

The range of action of the HPA axis is greatly expanded by their numerous peripheral targets, including the heart. Through a widely controlled hormonal cascade, the HPA axis, which is made up of the anterior pituitary, the adrenal cortex, and the paraventricular nucleus (PVN) of the hypothalamus, coordinates the body's reaction to stress. Corticotropin-releasing hormone (CRH) and arginine vasopressin (AVP) are released by hypothalamic neurons in response to a stress stimulus. These neurotransmitters cause the anterior pituitary to release adrenocorticotropic hormone (ACTH). The adrenal cortex then uses ACTH to promote the production and release of glucocorticoids, mainly cortisol [9].

CRH influences cardiovascular physiology both directly and indirectly. Indirect effects are mediated by the release of cortisol. Direct effects occur through receptors on cardiomyocytes and vascular walls (CRH-R2/CRH-R1): the CRHR2 receptor, predominantly expressed in the heart, mediates chronotropic and positive inotropic effects through activation of the cAMP signaling pathway [59]. The hormone–receptor interaction can trigger the release of peptides such as ANP and BNP. In addition, binding to CRH-R2 may promote an increase in coronary blood flow and directly participate in processes involved in cardiac remodeling, i.e., the structural changes that the heart enacts in response to chronic conditions such as hypertension or heart failure [9]. Furthermore, in other studies in animal models, it can be observed that central stimulation of the CRH system appears to increase blood pressure through CRHR1, while peripheral stimulation via CRHR2 tends to reduce blood pressure [59].

AVP regulates water balance and vascular tone through three receptors (V1aR, V1bR, V2R), with effects varying by tissue oxygenation and vascular region. ACTH may also directly influence vascular tone through MC2R on endothelial cells, independently of cortisol [9].

ACTH influences cardiovascular physiology mainly through the production of cortisol but can also have direct vascular effects. ACTH, in fact, acts directly on cardiovascular physiology by binding to MC2R receptors expressed on vascular endothelial cells. This interaction induces the local production of vasoactive mediators, modulating vascular tone and contributing to the regulation of blood pressure independently of cortisol via an autocrine/paracrine mechanism [9].

As already noted, each component of the HPA axis exerts direct effects on the cardiovascular system, however, it is undisputed that cortisol synthesis represents the most incisive of the final events put into play by the axis in the cardiovascular system.

Glucocorticoids influence cardiovascular physiology through genomic and non-genomic mechanisms. The effects defined as genomic are expressed through the activation of intracellular receptors. GR-mediated signaling is crucial for the maintenance of contractility and for the structural and functional organization of the sarcomere: it is closely linked to the expression and transcription of numerous genes involved in the encoding of contractile proteins and channel proteins responsible for the exchange of calcium ions between the cytoplasmic environment and the sarcoplasmic reticulum and in other key processes such as the inflammatory response [59]. By triggering intracellular signaling pathways like those mediated by PI3K and MAPK kinases, glucocorticoids can more quickly start non-genomic actions and enable instantaneous cardiac function adaptation to acute stimuli [59].

Additionally to GR, glucocorticoids can also activate mineralocorticoid receptors (MRs) in cardiac tissue [60]. This is made possible by the heart's low expression of the enzyme 11β-hydroxysteroid dehydrogenase type 2 (11β-HSD2). The previously mentioned enzyme catalyzes the conversion of physiologically active cortisol into its inactive metabolite cortisone, which has a low affinity for MRs, in mineralocorticoid-sensitive organs. In tissue environments where it is essential that glucocorticoids do not compete for MR activation, this enzyme is required. Glucocorticoids' interaction with MR tends to promote inflammation and fibrosis, processes involved in cardiac remodeling and dysfunction [61]. At the level of the vascular endothelium, glucocorticoids appear to be responsible for and to exert protective effects under conditions of acute stress due to the anti-inflammatory mechanisms and the regulation of vascular reactivity that they activate [62]. Chronic exposure to such stimuli, however, can impair vascular repair mechanisms, promoting chronic conditions of vascular dysfunction such as hypertension. This duality underlines the complexity of endocrine–cardiovascular interactions in cardiovascular physiology and the potential consequences of HPA axis dysfunction [60].

In addition to glucocorticoids and mineralocorticoids, adrenal medullary hormones also contribute significantly to cardiovascular regulation. Catecholamines (epinephrine and norepinephrine) exert their effects primarily through the activation of adrenergic receptors on cardiomyocytes, vascular smooth muscle cells, and endothelial cells. Binding to $\beta_1$-adrenergic receptors increases intracellular cAMP levels, leading to enhanced calcium influx, positive chronotropic and inotropic effects, and improved cardiac output. $\beta_2$-adrenergic stimulation promotes vasodilation in skeletal muscle and coronary arteries, optimizing blood flow distribution during stress [63]. Conversely, activation of $\alpha_1$-adrenergic receptors in the peripheral vasculature increases intracellular $IP_3$ and DAG, mobilizing calcium from the sarcoplasmic reticulum and causing vasoconstriction, thereby contributing to the maintenance of systemic vascular resistance and blood pressure [64]. Chronic catecholamine excess often results in sustained hypertension, concentric left ventricular hypertrophy, arrhythmias, and in some cases catecholamine-induced cardiomyopathy [65,66]. Conversely, catecholamine deficiency or secondary adrenal suppression following prolonged corticosteroid therapy impairs the cardiovascular stress response, leading to hypotension, reduced perfusion, and increased vulnerability during acute illness or surgery [67].

#### 4.2.3. Insulin's Roles in Cardiac Metabolism and Vascular Function

With effects that differ greatly depending on physiological conditions and insulin resistance, insulin plays a crucial role in the cardiovascular system through both direct and indirect mechanisms [29].

The heart is highly dependent on the availability of metabolic substrates and requires an enormous amount of ATP to maintain contraction and electrolytic activity. Both directly and indirectly, through their effects on peripheral tissues and the heart muscle, circulating insulin levels, which vary with dietary intake and circadian rhythms, shape cardiac metabolism [29]. Cardiac mitochondria occupy a significant portion of heart cells and are responsible for producing large amounts of energy (ATP). The heart mainly uses long-chain fatty acids as fuel, followed by glucose and lactate. This flexibility in metabolism depends on a continuous oxygen supply and is essential for meeting the heart's energy needs [68,69].

Studies on pediatric samples show a correlation between body mass and cardiac mass, suggesting that insulin modulates cardiac growth indirectly, via metabolic regulation linked to systemic nutritional status, and thus regardless of hemodynamic load [70]. This relationship is also confirmed by research conducted on adolescents suffering from anorexia nervosa, who show a significant reduction in cardiac mass that improves with nutritional recovery [71].

At the intracellular level, insulin binds to the IR, activating the PI3K/Akt pathway, with Akt1 mediating cardiac hypertrophy and normal postnatal growth. IGF-1, through its receptor and hybrid complexes with the IR, exerts similar effects on cardiac tissue, promoting cell survival and expansion [72]. Glucose uptake, mainly regulated by Akt2 via GLUT4, is crucial for cardiac metabolism [73].

The direct effects of insulin include stimulation of nitric oxide synthase (eNOS) in cells, increasing nitric oxide (NO) production, which promotes vasodilation and protects against atherosclerosis. In addition, insulin increases myocardial glucose uptake via GLUT4 and stimulates adaptive cardiomyocyte growth through mTOR signaling [74].

In insulin resistance, where PI3K/Akt signaling is disrupted while other pathways, such MAPK/ERK, remain active, the indirect effects become more evident. Activated in smooth muscle and vascular endothelial cells, this pathway generates endothelin-1 and stimulates cell proliferation, so promoting vasoconstriction and pathological vascular remodeling. This indirect mechanism is one of several maladaptive responses induced by hyperinsulinemia [74].

Among the effects mediated by hyperinsulinemia there is also the activation of the renin–angiotensin–aldosterone system (RAAS). Hyperinsulinemia activates the RAAS through various mechanisms: first, it directly stimulates renin secretion by juxtaglomerular cells and indirectly through the activity of the sympathetic nervous system [75]. Hyperinsulinism can also increase angiotensinogen synthesis, leading to higher levels of angiotensin II. Moreover, visceral obesity could be a last mechanism in charge of activating the system since it not only creates an environment of chronic inflammation and oxidative stress that prolongs system activation but also stimulates angiotensinogen synthesis. Chronic renin–angiotensin–aldosterone system (RAAS) activation promotes and aggravates congestive heart failure, systemic hypertension, and chronic kidney disease. A pro-fibrotic, pro-inflammatory, and hypertrophic environment resulting from too high levels of circulating and tissue angiotensin II (AngII) and aldosterone causes remodeling and dysfunction in renal and cardiovascular tissues [76].

Apart from the abovementioned consequences, loss of insulin control over lipolysis results in an increase in free fatty acids, which activate protein kinase C (PKC) and aggravate oxidative stress and vascular stiffness. Insulin's immunomodulating action is also compromised in conditions of insulin resistance: this compromises its capacity to reduce pro-inflammatory molecules in macrophages such MCP-1, so encouraging inflammation and atherogenesis [77]. In general, insulin protects the cardiovascular system under physiological conditions; in insulin resistance, indirect processes favoring vascular dysfunction, inflammation, and atherosclerosis predominate.

## 5. Cardiac Endocrine Function

Cardiac endocrine activity in children is predominantly mediated by the natriuretic peptides ANP and BNP, released mainly by atrial cardiomyocytes [73,74,78].

Their expression is developmentally regulated during fetal and postnatal life, responding to mechanical stress, inflammation, and neurohormonal input. The NPPA and NPPB genes encoding ANP and BNP are regulated tightly: NPPA is expressed in early cardiac development in both atria and ventricular regions and becomes more atrial-specific over time, influenced by transcription factors GATA4, NKX2-5, and TBX5 [79]. Although NPPA and NPPB tend to be co-regulated under mechanical strain such as pressure overload, BNP expression can also be upregulated independently by inflammatory cytokines, especially IL-1$\beta$ and TNF-$\alpha$, as observed in myocarditis [80]. In healthy conditions, natriuretic peptide secretion predominates from the atria, while ventricular production becomes clinically significant only in pathologic conditions such as heart failure [81].

These peptides exert their effects via three known receptors. NPR-A (NPR1) activates guanylyl cyclase to promote vasodilation, natriuresis, and anti-hypertrophic effects; NPR-B (NPR2) binds CNP and is more relevant to non-cardiac functions such as bone growth; and NPR-C (NPR3) serves primarily to clear circulating peptides [5].

BNP mediates potent anti-hypertrophic, anti-fibrotic, and anti-arrhythmic actions through paracrine and autocrine mechanisms [17]. BNP-deficient animal models develop ventricular hypertrophy and fibrosis even in the absence of hypertension, underscoring its role in preventing adverse remodeling [15]. Clinically, elevated NT-pro-BNP levels predict left ventricular hypertrophy independently of overt heart failure [7], though genetic variation in ANP and NPR-A may confer stronger susceptibility to hypertrophic remodeling than BNP itself [82–86]. BNP deficiency accelerates pro-fibrotic signaling via pathways involving noradrenaline, angiotensin II, and TGF-$\beta$ [83] and limits immunologic cell recruitment via NPR-A-mediated suppression of monocyte chemotaxis [6,84]. Clinically, higher BNP and NT-proBNP levels correlate with increased risk of atrial fibrillation, ventricular arrhythmias, and sudden cardiac death [80–83,87].

ANP and BNP also regulate blood pressure and fluid homeostasis by increasing capillary permeability and relaxing vascular smooth muscle [88]. While chronic loss of NPR-A does not significantly alter long-term blood pressure, acute endothelial deletion leads to hypertension, hypervolemia, and disrupted vascular permeability [89]. In heart failure patients, BNP infusion reduces arterial pressure and systemic resistance. Natriuresis and diuresis occur both in health and disease, but an increase in glomerular filtration rate (GFR) is seen predominantly in healthy subjects [90]. These renal effects arise from afferent arteriole dilation, efferent constriction, and reduced sodium reabsorption, though NPR-A expression seems limited to glomerular and vascular structures rather than tubules [87–89].

Furthermore, BNP appears to modulate the sympathetic nervous system: $\alpha$-adrenergic activation enhances peptide secretion while $\beta$-adrenergic stimulation suppresses it [91–94]. ANP has been shown to inhibit sympathetic nerve activity in various vascular beds in animal models, though human data remain inconsistent [91–98]. BNP infusion at low doses reduces cardiac sympathetic activity, with higher doses needed to suppress renal sympathetic output [99]. Finally, BNP suppresses RAAS activation via direct inhibition of renin and aldosterone, amplifying its natriuretic and diuretic effects [100].

## 6. Clinical Implications of Endocrine Disease-Related Cardiovascular Effects in Fetal and Pediatric Populations

Several endocrine conditions can alter cardiac function and development in fetal and pediatric age groups. The main ones are diabetes mellitus, obesity, thyroid disorders, and adrenal dysfunction.

### 6.1. Cardiac Involvement in Diabetes

Diabetes can negatively impact cardiac development and function both during pregnancy and in children affected by the disease.

In the fetal stage, maternal pregestational or gestational diabetes mellitus (GDM) disrupts normal cardiac morphogenesis as early as the first trimester [101]. This occurs through mechanisms such as oxidative stress, inflammation, and mitochondrial damage, which interfere with epigenetic regulation and molecular signaling pathways essential for heart formation [101].

Numerous epidemiological studies have demonstrated a strong and consistent association between maternal diabetes, both type 1 (T1D) and type 2 (T2D), and an increased risk of CHD in offspring.

Early reports [102], and more recent large-scale investigations, such as a Canadian population-based study involving nearly 2.3 million births between 2002 and 2010, have confirmed that the risk of CHD in infants born to mothers with pregestational diabetes is approximately 4–5%, compared to 1% in the general population [103].

This risk varies according to CHD subtype. The Baltimore-Washington Infant Study found an increased risk of laterality and cardiac looping defects (heterotaxy), outflow tract anomalies, and atrioventricular and membranous ventricular septal defects but no significant increase in left/right-sided obstructive defects, muscular ventricular septal defects, atrial septal defects, or persistent ductus arteriosus [104]. Additionally, late gestational exposure to maternal diabetes has been associated with hypertrophic cardiomyopathy [105].

The Canadian study further demonstrated that type 2 diabetes in mothers was associated with the highest risk for heterotaxy and left ventricular outflow tract obstructions, while type 1 diabetes was more strongly linked to conotruncal anomalies and atrioventricular septal defects [103]. Both types of diabetes also increased the risk of other CHDs, including right ventricular outflow tract obstructions and atrial/ventricular septal defects, though to a lesser extent.

Ghouse et al. [106] investigated whether prenatal exposure to maternal diabetes was associated with subtle alterations in neonatal left ventricular structure and function. Findings showed more pronounced changes in infants born to mothers with pre-existing diabetes compared to those with GDM, suggesting that, even in the absence of overt congenital heart defects, maternal diabetes, particularly when pre-existing, can subtly influence neonatal cardiac anatomy and function.

Gireada et al. [107] used Fetal HQ, a 2D speckle-tracking echocardiography technique, to assess whether fetuses of mothers with well-controlled GDM displayed altered cardiac shape and contractility; despite adequate glycemic control, 60% of fetuses in the GDM group exhibited at least one form of contractility abnormality (global, longitudinal, or transverse) in one or both ventricles.

The association between diabetes and CHD are supported by multiple other studies, which consistently report a 3- to 5-fold higher prevalence of CHD among infants exposed to pre-existing maternal diabetes [108–112].

In children diagnosed with diabetes, both T1D and T2D, various forms of cardiovascular involvement have been observed. In pediatric patients with T1D, subclinical myocardial dysfunction can emerge early, often without clinical symptoms [113]. Advanced imaging techniques have revealed reduced longitudinal and circumferential strain, altered ventricular remodeling, and early signs of myocardial fibrosis [114,115]. Cardiac autonomic neuropathy is also common in children with T1D, affecting more than 30% of patients in some cohorts, and is associated with poor glycemic variability and suboptimal adherence to therapy [113].

In pediatric T2D, which is becoming increasingly prevalent, the cardiovascular risk is compounded by additional metabolic abnormalities such as obesity, dyslipidemia, and insulin resistance. Adolescents with T2D had increased left ventricular mass and evidence of diastolic dysfunction compared to normoglycemic peers [116]. These structural and functional changes often occur within a few years of diagnosis and are compounded by endothelial dysfunction and early arterial stiffness [117].

Another important aspect concerns electrocardiographic (ECG) abnormalities. As reported by Galli-Tsinopoulou A et al. [118], T1DM youths have a sixfold increased risk for QT/QTc prolongation and should have regular follow-ups for cardiac autonomic dysfunction. A QTc length prolongation during the diagnosis of diabetic ketoacidosis and ketosis has been reported by Aygün D et al. [119]. Additionally, children with T1D are at high risk of impaired ventricular depolarization and repolarization [120]. The electrophysiological changes can precede detectable structural abnormalities and may serve as early indicators of arrhythmic risk [121].

### 6.2. *Cardiac Dysfunction and Childhood Obesity*

Severe obesity in children and adolescents is a significant global public health concern. According to current estimates, the prevalence of at least one cardiometabolic risk factor in children with severe obesity ranges from 67% to 86%. Obesity and insulin resistance in youth are associated with early signs of cardiac dysfunction.

Burden et al. [122] in a meta-analysis shows that obese children and adolescents exhibit increased left ventricular mass and cardiac wall thickening compared to their normal-weight peers. Impairments in both diastolic and systolic function are observed, even in the absence of clinical symptoms. The risk of developing heart disease in adulthood therefore appears to be already increased in childhood.

Childhood obesity is associated with changes in myocardial geometry and function indicating early onset of unfavorable alterations of the myocardium [123]. Erbs et al. [124], using cardiac MRI, observed that left atrial enlargement aligns with the pathophysiological model whereby obesity increases blood volume, cardiac output, and peripheral vascular resistance, ultimately leading to cardiac chamber dilatation, a condition that appears, at least at this stage, to be reversible with weight reduction.

Yi LF et al. [125] showed that school-age children with obesity have impaired autonomic nerve function, presenting with reduced vagal tone, which is particularly prominent in those with dyslipidemia; the more obese the children, the lower the vagal tone, which may increase the risks of cardiovascular diseases. In 2023, a cross-sectional observational study involving 70 children with a body mass index (BMI) greater than one standard deviation above the mean for Thai children was conducted at Naresuan University Hospital. The study found that cardiac functional abnormalities in childhood obesity were significantly positively correlated with BMI and several cardiac parameters, including increased ventricular wall thickness [126]. A cross-sectional study published in 2014 registered the heart rate during ten minutes in the supine position with spontaneous breathing [127]. Cardiac autonomic control was assessed by heart rate variability showing that obese normotensive children and adolescents present impairment of cardiac autonomic control [127].

As reported by Campos JO et al. [128], school-age obesity (BMI > 95th percentile) is associated with autonomic imbalances; in particular, obesity reduces vagal (parasympathetic) activity and relative sympathetic overdrive.

Similarly, another cross-sectional study evaluated cardiac autonomic function in obese school-age children (ages ~5–8 years) [125], showing an impaired autonomic nerve function, presenting with reduced vagal tone, which is particularly prominent in those with dyslipidemia.

The role of cortisol levels in cardiovascular regulation in children with obesity has also been hypothesized. Studies have reported that overweight or obese children exhibit significantly higher cortisone/creatinine ratios ($\beta$ = 1.26, 95% CI: 1.17–1.36) compared to children of normal weight, suggesting that cortisol, although not directly, may contribute to the development of hypertension in this population, thereby influencing cardiovascular risk [129].

### 6.3. Cardiac Impairment in Thyroid Disorders

Fetal hypothyroidism is associated with intrauterine growth restriction (IUGR), reduced cardiomyocyte proliferation, and impaired myocardial maturation. Affected neonates often exhibit altered chamber geometry, reduced systolic and diastolic function, and elevated resting heart rates by early childhood [5].

In congenital hypothyroidism, cardiovascular involvement is particularly concerning [130]. Thyroid hormone is required for normal cardiogenesis and fetal cardiac maturation. Infants born with untreated hypothyroidism may have cardiomegaly, reduced cardiac performance, and increased peripheral resistance [131]. Fortunately, with early neonatal screening and prompt initiation of levothyroxine therapy, these complications are largely preventable.

During the entire lifespan, the cardiovascular system is a key target of thyroid hormones, as they influence cardiac output, vascular resistance, myocardial contractility, heart rate, and autonomic function [50]. In children, disturbances in thyroid function, whether congenital or acquired, can result in significant cardiovascular consequences that may be transient or long-lasting depending on the timing, severity, and duration of the endocrine imbalance.

In hyperthyroidism, commonly due to Graves' disease in children and adolescents, excess T3 and T4 lead to an increase in resting heart rate, systolic blood pressure, and cardiac output [132]. The increased metabolic demand and sensitization of $\beta$-adrenergic receptors result in a hyperdynamic circulatory state [132]. Children may present with palpitations, restlessness, and exertional intolerance. Physical examination often reveals a hyperkinetic precordium, bounding pulses, and systolic flow murmurs [133]. Although atrial fibrillation is rare in pediatric populations, it has been reported in adolescents and may be associated with thyrotoxic cardiomyopathy, particularly in the presence of delayed diagnosis or underlying heart disease [134].

Cardiac imaging studies, including Doppler echocardiography, have shown increased left ventricular end-diastolic diameter, ejection fraction, and cardiac index in untreated pediatric hyperthyroidism [135]. There is also evidence of left ventricular hypertrophy in some cases, which is generally reversible with medical therapy (e.g., methimazole) or after definitive treatment such as radioiodine ablation or thyroidectomy [136]. With normalization of thyroid hormone levels, most functional and structural changes regress, highlighting the reversibility of cardiac involvement in the hyperthyroid state [136].

In contrast, hypothyroidism exerts depressive effects on the cardiovascular system [137]. The bradycardia and reduced myocardial contractility observed in hypothyroid children result in a low-output state, which may be clinically silent or present with fatigue, cold intolerance, and exercise intolerance [138]. Importantly, hypothyroidism increases systemic vascular resistance, leading to diastolic hypertension, a finding that can be overlooked unless blood pressure is carefully measured [138]. A frequent finding in moderate to severe hypothyroidism is pericardial effusion, which occurs due to altered capillary permeability and accumulation of protein-rich fluid in the pericardial sac [135,136]. While often asymptomatic, large effusions can cause muffled heart sounds and, in rare cases, hemodynamic compromise [139–141].

Even in the subclinical forms of thyroid dysfunction, where free T4 and T3 levels remain within normal limits but TSH is elevated (subclinical hypothyroidism), early signs of cardiovascular involvement have been documented in children [142,143]. These include increased carotid intima–media thickness (CIMT), a surrogate marker of early atherosclerosis, endothelial dysfunction, and alterations in heart rate variability, suggesting autonomic imbalance [140,141]. These vascular changes, though often subclinical, may signal a predisposition to future cardiovascular disease, particularly in children with additional risk factors such as obesity, sedentary lifestyle, or family history of dyslipidemia [144–146].

### 6.4. Cardiac Involvement in Growth Deficiency

During intrauterine life, fetuses affected by intrauterine growth restriction (IUGR) are frequently exposed to a state of chronic hypoxia due to placental insufficiency. This sustained low-oxygen environment has a profound impact on fetal development, particularly by downregulating the synthesis of IGF-I [46]. Low IGF signaling hinders cardiomyocyte proliferation, resulting in cardiomyocyte deficiency, myocardial thinning, and reduced stroke volume. Early and persistent alterations of myocardial structure and decreased cardiac function detected by echocardiography in fetal, neonatal, and juvenile patients with IUGR are reported [147]. Postnatally, these hearts are metabolically compromised, exhibit higher resting heart rates, and are at increased risk for hypertension and coronary disease [5].

During childhood, GH plays a fundamental role not only in linear growth and somatic development but also in the regulation of cardiovascular structure and function. Its biological actions are mediated both directly and through IGF-1, which exerts anabolic, metabolic, and vascular effects on various tissues, including the myocardium [48]. In children with isolated or combined GHD, several studies have highlighted the presence of cardiac alterations, even in the absence of clinical symptoms, emphasizing the importance of early recognition and monitoring of cardiovascular involvement [48].

Children with untreated GHD often exhibit a significant reduction in left ventricular (LV) mass, LV wall thickness, and end-diastolic volume when compared to age-matched healthy peers [148,149]. These findings have been consistently reported in echocardiographic and magnetic resonance imaging (MRI) studies [150]. GH and IGF-1 are known to stimulate protein synthesis in cardiomyocytes and contribute to cardiac myocyte hypertrophy during development. Their deficiency impairs myocardial growth and may lead to subclinical cardiac hypoplasia, with diminished preload and reduced myocardial compliance [48].

In addition to morphological changes, functional impairments have also been observed. Some studies have demonstrated reduced stroke volume, cardiac index, and systolic performance, particularly during exercise or stress echocardiography, indicating a decreased cardiovascular reserve [151].

Beyond myocardial structure and contractility, GH and IGF-1 also contribute to endothelial integrity and vascular tone regulation. Children with GHD often display signs of early vascular dysfunction, such as increased carotid intima–media thickness (CIMT) and reduced flow-mediated dilation, which may represent early markers of cardiovascular risk [152]. These vascular alterations are thought to be mediated by impaired nitric oxide production and increased oxidative stress in the absence of GH/IGF-1 activity [153].

Treatment with recombinant human growth hormone (rhGH) has shown favorable effects on both cardiac morphology and function. Several longitudinal studies have documented significant increases in LV mass index, improvements in ejection fraction, and normalization of diastolic parameters following GH therapy, typically within 6–12 months of initiation [148,149]. Moreover, GH replacement appears to enhance exercise capacity, likely due to improvements in myocardial contractility and oxygen delivery. These effects are most pronounced when treatment is started early in childhood, highlighting the importance of prompt diagnosis and intervention.

### 6.5. Cardiovascular Effects of Adrenal Gland Disorders

Both excessive and insufficient exposure to fetal glucocorticoids, whether due to maternal stress or antenatal corticosteroid therapy, can disrupt crucial steps in cardiac maturation. Prolonged maternal cortisol elevation induces fetal cardiomyocyte proliferation but also cardiomyocyte hypertrophy, disrupted myocardial architecture, mitochondrial dysfunction, and arrhythmias, resulting in increased perinatal mortality [44,45]. Postnatally, fetuses exposed to high levels of cortisol exhibit delayed heart rate acceleration, ECG changes, and impaired blood pressure regulation, suggesting failure to adapt to the hemodynamic stress of birth [45].

Excessive aldosterone production (hyperaldosteronism) is associated with an increased risk of atrial fibrillation (AF), as aldosterone promotes atrial enlargement and fibrosis through mechanisms independent of blood pressure [154,155]. This occurs via activation of mineralocorticoid receptors on cardiomyocytes and macrophages, increasing oxidative stress and promoting the transformation of fibroblasts into myofibroblasts [156]. Activation of the renin–angiotensin system (RAS) is also a risk factor for AF [157]. Angiotensin II induces atrial remodeling, fibrosis, and prolongation of P-wave duration, thereby increasing susceptibility to atrial fibrillation [157].

Similarly, androgens exert important regulatory effects on cardiac contractility. In patients with gonadal disorders, impaired cardiac function has been documented, with significant improvement following androgen administration. This effect has been particularly well described in Duchenne muscular dystrophy [158] but more generally applies to pediatric patients with delayed puberty, where androgen deficiency may compromise myocardial performance [159].

With regard to catecholamines, in the fetus, epinephrine and norepinephrine sustain heart rate, myocardial contractility, and vascular tone, ensuring adequate perinatal adaptation; deficiency (e.g., impaired adrenal development or defects in biosynthetic enzymes) may cause hypotension and poor cardiac output, whereas excess (e.g., maternal stress, placental dysfunction) predisposes to tachycardia, arrhythmias, and oxygen imbalance [160–162]. In pediatric age groups, catecholamine excess due to pheochromocytoma or paraganglioma leads to hypertension, ventricular hypertrophy, and arrhythmias, while insufficiency in the context of adrenal failure or iatrogenic suppression results in inadequate stress response, hypotension, and impaired cardiac performance [66,163].

## 7. Natriuretic Peptides as Clinical Biomarkers in Pediatric Disorders

### 7.1. BNP and NT-ProBNP in Pediatric Heart Failure

Heart failure (HF) in children is a condition with unique clinical and physiological characteristics, distinct from adult HF. It is defined as a clinical and pathophysiological syndrome resulting from ventricular dysfunction, volume or pressure overload, or a combination of both [164].

Pediatric HF is further complicated by its heterogenous etiology, clinical presentation, and compensatory mechanisms which are mainly age-dependent. Most importantly, unlike adults, where HF is often due to acquired myocardial disease, CHD represents the main cause of HF in children and especially in newborn patients. A pediatric classification for HF was proposed by Ross et al., which includes age-adapted parameters, such as feeding difficulties, failure to thrive, and exercise intolerance [165].

More recently, a more detailed scoring system designed for both children and adolescents was proposed by Connoly et al. [166], the New York University Pediatric Heart Failure Index. Unlike the Ross classification, the NYU-PHFI incorporates symptoms, necessity of medical therapy, and anatomical features such as the presence of single ventricle lesions.

Amid these clinical challenges, NT-proBNP has emerged as a diagnostic and prognostic biomarker. However, it is not yet included in the HF pediatric guidelines, due to the complex interpretation in neonates and infants. Indeed, one of the greatest challenges to overcome is the physiological pattern of BNP levels during the perinatal stage, with stability only weeks after birth [167].

In children with CHD, BNP and NT-proBNP have been extensively studied. NPs have been demonstrated to be reliable indicators of volume overload and their levels correlate with shunt and ventricular dilation [168].

Data also suggest a correlation between BNP levels, right ventricular dilation, and the severity of pulmonary regurgitation in patients with tetralogy of Fallot (TOF) [169]. Interestingly, in TOF, BNP levels also correlate with IP severity and RV end-diastolic volume, indicating that BNP may serve as a sensitive marker of RV stress and volume overload in certain CHD phenotypes [170].

NT-proBNP has also been investigated as a monitoring tool after surgery for CHD and, regardless of the presence of ventricular dysfunction, high blood levels of NT-proBNP were able to identify low cardiac output syndrome, prolonged mechanical ventilation, and ICU stay [171]. Additional studies have confirmed these findings [172], confirming that monitoring NT-proBNP levels in the perioperative stage can aid in early identification of complications.

Aside from CHD, most cases of HF in children are caused by cardiomyopathies. Periodic measurement of NPs is used in practice despite actual guideline recommendations. In is to be noted that, in the context of cardiomyopathies, some studies have identified that NPs are also a strong predictor of long-term outcome [173]. Also, a recent large multicenter registry analysis conducted by Schmitt et al. found that children with higher NP levels had a higher risk of death or heart transplant, with NT-proBNP showing a good performance in the stratification of high-risk patients [174].

### 7.2. Atrial Natriuretic Peptide in Heart Failure

ANP plays a critical regulatory role in HF, promoting vasodilation, natriuresis, diuresis, and inhibition of the renin–angiotensin–aldosterone system (RAAS) [175].

Although data on the role of ANP in pediatric patients are limited, existing studies consistently show that ANP levels are elevated in children with symptomatic heart failure compared to asymptomatic patients or healthy controls. Agnoletti et al. [176] reported mean plasma ANP levels of 232.5 $\pm$ 82.9 pg/mL in symptomatic children, compared to 48.4 $\pm$ 29.4 pg/mL in asymptomatic patients. This trend has also been observed in children with CHD [169,170].

Furthermore, ANP concentrations have also been shown to increase progressively with HF severity, showing consistency as a marker of severity in children with HF [177–179].

Patients diagnosed with cardiomyopathies exhibit the highest ANP levels among the reviewed studies, with median values reaching 431 pg/mL. This may reflect both increased peptide synthesis and reduced receptor sensitivity [180]. In contrast, patients with right ventricular disease show lower ANP levels, while those with congenital heart disease (CHD) display intermediate values, varying according to the anatomical defect and the significance of the shunt [181].

ANP has also attracted renewed interest in the context of pharmacological treatment of HF, specifically the use of neprilysin inhibitors such as sacubitril/valsartan. Neprilysin preferentially degrades ANP over BNP, so ANP plasma levels may provide a more dynamic reflection of therapeutic effects, an especially relevant consideration in pediatric trials such as PANORAMA-HF [182].

### 7.3. Brain Natriuretic Peptides in Pulmonary Hypertension

NPs have a central role as biomarkers in the diagnosis, risk assessment, and monitoring of pulmonary arterial hypertension. In fact, as already noted, NPs are reliable indicators of RV dysfunction, a hallmark of disease progression in PH. BNP and NT-proBNP can be used, together with other tools, to stratify disease severity and prognosis. Whereas low circulating levels are associated with low risk profile, elevated levels signal higher risk and clinical worsening [183]. BNP and NT-proBNP are also increasingly used in pediatric patients and they have been particularly investigated in NICU infants. In the work of Cuna et al., among 118 cases, BNP demonstrated a 100% sensitivity for detecting PH, thereafter confirmed by echocardiography [184]. A recent systematic review confirmed NPs have adequate diagnostic accuracy for PH and could be used to discriminate patients who should receive an echocardiogram [185]. An interesting study of Dasgupta et al. was conducted in order to evaluate NT-proBNP as a non-invasive marker for PH in children of different ages, comparing it against cardiac catheterization and echocardiography. NT-proBNP levels were significantly higher in the PH group and it correlated positively with pulmonary artery pressure and vascular resistance and negatively with pulmonary artery acceleration time and right ventricular function parameters [186].

Furthermore, many data have also established the role of BNP and NT-pro BNP as accurate markers of prognostic relevance in children. The SAUDIPH registry identified BNP as a significant marker in PH, showing that elevated BNP levels at diagnosis were independently associated with higher 1 year mortality. Therefore, data support the routine use of BNP at baseline and during follow-up to help identify high-risk patients who may benefit from intensified management strategies [187].

### 7.4. BNP in Inflammatory Syndromes

Kawasaki disease (KD) is vasculitis affecting primarily infants and is likely influenced by genetic predisposition and immune responses to infectious triggers and the most frequent cardiac complication is the development of coronary artery aneurysms (CAAs) [16]. Several studies explored BNP's predictive value for coronary lesions in KD and a recent meta-analysis supported the role of NPs as a diagnostic tool for artery lesions [188].

MIS-C is a hyperinflammatory response following mainly SARS-CoV-2 infection. While CAA is less common in MIS-C, HF is more common [52]. NPs are usually elevated in MIS-C and can assist in identifying myocardial involvement. In a large multicenter cohort, Ludwikowska et al. reported elevated NP values even in patients without reduced LVEF or echocardiographic abnormalities [18].

NT-proBNP levels are frequently elevated in pediatric patients with sepsis or septic shock, even in the absence of structural heart disease [18]. Age-adjusted NT-proBNP cut-offs have been proposed in order to improve diagnostic capacity. Sun et al. suggest a threshold of 5000 ng/L in infants and 3800 ng/L in adolescents for mortality prediction [189].

## 8. Limitations

This narrative review, while providing a broad and integrative overview of endocrine regulation during cardiac development and pediatric heart disease, has inherent limitations linked to its methodology. Narrative reviews do not follow systematic protocols for literature selection or data synthesis, which introduces potential selection bias and limits the reproducibility of the findings.

The existing literature on the endocrine regulation of cardiac development and pediatric heart disease, while rich in experimental insights, presents several notable limitations that hinder a comprehensive understanding of this complex interplay. First, most available data derive from animal models or in vitro systems, which, although informative, may not fully recapitulate the intricacies of human fetal and neonatal physiology. Translational gaps remain significant, particularly in the hormonal regulation of cardiomyocyte proliferation, maturation, and structural remodeling.

Secondly, longitudinal human studies are scarce. Most clinical data focus on postnatal hormone levels or cross-sectional assessments, limiting our ability to understand causal relationships between prenatal endocrine disruptions and later cardiovascular outcomes. Moreover, few studies have evaluated the long-term effects of altered fetal hormonal environments (e.g., maternal hypothyroidism, glucocorticoid overexposure) on cardiac function across the lifespan.

Another major limitation is the lack of standardized reference values for key endocrine biomarkers in neonates and children. Hormones such as BNP, NT-proBNP, IGF-I, and T3/T4 vary considerably with age, sex, and developmental stage, yet age-specific norms are often underreported or inconsistently applied across studies. This hampers clinical interpretation and limits the development of reliable diagnostic or prognostic tools in pediatrics.

Furthermore, many studies fail to account for biological variability, including sex differences, epigenetic regulation, and the influence of maternal–fetal factors (e.g., placental hormone metabolism, maternal diabetes, or obesity). These elements may significantly affect cardiac–endocrine interactions but are often not stratified or adequately addressed in available data. In addition, the interpretation of hormonal biomarkers in children is inherently limited by physiological variability across developmental stages. Age, sex, and pubertal status strongly influence circulating levels of TSH, IGF-I, sex steroids, and natriuretic peptides, making it difficult to establish universal thresholds. These developmental dynamics highlight the need for age- and sex-specific reference ranges when applying hormonal biomarkers in pediatric cardiovascular risk assessment.

Finally, there is limited exploration of multihormonal interactions. Most studies investigate single hormones in isolation, whereas fetal and postnatal cardiac physiology is regulated by a highly integrated hormonal network. Understanding how thyroid hormones, glucocorticoids, insulin, and growth factors interact synergistically or antagonistically remains an important but underexplored area of research.

To address these gaps, future studies should prioritize prospective, multicenter human cohorts, integrate multiomics and systems biology approaches, and aim for standardization of biomarker thresholds across developmental stages. Only through such efforts can the field advance toward personalized and preventive strategies in pediatric cardiovascular health.

## 9. Conclusions

Hormones regulate cardiac function through complex endocrine, paracrine, and autocrine pathways, influencing heart rate, myocardial contractility, vascular tone, and overall cardiac remodeling. Importantly, the heart is not only a passive target of hormonal regulation but also functions as an active endocrine organ, capable of synthesizing and secreting bioactive substances such as natriuretic peptides. These cardiokines contribute to the local and systemic modulation of cardiovascular homeostasis, thus playing a role in the autonomous regulation of cardiac function.

The synergistic interaction among hormonal signals, together with the maintenance of physiological homeostasis, is fundamental to ensure proper cardiac development and function, especially during growth. Early detection and targeted hormonal management during pregnancy are critical to mitigate these risks and support lifelong heart health. In pediatric populations, understanding the role of endocrine mechanisms in cardiac physiology is crucial, as alterations in hormonal signaling may lead to early-onset structural or functional heart abnormalities. Such dysregulations may act as silent precursors of cardiovascular disease later in life.

In clinical practice, selected hormonal biomarkers may provide valuable support for pediatric cardiovascular screening and follow-up. Table 2 summarizes the practical application of key hormonal biomarkers in pediatric cardiovascular medicine, highlighting their role in both screening (initial evaluation and risk identification) and follow-up (disease monitoring and management).

**Table 2.** Practical use of hormonal biomarkers in pediatric cardiovascular screening and follow-up.

| Biomarker | Target Population/Clinical Context | Screening (Initial Evaluation/Risk Identification) | Follow-Up (Monitoring/Management) |
|---|---|---|---|
| BNP/NT-proBNP | Children with congenital heart disease, cardiomyopathies, unexplained, heart failure (HF) symptoms | Rarely used for primary screening | Disease monitoring, perioperative assessment, HF follow-up |
| IGF-I | Children with growth hormone deficiency, intrauterine growth restriction, poor growth trajectory | Growth and cardiovascular risk assessment | Monitoring treatment response and progression |
| TSH, FT4 | Children with congenital or acquired thyroid disorders, arrhythmias, unexplained tachycardia/bradycardia, structural congenital heart disease | Routine cardiovascular risk evaluation | Monitoring disease progression or therapy |
| Cortisol | Newborns with maternal stress, antenatal corticosteroid exposure, suspected adrenal insufficiency | Evaluation when adrenal dysfunction is suspected | Monitoring if abnormalities detected or therapy started |
| Sex hormones (testosterone, estrogens) | Adolescents with delayed puberty, hypogonadism, neuromuscular diseases (e.g., Duchenne muscular dystrophy) | Evaluation of suspected pubertal or gonadal disorders | Follow-up in confirmed endocrine or cardiovascular involvement |
| Catecholamines | Children with suspected adrenal dysfunction, autonomic disorders, or tumors such as pheochromocytoma/paraganglioma | Assessment in cases of unexplained hypertension, tachycardia, or stress intolerance | Monitoring disease activity, treatment response, and recurrence in catecholamine-secreting tumors |

Gaining deeper insight into the molecular mechanisms that mediate hormone–heart crosstalk, as well as identifying age-specific patterns of cardiac endocrine response, may pave the way for the development of targeted preventive and therapeutic interventions. Ultimately, a comprehensive approach to the endocrine regulation of the heart in children will enhance early diagnosis, individualized care, and long-term cardiovascular outcomes in pediatric patients.

**Author Contributions:** Conceptualization, V.C., S.M. and G.Z.; methodology, V.C., S.M., F.P., G.F., G.C., M.E., B.B., C.G. and G.Z.; investigation, F.P., G.F., G.C., M.E., B.B. and C.G.; writing—original draft preparation, V.C., F.P., G.F., G.C., M.E., B.B. and C.G.; writing—review and editing, V.C. and S.M.; supervision, V.C., S.M. and G.Z. All authors have read and agreed to the published version of the manuscript.

**Funding:** This research received contribution from (1) the National Recovery and Resilience Plan (NRRP), Mission 4 Component 2 Investment 1.3—Call for tender No. 341 of 15 March 2022 of the Italian Ministry of University and Research funded by the European Union—NextGenerationEU. Award Number: Project code PE00000003, Concession Decree No. 1550 of 11 October 2022 adopted by the Italian Ministry of University and Research, CUP F13C22001210007; (2) PODiaCar Project 101128946-PODiaCar-EU4H-2022-PJ-3 co-funded by European Union.

**Institutional Review Board Statement:** Not applicable.

**Informed Consent Statement:** Not applicable.

**Data Availability Statement:** Not applicable.

**Conflicts of Interest:** The authors declare no conflicts of interest.

## References


1. Zaffran, S.; Frasch, M. Early signals in cardiac development. *Circ. Res.* **2002**, *91*, 457–469. [CrossRef]
2. Srivastava, D.; Olson, E.N. A genetic blueprint for cardiac development. *Nature* **2000**, *407*, 221–226. [CrossRef]
3. Buckingham, M.; Meilhac, S.; Zaffran, S. Building the mammalian heart from two sources of myocardial cells. *Nat. Rev. Genet.* **2005**, *6*, 826–835. [CrossRef]
4. Ventura-Clapier, R.; Piquereau, J.; Garnier, A.; Mericskay, M.; Lemaire, C.; Crozatier, B. Gender issues in cardiovascular diseases. Focus on energy metabolism. *Biochim. Biophys. Acta Mol. Basis Dis.* **2020**, *1866*, 165722. [CrossRef] [PubMed]
5. Chattergoon, N.N. Thyroid hormone signaling and consequences for cardiac development. *J. Endocrinol.* **2019**, *242*, T145–T160. [CrossRef] [PubMed]
6. Graham, N.; Huang, G.N. Endocrine Influence on Cardiac Metabolism in Development and Regeneration. *Endocrinology* **2021**, *162*, bqab081. [CrossRef] [PubMed]
7. Tan, L.; Bogush, N.; Naib, H.; Perry, J.; Calvert, J.W.; Martin, D.I.K.; Graham, R.M.; Naqvi, N.; Husain, A. Redox activation of JNK2$\alpha$2 mediates thyroid hormone-stimulated proliferation of neonatal murine cardiomyocytes. *Sci. Rep.* **2019**, *9*, 17731. [CrossRef]
8. Barjaktarovic, M.; Korevaar, T.I.M.; Gaillard, R.; de Rijke, Y.B.; Visser, T.J.; Jaddoe, V.W.V.; Peeters, R.P. Childhood thyroid function, body composition and cardiovascular function. *Eur. J. Endocrinol.* **2017**, *177*, 319–327. [CrossRef]
9. Burford, N.G.; Webster, N.A.; Cruz-Topete, D. Hypothalamic-Pituitary-Adrenal Axis Modulation of Glucocorticoids in the Cardiovascular System. *Int. J. Mol. Sci.* **2017**, *18*, 2150. [CrossRef]
10. Wang, K.C.W.; Zhang, L.; McMillen, I.C.; Botting, K.J.; Duffield, J.A.; Zhang, S.; Suter, C.M.; Brooks, D.A.; Morrison, J.L. Fetal growth restriction and the programming of heart growth and cardiac insulin-like growth factor 2 expression in the lamb. *J. Physiol.* **2011**, *589 Pt 19*, 4709–4722. [CrossRef]
11. Vetter, U.; Kupferschmid, C.; Lang, D.; Pentz, S. Insulin-like growth factors and insulin increase the contractility of neonatal rat cardiocytes in vitro. *Basic Res. Cardiol.* **1988**, *83*, 647–654. [CrossRef] [PubMed]
12. Sokou, R.; Lianou, A.; Lampridou, M.; Panagiotounakou, P.; Kafalidis, G.; Paliatsiou, S.; Volaki, P.; Tsantes, A.G.; Boutsikou, T.; Iliodromiti, Z.; et al. Neonates at Risk: Understanding the Impact of High-Risk Pregnancies on Neonatal Health. *Medicina* **2025**, *61*, 1077. [CrossRef] [PubMed]
13. Liang, J.; He, X.; Wang, Y. Cardiomyocyte proliferation and regeneration in congenital heart disease. *Pediatr. Discov.* **2024**, *2*, e2501. [CrossRef] [PubMed]
14. Upadhyay, R.K. Emerging risk biomarkers in cardiovascular diseases and disorders. *J. Lipids.* **2015**, *2015*, 971453. [CrossRef] [PubMed] [PubMed Central]
15. Holditch, S.J.; Schreiber, C.A.; Nini, R.; Tonne, J.M.; Peng, K.-W.; Geurts, A.; Jacob, H.J.; Burnett, J.C.; Cataliotti, A.; Ikeda, Y. B-Type Natriuretic Peptide Deletion Leads to Progressive Hypertension, Associated Organ Damage, and Reduced Survival: Novel Model for Human Hypertension. *Hypertension* **2015**, *66*, 199–210. [CrossRef]
16. McCrindle, B.W.; Rowley, A.H.; Newburger, J.W.; Burns, J.C.; Bolger, A.F.; Gewitz, M.; Baker, A.L.; Jackson, M.A.; Takahashi, M.; Shah, P.B.; et al. Diagnosis, Treatment, and Long-Term Management of Kawasaki Disease: A Scientific Statement for Health Professionals from the American Heart Association. *Circulation* **2017**, *135*, e927–e999. [CrossRef]
17. Goetze, J.P.; Bruneau, B.G.; Ramos, H.R.; Ogawa, T.; de Bold, M.K.; de Bold, A.J. Cardiac natriuretic peptides. *Nat. Rev. Cardiol.* **2020**, *17*, 698–717. [CrossRef]
18. Ludwikowska, K.M.; Tokarczyk, M.; Paleczny, B.; Tracewski, P.; Szenborn, L.; Kusa, J. Clinical Significance of B-Type Natriuretic Peptide and N-Terminal Pro-B-Type Natriuretic Peptide in Pediatric Patients: Insights into Their Utility in the Presence or Absence of Pre-Existing Heart Conditions. *Int. J. Mol. Sci.* **2024**, *25*, 8781. [CrossRef]
19. Kloesel, B.; DiNardo, J.A.; Body, S.C. Cardiac Embryology and Molecular Mechanisms of Congenital Heart Disease: A Primer for Anesthesiologists. *Anesth. Analg.* **2016**, *123*, 551–569. [CrossRef]
20. Li, Y.; Du, J.; Deng, S.; Liu, B.; Jing, X.; Yan, Y.; Liu, Y.; Wang, J.; Zhou, X.; She, Q. The molecular mechanisms of cardiac development and related diseases. *Signal Transduct. Target. Ther.* **2024**, *9*, 368. [CrossRef]
21. Sun, H.; Saeedi, P.; Karuranga, S.; Pinkepank, M.; Ogurtsova, K.; Duncan, B.B.; Stein, C.; Basit, A.; Chan, J.C.; Mbanya, J.C.; et al. IDF Diabetes Atlas: Global, regional and country-level diabetes prevalence estimates for 2021 and projections for 2045. *Diabetes Res. Clin. Pract.* **2022**, *183*, 109119, Erratum in *Diabetes Res. Clin. Pract.* **2023**, *204*, 110945. https://doi.org/10.1016/j.diabres.2023.110945. [CrossRef] [PubMed] [PubMed Central]
22. Katano, W.; Moriyama, Y.; Takeuchi, J.K.; Koshiba-Takeuchi, K. Cardiac septation in heart development and evolution. *Dev. Growth Differ.* **2019**, *61*, 114–123. [CrossRef]

23. Galow, A.M.; Brenmoehl, J.; Hoeflich, A. Synergistic effects of hormones on structural and functional maturation of cardiomyocytes and implications for heart regeneration. *Cell. Mol. Life Sci.* **2023**, *80*, 240. [CrossRef]

24. Bianco, A.C.; Dumitrescu, A.; Gereben, B.; O Ribeiro, M.; Fonseca, T.L.; Fernandes, G.W.; Bocco, B.M.L.C. Paradigms of Dynamic Control of Thyroid Hormone Signaling. *Endocr. Rev.* **2019**, *40*, 1000–1047. [CrossRef]
25. Amanollahi, R.; Holman, S.L.; Bertossa, M.R.; Meakin, A.S.; Thornburg, K.L.; McMillen, I.C.; Wiese, M.D.; Lock, M.C.; Morrison, J.L. Ontogeny of Fetal Cardiometabolic Pathways: The Potential Role of Cortisol and Thyroid Hormones in Driving the Transition from Preterm to Near-Term Heart Development in Sheep. *J. Cardiovasc. Dev. Dis.* **2025**, *12*, 36. [CrossRef]
26. CCruz-Topete, D.; Oakley, R.H.; Cidlowski, J.A. Glucocorticoid Signaling and the Aging Heart. *Front. Endocrinol.* **2020**, *11*, 347. [CrossRef]
27. Song, R.; Hu, X.-Q.; Zhang, L. Glucocorticoids and programming of the microenvironment in heart. *J. Endocrinol.* **2019**, *242*, T121–T133. [CrossRef]
28. De Bosscher, K.; Beck, I.M.; Dejager, L.; Bougarne, N.; Gaigneaux, A.; Chateauvieux, S.; Ratman, D.; Bracke, M.; Tavernier, J.; Berghe, W.V.; et al. Selective modulation of the glucocorticoid receptor can distinguish between transrepression of NF-κB and AP-1. *Cell. Mol. Life Sci.* **2014**, *71*, 143–163. [CrossRef]
29. Abel, E.D. Insulin signaling in the heart. American journal of physiology. *Endocrinol. Metab.* **2021**, *321*, E130–E145. [CrossRef]
30. Ayrout, M.; Simon, V.; Bernard, V.; Binart, N.; Cohen-Tannoudji, J.; Lombès, M.; Chauvin, S. A novel non genomic glucocorticoid signaling mediated by a membrane palmitoylated glucocorticoid receptor cross talks with GnRH in gonadotrope cells. *Sci. Rep.* **2017**, *7*, 1537. [CrossRef]
31. Knowlton, A.A.; Lee, A.R. Estrogen and the cardiovascular system. *Pharmacol. Ther.* **2012**, *135*, 54–70. [CrossRef]
32. Mattingly, K.A.; Ivanova, M.M.; Riggs, K.A.; Wickramasinghe, N.S.; Barch, M.J.; Klinge, C.M. Estradiol Stimulates Transcription of Nuclear Respiratory Factor-1 and Increases Mitochondrial Biogenesis. *Mol. Endocrinol.* **2008**, *22*, 609–622. [CrossRef]
33. Ventura-Clapier, R.; Moulin, M.; Piquereau, J.; Lemaire, C.; Mericskay, M.; Veksler, V.; Garnier, A. Mitochondria: A central target for sex differences in pathologies. *Clin. Sci.* **2017**, *131*, 803–822. [CrossRef]
34. El Ghorayeb, N.; Bourdeau, I.; Lacroix, A. Role of ACTH and Other Hormones in the Regulation of Aldosterone Production in Primary Aldosteronism. *Front. Endocrinol.* **2016**, *7*, 72. [CrossRef]
35. Epstein, M. Aldosterone and mineralocorticoid receptor signaling as determinants of cardiovascular and renal injury: An extraordinary paradigm shift. *Kidney Int. Suppl.* **2022**, *12*, 1–6. [CrossRef]
36. Yamashita, K.; Yamatabe, T.; Abe, H.; Yokoyama, M. Influence of systemic vascular resistance on cardiac output measured by new non-invasive cardiac output monitor. *Crit. Care* **2010**, *14* (Suppl. S1), P104. [CrossRef]
37. Casey, M.L.; Howell, M.L.; Winkel, C.A.; Simpson, E.R.; Macdonald, P.C. Deoxycorticosterone Sulfate Biosynthesis in Human Fetal Kidney. *J. Clin. Endocrinol. Metab.* **1981**, *53*, 990–996. [CrossRef]
38. Comline, R.S.; Silver, M. The development of the adrenal medulla of the foetal and new-born calf. *J. Physiol.* **1966**, *183*, 305–340. [CrossRef]
39. Dongdem, J.T.; Etornam, A.E.; Beletaa, S.; Alidu, I.; Kotey, H.; Wezena, C.A. The $\beta_3$-Adrenergic Receptor: Structure, Physiopathology of Disease, and Emerging Therapeutic Potential. *Adv. Pharmacol. Pharm. Sci.* **2024**, *2024*, 2005589. [CrossRef]
40. Lymperopoulos, A.; Rengo, G.; Koch, W.J. Adrenergic nervous system in heart failure: Pathophysiology and therapy. *Circ. Res.* **2013**, *113*, 739–753. [CrossRef]
41. Bode, C.; Preissl, S.; Hein, L.; Lother, A. Catecholamine treatment induces reversible heart injury and cardiomyocyte gene expression. *Intensive Care Med. Exp.* **2024**, *12*, 48. [CrossRef]
42. Hooper, S.B.; Te Pas, A.B.; Lang, J.; van Vonderen, J.J.; Roehr, C.C.; Kluckow, M.; Gill, A.W.; Wallace, E.M.; Polglase, G.R. Cardiovascular transition at birth: A physiological sequence. *Pediatr. Res.* **2015**, *77*, 608–614. [CrossRef] [PubMed]
43. Short, B.L.; Van Meurs, K.; Evans, J.R.; Cardiology Group. Summary proceedings from the cardiology group on cardiovascular instability in preterm infants. *Pediatrics* **2006**, *117 Pt 2*, S34–S39. [CrossRef] [PubMed]
44. Antolic, A.; Wood, C.E.; Keller-Wood, M. Chronic maternal hypercortisolemia in late gestation alters fetal cardiac function at birth. *Am. J. Physiol. Regul. Integr. Comp. Physiol.* **2018**, *314*, R342–R352. [CrossRef] [PubMed]
45. Antolic, A.; Li, M.; Richards, E.M.; Curtis, C.W.; Wood, C.E.; Keller-Wood, M. Mechanisms of in utero cortisol effects on the newborn heart revealed by transcriptomic modeling. *Am. J. Physiol. Regul. Integr. Comp. Physiol.* **2019**, *316*, R323–R337. [CrossRef]
46. Troncoso, R.; Ibarra, C.; Vicencio, J.M.; Jaimovich, E.; Lavandero, S. New insights into IGF-1 signaling in the heart. *Trends Endocrinol. Metab.* **2014**, *25*, 128–137. [CrossRef]
47. Caicedo, D.; Díaz, O.; Devesa, P.; Devesa, J. Growth Hormone (GH) and Cardiovascular System. *Int. J. Mol. Sci.* **2018**, *19*, 290. [CrossRef]
48. Isgaard, J.; Arcopinto, M.; Karason, K.; Cittadini, A. GH and the cardiovascular system: An update on a topic at heart. *Endocrine* **2015**, *48*, 25–35. [CrossRef]
49. Giammanco, M.; Di Liegro, C.M.; Schiera, G.; Di Liegro, I. Genomic and Non-Genomic Mechanisms of Action of Thyroid Hormones and Their Catabolite 3,5-Diiodo-L-Thyronine in Mammals. *Int. J. Mol. Sci.* **2020**, *21*, 4140. [CrossRef] [PubMed] [PubMed Central]

50. Yamakawa, H.; Kato, T.S.; Noh, J.Y.; Yuasa, S.; Kawamura, A.; Fukuda, K.; Aizawa, Y. Thyroid Hormone Plays an Important Role in Cardiac Function: From Bench to Bedside. *Front. Physiol.* **2021**, *12*, 606931. [CrossRef]
51. Carrillo, E.D.; Alvarado, J.A.; Hernández, A.; Lezama, I.; García, M.C.; Sánchez, J.A. Thyroid Hormone Upregulates Cav1.2 Channels in Cardiac Cells via the Downregulation of the Channels' β4 Subunit. *Int. J. Mol. Sci.* **2024**, *25*, 10798. [CrossRef]
52. Belhadjer, Z.; Méot, M.; Bajolle, F.; Khraiche, D.; Legendre, A.; Abakka, S.; Auriau, J.; Grimaud, M.; Oualha, M.; Beghetti, M.; et al. Acute Heart Failure in Multisystem Inflammatory Syndrome in Children in the Context of Global SARS-CoV-2 Pandemic. *Circulation* **2020**, *142*, 429–436. [CrossRef]
53. Soetedjo, N.N.M.; Agustini, D.; Permana, H. The impact of thyroid disorder on cardiovascular disease: Unraveling the connection and implications for patient care. *IJC Heart Vasc.* **2024**, *55*, 101536. [CrossRef]
54. Banu Rupani, N.; Alijanpour, M.; Babazadeh, K.; Hajian-Tilaki, K.; Moadabdoost, F. Effect of levothyroxine on cardiac function in children with subclinical hypothyroidism: A quasi-experimental study. *Casp. J. Intern. Med.* **2019**, *10*, 332–338. [CrossRef]
55. Haas, N.A.; Camphausen, C.K.; Kececioglu, D. Clinical review: Thyroid hormone replacement in children after cardiac surgery—Is it worth a try? *Crit. Care* **2006**, *10*, 213. [CrossRef]
56. Flores, S.; Loomba, R.S.; Checchia, P.A.; Graham, E.M.; Bronicki, R.A. Thyroid Hormone (Triiodothyronine) Therapy in Children After Congenital Heart Surgery: A Meta-Analysis. *Semin. Thorac. Cardiovasc. Surg.* **2020**, *32*, 87–95. [CrossRef] [PubMed]
57. Radman, M.; Portman, M.A. Thyroid Hormone in the Pediatric Intensive Care Unit. *J. Pediatr. Intensive Care* **2016**, *5*, 154–161. [CrossRef] [PubMed]
58. Sahoo, B.; Dabas, A.; Goswami, B.; Agarwal, A.; Kurian, S. Non-thyroidal Illness in Children with Congestive Heart Failure. *J. Clin. Res. Pediatr. Endocrinol.* **2024**, *16*, 192–199. [CrossRef] [PubMed]
59. Takefuji, M.; Murohara, T. Corticotropin-Releasing Hormone Family and Their Receptors in the Cardiovascular System. *Circ. J.* **2019**, *83*, 261–266. [CrossRef]
60. Oakley, R.H.; Cidlowski, J.A. Glucocorticoid signaling in the heart: A cardiomyocyte perspective. *J. Steroid Biochem. Mol. Biol.* **2015**, *153*, 27–34. [CrossRef]
61. Omori, Y.; Mano, T.; Ohtani, T.; Sakata, Y.; Takeda, Y.; Tamaki, S.; Tsukamoto, Y.; Miwa, T.; Yamamoto, K.; Komuro, I. Glucocorticoids Induce Cardiac Fibrosis via Mineralocorticoid Receptor in Oxidative Stress: Contribution of Elongation Factor Eleven-Nineteen Lysine-Rich Leukemia (ELL). *Yonago Acta Medica* **2014**, *57*, 109–116.
62. Coutinho, A.E.; Chapman, K.E. The anti-inflammatory and immunosuppressive effects of glucocorticoids, recent developments and mechanistic insights. *Mol. Cell. Endocrinol.* **2011**, *335*, 2–13. [CrossRef]
63. Motiejunaite, J.; Amar, L.; Vidal-Petiot, E. Adrenergic receptors and cardiovascular effects of catecholamines. *Ann. D'endocrinologie* **2021**, *82*, 193–197. [CrossRef]
64. Parichatikanond, W.; Duangrat, R.; Kurose, H.; Mangmool, S. Regulation of β-Adrenergic Receptors in the Heart: A Review on Emerging Therapeutic Strategies for Heart Failure. *Cells* **2024**, *13*, 1674. [CrossRef] [PubMed]
65. Varghese, R.T.; John, A.M.; Paul, T.V. Catecholamine induced cardiomyopathy in pheochromocytoma. *Indian J. Endocrinol. Metab.* **2013**, *17*, 733–735. [CrossRef] [PubMed]
66. Y-Hassan, S.; Falhammar, H. Cardiovascular Manifestations and Complications of Pheochromocytomas and Paragangliomas. *J. Clin. Med.* **2020**, *9*, 2435. [CrossRef] [PubMed]
67. Prigent, H.; Maxime, V.; Annane, D. Science review: Mechanisms of impaired adrenal function in sepsis and molecular actions of glucocorticoids. *Crit. Care* **2004**, *8*, 243–252. [CrossRef]
68. Ye, W.; Han, K.; Xie, M.; Li, S.; Chen, G.; Wang, Y.; Li, T. Mitochondrial energy metabolism in diabetic cardiomyopathy: Physiological adaption, pathogenesis, and therapeutic targets. *Chin. Med. J.* **2024**, *137*, 936–948. [CrossRef]
69. Lin, J.; Duan, J.; Wang, Q.; Xu, S.; Zhou, S.; Yao, K. Mitochondrial Dynamics and Mitophagy in Cardiometabolic Disease. *Front. Cardiovasc. Med.* **2022**, *9*, 917135. [CrossRef]
70. DeBosch, B.J.; Muslin, A.J. Insulin signaling pathways and cardiac growth. *J. Mol. Cell. Cardiol.* **2008**, *44*, 855–864. [CrossRef]
71. Barakat, S.; McLean, S.A.; Bryant, E.; Le, A.; Marks, P.; National Eating Disorder Research Consortium; Touyz, S.; Maguire, S. Risk factors for eating disorders: Findings from a rapid review. *J. Eat. Disord.* **2023**, *11*, 8. [CrossRef]
72. Clerk, A.; Sugden, P.H. The insulin receptor family in the heart: New light on old insights. *Biosci. Rep.* **2022**, *42*, BSR20221212. [CrossRef]
73. Shao, D.; Tian, R. Glucose Transporters in Cardiac Metabolism and Hypertrophy. *Compr. Physiol.* **2015**, *6*, 331–351. [CrossRef]
74. Caturano, A.; Galiero, R.; Vetrano, E.; Sardu, C.; Rinaldi, L.; Russo, V.; Monda, M.; Marfella, R.; Sasso, F.C. Insulin-Heart Axis: Bridging Physiology to Insulin Resistance. *Int. J. Mol. Sci.* **2024**, *25*, 8369. [CrossRef] [PubMed]
75. Libianto, R.; Batu, D.; MacIsaac, R.J.; Cooper, M.E.; Ekinci, E.I. Pathophysiological Links Between Diabetes and Blood Pressure. *Can. J. Cardiol.* **2018**, *34*, 585–594. [CrossRef] [PubMed]
76. Ames, M.K.; Atkins, C.E.; Pitt, B. The renin-angiotensin-aldosterone system and its suppression. *J. Vet. Intern. Med.* **2019**, *33*, 363–382. [CrossRef] [PubMed]

77. Geraldes, P.; King, G.L. Activation of protein kinase C isoforms and its impact on diabetic complications. *Circ. Res.* **2010**, *106*, 1319–1331. [CrossRef]
78. Sarzani, R.; Allevi, M.; Di Pentima, C.; Schiavi, P.; Spannella, F.; Giulietti, F. Role of Cardiac Natriuretic Peptides in Heart Structure and Function. *Int. J. Mol. Sci.* **2022**, *23*, 14415. [CrossRef]
79. de Bold, A.; Borenstein, H.; Veress, A.; Sonnenberg, H. A rapid and potent natriuretic response to intravenous injection of atrial myocardial extract in rats. *Life Sci.* **1981**, *28*, 89–94. [CrossRef]
80. Ogawa, T.; de Bold, A.J. Uncoordinated regulation of atrial natriuretic factor and brain natriuretic peptide in lipopolysaccharide-treated rats. *Biomarkers Biochem. Indic. Expo. Response Susceptibility Chem.* **2012**, *17*, 140–149. [CrossRef]
81. Murakami, Y.; Shimada, T.; Inoue, S.; Shimizu, H.; Ohta, Y.; Katoh, H.; Nakamura, K.; Ishibashi, Y. New insights into the mechanism of the elevation of plasma brain natriuretic polypeptide levels in patients with left ventricular hypertrophy. *Can. J. Cardiol.* **2002**, *18*, 1294–1300. [PubMed]
82. Bogush, N.; Tan, L.; Naib, H.; Faizullabhoy, E.; Calvert, J.W.; Iismaa, S.E.; Gupta, A.; Ramchandran, R.; Martin, D.I.K.; Graham, R.M.; et al. DUSP5 expression in left ventricular cardiomyocytes of young hearts regulates thyroid hormone (T3)-induced proliferative ERK1/2 signaling. *Sci. Rep.* **2020**, *10*, 21918. [CrossRef] [PubMed]
83. D'uVa, G.; Aharonov, A.; Lauriola, M.; Kain, D.; Yahalom-Ronen, Y.; Carvalho, S.; Weisinger, K.; Bassat, E.; Rajchman, D.; Yifa, O.; et al. ERBB2 triggers mammalian heart regeneration by promoting cardiomyocyte dedifferentiation and proliferation. *Nat. Cell Biol.* **2015**, *17*, 627–638. [CrossRef] [PubMed]
84. Glezeva, N.; Collier, P.; Voon, V.; Ledwidge, M.; McDonald, K.; Watson, C.; Baugh, J. Attenuation of monocyte chemotaxis--a novel anti-inflammatory mechanism of action for the cardio-protective hormone B-type natriuretic peptide. *J. Cardiovasc. Transl. Res.* **2013**, *6*, 545–557. [CrossRef]
85. Huang, L.; Huang, L.; Yu, J.; Wu, X.; Zhao, J. An association between N-terminal pro-brain natriuretic protein level and risk of left ventricular hypertrophy in patients without heart failure. *Exp. Ther. Med.* **2020**, *19*, 3259–3266. [CrossRef]
86. Rubattu, S.; Bigatti, G.; Evangelista, A.; Lanzani, C.; Stanzione, R.; Zagato, L.; Manunta, P.; Marchitti, S.; Venturelli, V.; Bianchi, G.; et al. Association of atrial natriuretic peptide and type a natriuretic peptide receptor gene polymorphisms with left ventricular mass in human essential hypertension. *J. Am. Coll. Cardiol.* **2006**, *48*, 499–505. [CrossRef]
87. Pye, M.P.; Cobbe, S.M. Mechanisms of ventricular arrhythmias in cardiac failure and hypertrophy. *Cardiovasc. Res.* **1992**, *26*, 740–750. [CrossRef]
88. Wang, T.J.; Larson, M.G.; Levy, D.; Benjamin, E.J.; Leip, E.P.; Omland, T.; Wolf, P.A.; Vasan, R.S. Plasma natriuretic peptide levels and the risk of cardiovascular events and death. *N. Engl. J. Med.* **2004**, *350*, 655–663. [CrossRef]
89. Patton, K.K.; Ellinor, P.T.; Heckbert, S.R.; Christenson, R.H.; DeFilippi, C.; Gottdiener, J.S.; Kronmal, R.A. N-terminal pro-B-type natriuretic peptide is a major predictor of the development of atrial fibrillation: The Cardiovascular Health Study. *Circulation* **2009**, *120*, 1768–1774. [CrossRef]
90. Yamauchi, T.; Sakata, Y.; Miura, M.; Onose, T.; Tsuji, K.; Abe, R.; Oikawa, T.; Kasahara, S.; Sato, M.; Nochioka, K.; et al. Prognostic Impact of Atrial Fibrillation and New Risk Score of Its Onset in Patients at High Risk of Heart Failure—A Report From the CHART-2 Study. *Circ. J. Off. J. Jpn. Circ. Soc.* **2017**, *81*, 185–194. [CrossRef]
91. Potter, L.R.; Yoder, A.R.; Flora, D.R.; Antos, L.K.; Dickey, D.M. Natriuretic peptides: Their structures, receptors, physiologic functions and therapeutic applications. In *cGMP: Generators, Effectors and Therapeutic Implications*; Handbook of Experimental Pharmacology; Springer: Berlin/Heidelberg, Germany, 2009; pp. 341–366. [CrossRef]
92. Scott, P.A.; Barry, J.; Roberts, P.R.; Morgan, J.M. Brain natriuretic peptide for the prediction of sudden cardiac death and ventricular arrhythmias: A meta-analysis. *Eur. J. Heart Fail.* **2009**, *11*, 958–966. [CrossRef] [PubMed]
93. McGrath, M.F.; de Bold, M.L.; de Bold, A.J. The endocrine function of the heart. *Trends Endocrinol. Metab. TEM* **2005**, *16*, 469–477. [CrossRef] [PubMed]
94. Okamoto, R.; Ali, Y.; Hashizume, R.; Suzuki, N.; Ito, M. BNP as a Major Player in the Heart-Kidney Connection. *Int. J. Mol. Sci.* **2019**, *20*, 3581. [CrossRef] [PubMed]
95. Luchner, A.; Schunkert, H. Interactions between the sympathetic nervous system and the cardiac natriuretic peptide system. *Cardiovasc. Res.* **2004**, *63*, 443–449. [CrossRef]
96. Maack, T. Role of atrial natriuretic factor in volume control. *Kidney Int.* **1996**, *49*, 1732–1737. [CrossRef]
97. Heinl, E.S.; Broeker, K.A.; Lehrmann, C.; Heydn, R.; Krieger, K.; Ortmaier, K.; Tauber, P.; Schweda, F. Localization of natriuretic peptide receptors A, B, and C in healthy and diseased mouse kidneys. *Pflug. Arch. Eur. J. Physiol.* **2023**, *475*, 343–360. [CrossRef]
98. Jensen, B.L. Natriuretic peptide receptors in the kidneys are predominantly vascular and not tubular. *Pflug. Arch. Eur. J. Physiol.* **2023**, *475*, 289–290. [CrossRef]
99. Barrett, C.J.; Schultz, H.D. Sympathoinhibitory effects of atrial natriuretic peptide in rats with heart failure. *J. Card. Fail.* **1999**, *5*, 316–323. [CrossRef]
100. Oparil, S.; Chen, Y.F.; Wang, R.P. Atrial natriuretic peptide inhibits sympathetic outflow in NaCl-sensitive spontaneously hypertensive rats. *J. Hypertens.* **1991**, *9*, 1177–1185.

101. Basu, M.; Garg, V. Maternal hyperglycemia and fetal cardiac development: Clinical impact and underlying mechanisms. *Birth Defects Res.* **2018**, *110*, 1504–1516. [CrossRef] [PubMed] [PubMed Central]
102. Rowland, T.W.; Hubbell, J.P., Jr.; Nadas, A.S. Congenital heart disease in infants of diabetic mothers. *J. Pediatr.* **1973**, *83*, 815–820. [CrossRef]
103. Liu, S.; Joseph, K.S.; Lisonkova, S.; Rouleau, J.; Van den Hof, M.; Sauve, R.; Kramer, M.S.; Canadian Perinatal Surveillance System (Public Health Agency of Canada). Association between maternal chronic conditions and congenital heart defects: A population-based cohort study. *Circulation* **2013**, *128*, 583–589. [CrossRef]
104. Gillette, P.C. Genetic and environmental risk factors of major cardiovascular malformations: The baltimore–washington infant study: 1981–1989 edited by Charlotte Ferencz, Adolfo Correa–Villasenor, Christopher A. Loffredo, P. David Wilson, Futura Publishing Company, Inc., Armonk, N.Y. (1998) 463 pages, illustrated, $95.00 ISBN: 1044–4157. *Clin. Cardiol.* **2009**, *21*, 867–868. [CrossRef] [PubMed Central]
105. Loffredo, C.A.; Wilson, P.D.; Ferencz, C. Maternal diabetes: An independent risk factor for major cardiovascular malformations with increased mortality of affected infants. *Teratology* **2001**, *64*, 98–106. [CrossRef] [PubMed]
106. Ghouse, J.; Hansson, M.; Vøgg, R.O.B.; Sillesen, A.S.; Pærregaard, S.; Raja, A.A.; Vejlstrup, N.; Frikke-Schmidt, R.; Øyen, N.; Kulkarni, A.; et al. Maternal Diabetes and Cardiac Left Ventricular Structure and Function in the Infant: A Copenhagen Baby Heart Study. *Diabetes Care* **2024**, *47*, 2230–2238. [CrossRef] [PubMed]
107. Gireadă, R.; Socolov, D.; Mihălceanu, E.; Lazăr, I.T.; Luca, A.; Matasariu, R.; Ursache, A.; Bujor, I.; Gireadă, T.; Boiculese, V.L.; et al. Evaluation of Fetal Cardiac Geometry and Contractility in Gestational Diabetes Mellitus by Two-Dimensional Speckle-Tracking Technology. *Diagnostics* **2022**, *12*, 2053. [CrossRef] [PubMed] [PubMed Central]
108. Marchincin, S.L.; Howley, M.M.; Van Zutphen, A.R.; Fisher, S.C.; Nestoridi, E.; Tinker, S.C.; Browne, M.L.; National Birth Defects. Prevention Study Risk of birth defects by pregestational type 1 or type 2 diabetes: National Birth Defects Prevention Study, 1997–2011. *Birth Defects Res.* **2023**, *115*, 56–66. [CrossRef]
109. Wren, C.; Birrell, G.; Hawthorne, G. Cardiovascular malformations in infants of diabetic mothers. *Heart* **2003**, *89*, 1217–1220. [CrossRef]
110. Simeone, R.M.; Devine, O.J.; Marcinkevage, J.A.; Gilboa, S.M.; Razzaghi, H.; Bardenheier, B.H.; Sharma, A.J.; Honein, M.A. Diabetes and congenital heart defects: A systematic review, meta-analysis, and modeling project. *Am. J. Prev. Med.* **2015**, *48*, 195–204. [CrossRef]
111. Øyen, N.; Diaz, L.J.; Leirgul, E.; Boyd, H.A.; Priest, J.; Mathiesen, E.R.; Quertermous, T.; Wohlfahrt, J.; Melbye, M. Prepregnancy Diabetes and Offspring Risk of Congenital Heart Disease: A Nationwide Cohort Study. *Circulation* **2016**, *133*, 2243–2253. [CrossRef]
112. Ramos-Arroyo, M.A.; Rodriguez-Pinilla, E.; Cordero, J.F. Maternal diabetes: The risk for specific birth defects. *Eur. J. Epidemiol.* **1992**, *8*, 503–508. [CrossRef]
113. Gözüküçük, D.; İleri, B.A.; Başkan, S.K.; Öztarhan, E.; Güller, D.; Önal, H.; Öztarhan, K. Evaluation of cardiac autonomic dysfunctions in children with type 1 diabetes mellitus. *BMC Pediatr.* **2024**, *24*, 229. [CrossRef]
114. Bjornstad, P.; Truong, U.; Pyle, L.; Dorosz, J.L.; Cree-Green, M.; Baumgartner, A.; Coe, G.; Regensteiner, J.G.; Reusch, J.E.; Nadeau, K.J. Youth with type 1 diabetes have worse strain and less pronounced sex differences in early echocardiographic markers of diabetic cardiomyopathy compared to their normoglycemic peers: A RESistance to InSulin in Type 1 ANd Type 2 diabetes (RESISTANT) Study. *J. Diabetes Its Complicat.* **2016**, *30*, 1103–1110. [CrossRef]
115. Ng, A.C.; Auger, D.; Delgado, V.; van Elderen, S.G.; Bertini, M.; Siebelink, H.M.; van der Geest, R.J.; Bonetti, C.; van der Velde, E.T.; de Roos, A.; et al. Association between diffuse myocardial fibrosis by cardiac magnetic resonance contrast-enhanced $T_1$ mapping and subclinical myocardial dysfunction in diabetic patients: A pilot study. *Circ. Cardiovasc. Imaging* **2012**, *5*, 51–59. [CrossRef] [PubMed]
116. TODAY Study Group. Longitudinal Changes in Cardiac Structure and Function from Adolescence to Young Adulthood in Participants With Type 2 Diabetes Mellitus: The TODAY Follow-Up Study. *Circ. Heart Fail.* **2020**, *13*, e006685. [CrossRef] [PubMed]
117. Shah, A.S.; Isom, S.; Dabelea, D.; D'Agostino, R.; Dolan, L.M., Jr.; Wagenknecht, L.; Imperatore, G.; Saydah, S.; Liese, A.D.; Lawrence, J.M.; et al. A cross sectional study to compare cardiac structure and diastolic function in adolescents and young adults with youth-onset type 1 and type 2 diabetes: The SEARCH for Diabetes in Youth Study. *Cardiovasc. Diabetol.* **2021**, *20*, 136. [CrossRef] [PubMed]
118. Galli-Tsinopoulou, A.; Chatzidimitriou, A.; Kyrgios, I.; Rousso, I.; Varlamis, G.; Karavanaki, K. Children and adolescents with type 1 diabetes mellitus have a sixfold greater risk for prolonged QTc interval. *J. Pediatr. Endocrinol. Metab.* **2014**, *27*, 237–243. [CrossRef]
119. Aygün, D.; Aygün, F.; Nişli, K.; Baş, F.; Çıtak, A. Electrocardiographic changes in children with diabetic ketoacidosis and ketosis. *Turk Pediatri Ars.* **2017**, *52*, 194–201. [CrossRef]

120. Ertaş, K.; Gül, Ö.; Yıldırım, R.; Özalkak, Ş. Evaluation of cardiac electrophysiological balance index in children diagnosed with type 1 diabetes mellitus. *Cardiol. Young* **2025**, *35*, 593–599. [CrossRef]
121. Türe, M.; Akın, A.; Unal, E.; Kan, A.; Savaş, S. Electrocardiographic data of children with type 1 diabetes mellitus. *Cardiol. Young* **2022**, *32*, 106–110. [CrossRef]
122. Burden, S.; Weedon, B.; Whaymand, L.; Rademaker, J.; Dawes, H.; Jones, A. The effect of overweight/obesity on diastolic function in children and adolescents: A meta-analysis. *Clin. Obes.* **2021**, *11*, e12476. [CrossRef]
123. Mangner, N.; Scheuermann, K.; Winzer, E.; Wagner, I.; Hoellriegel, R.; Sandri, M.; Zimmer, M.; Mende, M.; Linke, A.; Kiess, W.; et al. Childhood obesity: Impact on cardiac geometry and function. *JACC Cardiovasc. Imaging* **2014**, *7*, 1198–1205. [CrossRef] [PubMed]
124. Erbs, S.; Broniecki, H.; Scheuermann, K.; Winzer, E.; Adam, J.; Spielau, U.; Woitek, F.; Sandri, M.; Zimmer, M.; Besler, C.; et al. Impact of Weight Reduction During Adolescence on Parameters of Cardiac Geometry and Function in Obese Children. *JACC Cardiovasc. Imaging* **2018**, *11*, 1915–1917. [CrossRef] [PubMed]
125. Yi, L.F.; Wen, H.X.; Huang, X.L.; Qiu, M.; Cao, X.X. Cardiac autonomic nerve function in obese school-age children. *Chin. J. Contemp. Pediatr.* **2017**, *19*, 524–528. [CrossRef]
126. Nimpum, D.; Jittham, W. Echocardiographic assessment of cardiac function abnormalities and related risk factors in Thai overweight and obese children. *Ann. Pediatr. Cardiol.* **2023**, *16*, 413–421. [CrossRef]
127. Freitas, I.M.; Miranda, J.A.; Mira, P.A.; Lanna, C.M.; Lima, J.R.; Laterza, M.C. Cardiac autonomic dysfunction in obese normotensive children and adolescents. *Rev. Paul. Pediatr.* **2014**, *32*, 244–249. [CrossRef]
128. Campos, J.O.; Barros, M.A.V.; Oliveira, T.L.P.S.A.; Nobre, I.G.; de Morais, A.S.; Santos, M.A.M.; Leandro, C.G.; Costa-Silva, J.H. Cardiac autonomic dysfunction in school age children with overweight and obesity. *Nutr. Metab. Cardiovasc. Dis. NMCD* **2022**, *32*, 2410–2417. [CrossRef]
129. Wirix, A.J.; Finken, M.J.; von Rosenstiel-Jadoul, I.A.; Heijboer, A.C.; Nauta, J.; Groothoff, J.W.; Chinapaw, M.J.; Kist-van Holthe, J.E. Is There an Association Between Cortisol and Hypertension in Overweight or Obese Children? *J. Clin. Res. Pediatr. Endocrinol.* **2017**, *9*, 344–349. [CrossRef]
130. Scavone, M.; Tallarico, V.; Stefanelli, E.; Parisi, F.; De Sarro, R.; Salpietro, C.; Ceravolo, G.; Sestito, S.; Pensabene, L.; Chimenz, R.; et al. Cardiac malformations in children with congenital hypothyroidism. *J. Biol. Regul. Homeost. Agents* **2020**, *34* (Suppl. S2), 91–97.
131. Soto-Rivera, C.L.; Schwartz, S.M.; Sawyer, J.E.; Macrae, D.J.; Agus, M.S. Endocrinologic Diseases in Pediatric Cardiac Intensive Care. *Pediatr. Crit. Care Med.* **2016**, *17* (Suppl. S1), S296–S301. [CrossRef]
132. Metwalley, K.A.; Farghaly, H.S. Graves' Disease in Children: An Update. *Clin. Med. Insights Endocrinol. Diabetes* **2023**, *16*, 11795514221150615. [CrossRef]
133. Léger, J.; Carel, J.C. Hyperthyroidism in childhood: Causes, when and how to treat. *J. Clin. Res. Pediatr. Endocrinol.* **2013**, *5* (Suppl. S1), 50–56. [CrossRef]
134. Subramonian, D.; Wu, Y.J.; Amed, S.; Sanatani, S. Hyperthyroidism with Atrial Fibrillation in Children: A Case Report and Review of the Literature. *Front. Endocrinol.* **2021**, *12*, 689497. [CrossRef]
135. Petrovic Djordjevic, I.; Petrovic, J.; Radomirovic, M.; Petrovic, S.; Biorac, B.; Jemuovic, Z.; Tesic, M.; Trifunovic Zamaklar, D.; Nedeljkovic, I.; Nedeljkovic Beleslin, B.; et al. Impairment of Left Ventricular Function in Hyperthyroidism Caused by Graves' Disease: An Echocardiographic Study. *J. Clin. Med.* **2024**, *13*, 7348. [CrossRef]
136. Park, S.K.; Ryoo, J.H.; Kang, J.G.; Jung, J.Y. Association of left ventricular hypertrophy with the level of thyroid hormone in euthyroid state. *J. Endocrinol. Investig.* **2021**, *44*, 111–117. [CrossRef] [PubMed]
137. Zúñiga, D.; Balasubramanian, S.; Mehmood, K.T.; Al-Baldawi, S.; Zúñiga Salazar, G. Hypothyroidism and Cardiovascular Disease: A Review. *Cureus* **2024**, *16*, e52512. [CrossRef]
138. Udovcic, M.; Pena, R.H.; Patham, B.; Tabatabai, L.; Kansara, A. Hypothyroidism and the Heart. *Methodist DeBakey Cardiovasc. J.* **2017**, *13*, 55–59. [CrossRef] [PubMed]
139. Sahoo, R.; Sirohi, T.; Talapa, R.; Jamir, I. Case series of unusual cases of hypothyroidism-induced pericardial effusion. *J. Fam. Med. Prim. Care* **2024**, *13*, 2776–2779. [CrossRef] [PubMed]
140. Karki, S.; Rayamajhi, R.J.; Shikhrakar, S.; Shahi, S.; Dhakal, B.; Khadka, M. Pericardial effusion in hypothyroidism: A case report. *Ann. Med. Surg.* **2021**, *72*, 102999. [CrossRef]
141. Ebrahimi, P.; Taheri, H.; Bahiraie, P.; Rader, F.; Siegel, R.J.; Mandegar, M.H.; Hosseini, K.; Shahid, F. Incidence of secondary pericardial effusions associated with different etiologies: A comprehensive review of literature. *J. Cardiothorac. Surg.* **2025**, *20*, 141. [CrossRef]
142. Cojić, M.; Cvejanov-Kezunović, L. Subclinical Hypothyroidism—Whether and When to Start Treatment? *Open Access Maced. J. Med. Sci.* **2017**, *5*, 1042–1046. [CrossRef]
143. Radetti, G.; Renzullo, L.; Gottardi, E.; D'Addato, G.; Messner, H. Altered thyroid and adrenal function in children born at term and preterm, small for gestational age. *J. Clin. Endocrinol. Metab.* **2004**, *89*, 6320–6324. [CrossRef]

144. Biondi, B.; Palmieri, E.A.; Lombardi, G.; Fazio, S. Subclinical hypothyroidism and cardiac function. *Thyroid. Off. J. Am. Thyroid Assoc.* **2002**, *12*, 505–510. [CrossRef]
145. Brienza, C.; Grandone, A.; Di Salvo, G.; Corona, A.M.; Di Sessa, A.; Pascotto, C.; Calabrò, R.; Toraldo, R.; Perrone, L.; del Giudice, E.M. Subclinical hypothyroidism and myocardial function in obese children. *Nutr. Metab. Cardiovasc. Dis.* **2013**, *23*, 898–902. [CrossRef]
146. Danzi, S.; Klein, I. Thyroid hormone and the cardiovascular system. *Minerva Endocrinol.* **2004**, *29*, 139–150. [CrossRef]
147. Menendez-Castro, C.; Rascher, W.; Hartner, A. Intrauterine growth restriction-impact on cardiovascular diseases later in life. *Mol. Cell. Pediatr.* **2018**, *5*, 4. [CrossRef]
148. Ozdemir, O.; Abaci, A.; Hizli, S.; Razi, C.H.; Akelma, A.Z.; Kocak, M.; Kislal, F.M. Cardiac functions in children with growth hormone deficiency before and during growth hormone-replacement therapy. *Pediatr. Cardiol.* **2011**, *32*, 766–771. [CrossRef] [PubMed]
149. Alkan, F.; Ersoy, B.; Kızılay, D.O.; Coskun, S. Cardiac functions in children with growth hormone deficiency: Effects of one year of GH replacement therapy. *Growth Horm. IGF* **2021**, *60–61*, 101432. [CrossRef] [PubMed]
150. Thomas, J.D.; Dattani, A.; Zemrak, F.; Burchell, T.; Akker, S.A.; Gurnell, M.; Grossman, A.B.; Davies, L.C.; Korbonits, M. Characterisation of myocardial structure and function in adult-onset growth hormone deficiency using cardiac magnetic resonance. *Endocrine* **2016**, *54*, 778–787. [CrossRef] [PubMed]
151. Minczykowski, A.; Gryczynska, M.; Ziemnicka, K.; Czepczynski, R.; Sowinski, J.; Wysocki, H. The influence of growth hormone (GH) therapy on cardiac performance in patients with childhood onset GH deficiency. *Growth Horm. IGF Res.* **2005**, *15*, 156–164. [CrossRef]
152. Improda, N.; Moracas, C.; Mattace Raso, G.; Valente, V.; Crisci, G.; Lorello, P.; Di Mase, R.; Salerno, M.; Capalbo, D. Vascular Function and Intima-Media Thickness in Children and Adolescents with Growth Hormone Deficiency: Results from a Prospective Case-Control Study. *Horm. Res. Paediatr.* **2024**, *97*, 140–147. [CrossRef]
153. Kościuszko, M.; Buczyńska, A.; Hryniewicka, J.; Jankowska, D.; Adamska, A.; Siewko, K.; Jacewicz-Święcka, M.; Zaniuk, M.; Krętowski, A.J.; Popławska-Kita, A. Early Cardiovascular and Metabolic Benefits of rhGH Therapy in Adult Patients with Severe Growth Hormone Deficiency: Impact on Oxidative Stress Parameters. *Int. J. Mol. Sci.* **2025**, *26*, 5434. [CrossRef]
154. Aguilar, M.; Rose, R.A.; Takawale, A.; Nattel, S.; Reilly, S. New aspects of endocrine control of atrial fibrillation and possibilities for clinical translation. *Cardiovasc. Res.* **2021**, *117*, 1645–1661. [CrossRef]
155. Fauchier, L.; de Groote, P. Atrial fibrillation and renin-angiotensin-aldosterone system: Believe it or not. *Europace* **2011**, *13*, 297–298. [CrossRef] [PubMed]
156. Jiang, W.; Xiong, Y.; Li, X.; Yang, Y. Cardiac Fibrosis: Cellular Effectors, Molecular Pathways, and Exosomal Roles. *Front. Cardiovasc. Med.* **2021**, *8*, 715258. [CrossRef] [PubMed]
157. Nair, G.M.; Nery, P.B.; Redpath, C.J.; Birnie, D.H. The Role of Renin Angiotensin System In Atrial Fibrillation. *J. Atr. Fibrillation* **2014**, *6*, 972. [PubMed]
158. Sodero, G.; Cipolla, C.; Rigante, D.; Arzilli, F.; Mercuri, E.M. Pubertal induction therapy in pediatric patients with Duchenne muscular dystrophy. *J. Pediatr. Endocrinol. Metab.* **2025**, *38*, 781–787. [CrossRef]
159. Peck, M.; Connelly, P.; Lucas-Herald, A.K. Cardiometabolic outcomes of early onset hypogonadism in males. *Best Pract. Res. Clin. Endocrinol. Metab.* **2025**. *advance online publication*. [CrossRef]
160. Johansson, S.; Norman, M.; Legnevall, L.; Dalmaz, Y.; Lagercrantz, H.; Vanpée, M. Increased catecholamines and heart rate in children with low birth weight: Perinatal contributions to sympathoadrenal overactivity. *J. Intern. Med.* **2007**, *261*, 480–487. [CrossRef]
161. Ream, M.A.; Chandra, R.; Peavey, M.; Ray, A.M.; Roffler-Tarlov, S.; Kim, H.G.; Wetsel, W.C.; Rockman, H.A.; Chikaraishi, D.M. High oxygen prevents fetal lethality due to lack of catecholamines. *Am. J. Physiol. Regul. Integr. Comp. Physiol.* **2008**, *295*, R942–R953. [CrossRef]
162. Szatko, A.; Glinicki, P.; Gietka-Czernel, M. Pheochromocytoma/paraganglioma-associated cardiomyopathy. *Front. Endocrinol.* **2023**, *14*, 1204851. [CrossRef]
163. Heidarpour, M.; Zare, M.M.; Armani, S.; Torkashvan, H.; Mazaheri-Tehrani, S.; Shafie, D. Acute Heart Failure as a First Manifestation of Primary Adrenal Insufficiency: Highly Lethal If Not Diagnosed! *Case Rep. Endocrinol.* **2024**, *2024*, 5759629. [CrossRef] [PubMed]
164. Kirk, R.; Dipchand, A.I.; Rosenthal, D.N.; Addonizio, L.; Burch, M.; Chrisant, M.; Dubin, A.; Everitt, M.; Gajarski, R.; Mertens, L.; et al. The International Society for Heart and Lung Transplantation Guidelines for the management of pediatric heart failure: Executive summary. *J. Heart Lung Transplant.* **2014**, *33*, 888–909. [CrossRef] [PubMed]
165. Ross, R.D. The Ross classification for heart failure in children after 25 years: A review and an age-stratified revision. *Pediatr. Cardiol.* **2012**, *33*, 1295–1300. [CrossRef] [PubMed]

166. Connolly, D.; Rutkowski, M.; Auslender, M.; Artman, M. The New York University Pediatric Heart Failure Index: A new method of quantifying chronic heart failure severity in children. *J. Pediatr.* **2001**, *138*, 644–648. [CrossRef]
167. Cantinotti, M.; Storti, S.; Ripoli, A.; Zyw, L.; Crocetti, M.; Assanta, N.; Murzi, B.; Clerico, A. Diagnostic accuracy of B-type natriuretic hormone for congenital heart disease in the first month of life. *Clin. Chem. Lab. Med.* **2010**, *48*, 1333–1338. [CrossRef]
168. Holmgren, D.; Westerlind, A.; Lundberg, P.A.; Wåhlander, H. Increased plasma levels of natriuretic peptide type B and A in children with congenital heart defects with left compared with right ventricular volume overload or pressure overload. *Clin. Physiol. Funct. Imaging* **2005**, *25*, 263–269. [CrossRef]
169. Sandeep, B.; Huang, X.; Li, Y.; Wang, X.; Mao, L.; Kan, Y.; Xiong, D.; Gao, K.; Zongwei, X. Evaluation of right ventricle pulmonary artery coupling on right ventricular function in post operative tetralogy of Fallot patients underwent for pulmonary valve replacement. *J. Cardiothorac. Surg.* **2020**, *15*, 241. [CrossRef]
170. Tatani, S.B.; Carvalho, A.C.; Andriolo, A.; Rabelo, R.; Campos, O.; Moises, V.A. Echocardiographic parameters and brain natriuretic peptide in patients after surgical repair of tetralogy of Fallot. *Echocardiography* **2010**, *27*, 442–447. [CrossRef]
171. Afshani, N.; Schülein, S.; Biccard, B.M.; Thomas, J.M. Clinical utility of B-type natriuretic peptide (NP) in pediatric cardiac surgery—A systematic review. *Paediatr. Anaesth.* **2015**, *25*, 115–126. [CrossRef]
172. Lin, F.; Zheng, L.; Cui, Y.; Chen, W.; Gupta, R.K.; Li, H.; Chen, X.; Xia, H.; Liang, H. Prognostic value of perioperative NT-proBNP after corrective surgery for pediatric congenital heart defects. *BMC Pediatr.* **2019**, *19*, 497. [CrossRef]
173. Palm, J.; Holdenrieder, S.; Hoffmann, G.; Hörer, J.; Shi, R.; Klawonn, F.; Ewert, P. Predicting Major Adverse Cardiovascular Events in Children with Age-Adjusted NT-proBNP. *J. Am. Coll. Cardiol.* **2021**, *78*, 1890–1900. [CrossRef] [PubMed]
174. Schmitt, W.; Diedrich, C.; Hamza, T.H.; Meyer, M.; Eissing, T.; Breitenstein, S.; Rossano, J.W.; Lipshultz, S.E. NT-proBNP for Predicting All-Cause Death and Heart Transplant in Children and Adults with Heart Failure. *Pediatr. Cardiol.* **2025**, *46*, 694–703. [CrossRef] [PubMed]
175. Volpe, M.; Carnovali, M.; Mastromarino, V. The natriuretic peptides system in the pathophysiology of heart failure: From molecular basis to treatment. *Clin. Sci.* **2016**, *130*, 57–77. [CrossRef] [PubMed] [PubMed Central]
176. Agnoletti, G.; Panzali, A.F.; Ferrari, R. Levels of atrial natriuretic factor in the plasma and urine of patients with and without heart failure. *Int. J. Cardiol.* **1993**, *40*, 179–182. [CrossRef]
177. Cantinotti, M.; Walters, H.L.; Crocetti, M.; Marotta, M.; Murzi, B.; Clerico, A. BNP in children with congenital cardiac disease: Is there now sufficient evidence for its routine use? *Cardiol. Young* **2015**, *25*, 424–437. [CrossRef]
178. Westerlind, A.; Wåhlander, H.; Berggren, H.; Lundberg, P.A.; Holmgren, D. Plasma levels of natriuretic peptide type B and A in children with heart disease with different types of cardiac load or systolic dysfunction. *Clin. Physiol. Funct. Imaging* **2008**, *28*, 277–284. [CrossRef]
179. Kotby, A.A.; Taman, K.H.; Sedky, H.T.; Habeeb, N.M.; El-Hadidi, E.S.; Yosseif, H.S. Atrial natriuretic peptide as a marker of heart failure in children with left ventricular volume overload. *J. Paediatr. Child Health* **2013**, *49*, 43–47. [CrossRef]
180. Ghosh, N.; Haddad, H. Atrial natriuretic peptides in heart failure: Pathophysiological significance, diagnostic and prognostic value. *Can. J. Physiol. Pharmacol.* **2011**, *89*, 587–591. [CrossRef]
181. Uner, A.; Doğan, M.; Ay, M.; Acar, C. The evaluation of serum N-terminal prohormone brain-type natriuretic peptide, troponin-I, and high-sensitivity C-reactive protein levels in children with congenital heart disease. *Hum. Exp. Toxicol.* **2014**, *33*, 1158–1166. [CrossRef]
182. Shaddy, R.; Burch, M.; Kantor, P.F.; Solar-Yohay, S.; Garito, T.; Zhang, S.; Kocun, M.; Mao, C.; Cilliers, A.; Wang, X.; et al. PANORAMA-HF Investigators Sacubitril/Valsartan in Pediatric Heart Failure (PANORAMA-HF): A Randomized, Multicenter, Double-Blind Trial. *Circulation* **2024**, *150*, 1756–1766. [CrossRef]
183. Benza, R.L.; Miller, D.P.; Gomberg-Maitland, M.; Frantz, R.P.; Foreman, A.J.; Coffey, C.S.; Frost, A.; Barst, R.J.; Badesch, D.B.; Elliott, C.G.; et al. Predicting survival in pulmonary arterial hypertension: Insights from the Registry to Evaluate Early and Long-Term Pulmonary Arterial Hypertension Disease Management (REVEAL). *Circulation* **2010**, *122*, 164–172. [CrossRef]
184. Cuna, A.; Kandasamy, J.; Sims, B. B-type natriuretic peptide and mortality in extremely low birth weight infants with pulmonary hypertension: A retrospective cohort analysis. *BMC Pediatr.* **2014**, *14*, 68. [CrossRef]
185. Xiong, T.; Kulkarni, M.; Gokulakrishnan, G.; Shivanna, B.; Pammi, M. Natriuretic peptides in bronchopulmonary dysplasia: A systematic review. *J. Perinatol.* **2020**, *40*, 607–615. [CrossRef] [PubMed]
186. Dasgupta, S.; Bettermann, E.; Kelleman, M.; Kanaan, U.; Sachdeva, R.; Petit, C.; Kim, D.; Vincent, R.; Bauser-Heaton, H. N-terminal pro-B-type-natriuretic peptide as a screening tool for pulmonary hypertension in the paediatric population. *Cardiol. Young* **2021**, *31*, 1595–1607. [CrossRef] [PubMed]
187. Aldalaan, A.M.; Saleemi, S.A.; Weheba, I.; Abdelsayed, A.; Aleid, M.M.; Alzubi, F.; Zaytoun, H.; Alharbi, N. Pulmonary hypertension in Saudi Arabia: First data from the SAUDIPH registry with a focus on pulmonary arterial hypertension. *Respirology* **2021**, *26*, 92–101. [CrossRef]

188. Li, Y.; Zheng, C.; Cheng, S.; Chu, L.; Chen, Z.; Liu, G. The relationship between N-terminal pro-brain natriuretic and coronary artery lesions in Kawasaki disease: A meta-analysis. *PLoS ONE* **2025**, *20*, e0324552. [CrossRef] [PubMed]
189. Sun, Y.N.; Lin, C.W.; Tang, W.; Zhang, J.P. Age-Stratified Cut-Off Values for Serum Levels of N-Terminal ProB-Type Natriuretic Peptide and Mortality from Sepsis in Children Under Age 18 Years: A Retrospective Study from a Single Center. *Med. Sci. Monit.* **2021**, *27*, e933400. [CrossRef]

biomedicines

MDPI

Article

# A Four-Phenotype Model for Risk Stratification in Heart Failure with Preserved and Mildly Reduced Ejection Fraction: The Role of Sex and Diabetes

**Flavia-Mihaela Stoiculescu [1,2], Diana-Ruxandra Hădăreanu [2,3,*], Călin-Dinu Hădăreanu [1,4], Maria-Livia Iovănescu [2,3], Georgică-Costinel Târtea [2,3], Ionuț Donoiu [2,3], Petre-Alexandru Cojocaru [3], Sebastian Militaru [2,3], Octavian Istrătoaie [2,3] and Cristina Florescu [3,5]**

1 Doctoral School, University of Medicine and Pharmacy of Craiova, 200349 Craiova, Romania
2 Department of Cardiology, Emergency Clinical County Hospital of Craiova, 200642 Craiova, Romania
3 Department of Cardiology, University of Medicine and Pharmacy of Craiova, 200349 Craiova, Romania; cristina.t.florescu@umfcv.ro (C.F.)
4 Department of Cardiovascular Surgery, Emergency Clinical County Hospital of Craiova, 200642 Craiova, Romania
5 Department of Cardiology, Filantropia Municipal Hospital Craiova, 200516 Craiova, Romania
* Correspondence: diana.hadareanu@umfcv.ro

**Abstract**

**Background/Objectives**: Sex and diabetes are important determinants of risk in heart failure with mildly reduced and preserved ejection fraction (HFmrEF/HFpEF), yet their combined effects have not been systematically evaluated. This study examined how sex–diabetes phenotypes influence clinical characteristics and the risk of heart failure rehospitalization. **Methods**: We retrospectively analyzed 1018 HFmrEF/HFpEF patients (2019–2023), classified into four sex–diabetes phenotypes, and performed group comparisons. The primary endpoint was heart failure rehospitalization. **Results**: Over a mean follow-up of 1463 $\pm$ 496 days, 307 patients (30.1%) were rehospitalized for heart failure decompensation. The four phenotypes differed significantly in age, renal function, LV mass, LV dimensions, glycemia, and comorbidity burden (all $p < 0.05$). Men—particularly those with diabetes—had greater structural remodeling and higher prevalence of smoking, hypercholesterolemia, and atrial fibrillation. In univariate analysis, male sex, diabetes, smoking, NYHA class, lower TAPSE, and lower LVEF were associated with increased risk of rehospitalization. After adjustment for LVEF and NYHA class, male sex (HR 1.28; $p = 0.035$) and diabetes (HR 1.28; $p = 0.036$) remained independent predictors. Kaplan–Meier curves demonstrated a clear gradient in event-free survival (log-rank $p = 0.015$), with women without diabetes showing the best prognosis and diabetic men the worst. **Conclusions**: Sex and diabetes interact to define distinct risk profiles in HFmrEF/HFpEF. Women without diabetes represent a low-risk phenotype, whereas diabetic men exhibit the highest risk of recurrent heart failure decompensation. These findings support incorporating sex–diabetes phenotyping into routine risk stratification and personalized management.

**Keywords:** heart failure; heart failure with preserved ejection fraction; heart failure with mildly reduced ejection fraction; sex differences; diabetes mellitus; risk stratification; rehospitalization; prognosis

## 1. Introduction

Heart failure (HF) with mildly reduced (HFmrEF) or preserved ejection fraction (HFpEF) accounts for over half of the global HF burden and continues to rise due to population aging and increasing cardiometabolic disease. Despite its prevalence, HFmrEF/HFpEF remains a heterogeneous syndrome with limited therapeutic options and substantial risk of recurrent decompensation [1]. Identifying factors that modify risk within this population is therefore essential.

Sex strongly shapes HFmrEF/HFpEF pathophysiology and outcomes. Large epidemiologic studies have shown that sex differences influence HF risk across the lifespan, with women experiencing lower lifetime HF mortality but greater symptom burden and a higher prevalence of comorbidities such as obesity, hypertension, and diabetes [2]. Women with HFpEF typically exhibit smaller left ventricular (LV) cavities, concentric remodeling, greater diastolic stiffness, and more severe vascular–ventricular coupling abnormalities [3]. Despite these structural disadvantages, women consistently demonstrate better survival than men in major HFpEF cohorts, despite reporting worse symptoms and functional impairment [4]. In contrast, men present with more ischemic heart disease, greater myocardial fibrosis, higher pulmonary vascular resistance, and worse right ventricular function, factors associated with poorer prognosis [5].

Diabetes mellitus is another major determinant of HF risk and progression. Mechanistically, diabetes promotes microvascular inflammation, endothelial dysfunction, oxidative stress, and myocardial stiffening, key processes implicated in HFmrEF/HFpEF pathophysiology [6]. Epidemiologically, diabetes increases HF risk in both sexes but confers a significantly higher relative excess risk in women, as demonstrated in a meta-analysis of 47 cohorts and over 12 million individuals [7]. Conversely, men with diabetes carry a higher absolute burden of HF events due to greater ischemic and arrhythmic vulnerability [8].

Despite robust evidence that sex and diabetes independently influence HF outcomes, their combined interaction in HFmrEF/HFpEF remains inadequately characterized. Most prior studies have examined sex or diabetes in isolation, without evaluating whether specific sex–diabetes combinations define distinct prognostic endotypes. This gap is clinically relevant, as early mortality is higher in HFmrEF men even when baseline characteristics are matched, whereas HF hospitalization rates often depend more on comorbidity clusters—particularly diabetes—than on sex alone [9].

Therefore, the objective of this study was to investigate the prognostic impact of sex, diabetes, and their interaction on HF rehospitalization in HFmrEF/HFpEF. By identifying whether specific sex–diabetes combinations define distinct clinical risk profiles, this study aims to clarify a currently underexplored source of heterogeneity in HFmrEF/HFpEF and support more individualized approaches to risk stratification.

## 2. Materials and Methods

### 2.1. Study Design and Patient Selection

We conducted a retrospective cohort study using all consecutive discharges from the Clinical Emergency County Hospital of Craiova, Romania, between 1 January 2019 and 31 May 2023. During this interval, the charts from 15,160 consecutive inpatients were analyzed. Individuals were eligible if the index hospitalization resulted in a new diagnosis of HFmrEF or HFpEF, defined according to contemporary guideline criteria [1]. HFmrEF was defined by the presence of HF symptoms and/or signs, a LVEF between 41% and 49%, and additional evidence of cardiac structural or functional abdormalities or elevated natriuretic peptides. HFpEF was defined by the presence of HF symptoms and/or signs, a LVEF above 50%, and evidence of structural or functional abnormalities (such as LV hypertrophy, diastolic dysfunction, left atrial enlargement). Patients were excluded if any

https://doi.org/10.3390/biomedicines14010173

of the following conditions were present during the index admission: (1) acute coronary syndrome, (2) acute pulmonary embolism, (3) sustained ventricular tachyarrhythmia or resuscitated cardiac arrest, (4) HF with reduced ejection fraction, (5) second- or third-degree atrioventricular block, (6) missing or incomplete clinical documentation, or (7) insufficient follow-up duration (<1 year) or unavailable follow-up data. The study received approval from the institutional ethics committee (Approval no. 86/19.02.2024). Because of its retrospective nature, the requirement for informed consent was waived, and all patient identifiers were removed before analysis.

*2.2. Data Collection and Assessment*

Baseline demographic, clinical, and therapeutic information was retrieved from electronic medical records, including age, sex, heart rhythm, vital signs, New York Heart Association (NYHA) functional class at discharge, cardiovascular risk factors, medication use, and prior medical history.

Laboratory parameters collected at discharge included hemoglobin, hepatic enzymes, serum creatinine, serum sodium and potassium, and NT-proBNP. Estimated glomerular filtration rate (eGFR) was calculated using the CKD-EPI equation.

Comprehensive echocardiographic evaluation was performed in accordance with the recommendations of the European Association of Cardiovascular Imaging [10]. Measurements included left and right ventricular and atrial dimensions, LVEF using the Simpson biplane method, E/A ratio, tricuspid annular plane systolic excursion (TAPSE), tricuspid regurgitation peak velocity, and systolic pulmonary artery pressure, derived from the tricuspid regurgitation jet and inferior vena cava indices. LV mass was calculated with the Devereux formula, and relative wall thickness was obtained as: 2 $\times$ posterior wall thickness/LV end-diastolic diameter.

As echocardiographic data was obtained from archive clinical reports, in which left atrial volume index was not consistently available, to enusre internal validity and minimize missing data, left atrial diameter was used as a surrogate.

Echocardiographic examinations in our study were performed as part of routine clinical care by board-certified cardiologists with echocardiography accreditation according to national and European training standards. However, because of the retrospective design and reliance on existing clinical records, formal intra- and inter-observer variability assessments were not available.

Conventional risk factors were recorded as follows: smoking status (current or former), dyslipidemia (LDL-cholesterol > 130 mg/dL or lipid-lowering treatment), hypertension (BP $\geq$ 140/90 mmHg or antihypertensive therapy), type 2 diabetes mellitus, and chronic kidney disease.

Ischemic heart disease was determined based on clinical evaluation and available coronary imaging. Most patients with suspected coronary disease underwent invasive coronary angiography; coronary CT angiography was rarely used. Ischemic etiology was assigned when at least one of the following was documented: significant obstructive stenosis (>70% in a major epicardial artery or >50% in the left main), prior myocardial infarction, or prior revascularization. Patients with previous percutaneous intervention but no infarction were also considered to have ischemic HF. In patients without angiography and low clinical suspicion, ischemia may have remained unconfirmed. Valvular heart disease was defined as moderate or severe stenosis or regurgitation. Congenital heart disease encompassed any congenital structural cardiac abnormality recorded in the medical history (e.g., atrial or ventricular septal defects, bicuspid aortic valve, Ebstein anomaly, atrioventricular canal defects, or repaired tetralogy of Fallot).

https://doi.org/10.3390/biomedicines14010173

### 2.3. Follow-Up and Study Endpoint

The primary endpoint was first rehospitalization for heart failure decompensation. Rehospitalization events were identified based on hospital discharge diagnoses recorded in the institutional electronic medical records. For patients followed outside the index hospital, follow-up information was supplemented by structured outpatient visits and telephone contact when available. Independent adjudication of events was not performed due to the retrospective study design. Mortality status was verified using each individual's national identification number. The chosen endpoint was first rehospitalization for HF decompensation, defined as an unplanned hospital admission requiring intravenous diuretics, vasodilators, or inotropes for HF worsening. For patients without events, follow-up was censored at the date of last contact. For those with multiple HF readmissions, only the first event was used in time-to-event analyses.

### 2.4. Statistical Analysis

The normality of continuous variables was assessed using the Shapiro–Wilk test. Continuous data are reported as means $\pm$ standard deviations (SDs), while categorical variables are expressed as counts and percentages.

Variables were compared across the four sex–diabetes groups. For normally distributed continuous variables, one-way ANOVA was applied together with Levene's test for homogeneity of variances; when variances were equal, post hoc pairwise comparisons were performed with Tukey's test, and when variances were unequal, the Games–Howell procedure was used. Categorical variables were compared using the Pearson chi-square test. For each contingency table, SPSS-generated adjusted standardized residuals were examined to identify which cells contributed to statistically significant differences. Residuals with an absolute value $\geq$ 1.96 were considered significant at $p < 0.05$, $\geq$2.58 at $p < 0.01$, and $\geq$3.29 at $p < 0.001$. Positive residuals indicated that the observed frequency was higher than expected under the null hypothesis, while negative residuals indicated a lower-than-expected frequency. This method allowed post hoc identification of pairwise group differences without conducting multiple chi-square tests. Variables with missing observations were analyzed according to the number of valid cases reported by SPSS.

Univariable Cox proportional hazards models were used to explore associations between baseline clinical, biochemical, and echocardiographic parameters and the study endpoints. Variables included in the multivariable Cox model were selected a priori based on clinical relevance rather than on univariate statistical significance. This approach avoids confounder omission and ensures appropriate adjustment when evaluating the independent prognostic contribution of sex and diabetes.

Kaplan–Meier curves were generated to assess event-free survival, and differences were evaluated using the log-rank test.

A two-sided $p$ value < 0.05 was considered statistically significant. Analyses were performed using available-case data for each variable, as the extent of missingness varied across parameters. All analyses were conducted using SPSS (version 23).

## 3. Results

### 3.1. Study Population and Baseline Characteristics

A total of 1018 patients represented the final study population. The mean follow-up duration was 1463 $\pm$ 496 days (approximately 4 years). During follow-up, 307 patients (30.1%) met the primary endpoint of rehospitalization for heart failure decompensation.

The cohort had a mean age of 74 $\pm$ 11 years, and comorbidity burden was high: hypertension (79%), hypercholesterolemia (65%), atrial fibrillation (48%), type 2 diabetes

https://doi.org/10.3390/biomedicines14010173

mellitus (36.7%), chronic kidney disease (34%), and a history of smoking (14%). Most patients were in NYHA class II–III, with a mean LVEF of 49 ± 6%.

Across the four sex–diabetes phenotypes, several significant differences were found between baseline characteristics (Table 1). Women without diabetes were the oldest group, whereas diabetic men were the youngest ($p < 0.001$). Diabetic patients—both men and women—had significantly higher blood glucose levels compared with their non-diabetic counterparts ($p < 0.001$). Men (with and without diabetes) exhibited larger LV dimensions and greater LV mass than women, with all pairwise comparisons showing statistically significant differences ($p < 0.001$). Women without diabetes had higher eGFR values compared with both male groups and diabetic women ($p = 0.001$). Finally, diabetic men had the shortest time to first rehospitalization compared with women without diabetes and non-diabetic men ($p = 0.012$).

**Table 1.** Baseline characteristics across the four sex-diabetes phenotypes.

| Variable | All Patients (N = 1018) | Non-Diabetic Women (G1, N = 334) | Diabetic Women (G2, N = 167) | Non-Diabetic Men (G3, N = 310) | Diabetic Men (G4, N = 207) | ANOVA *p*-Value |
|---|---|---|---|---|---|---|
| Age (years) | 74 ± 11 | 76 ± 11 | 75 ± 9 | 73 ± 13 | 71 ± 10 | <0.001 [1] |
| Heart rate (bpm) | 73 ± 14 | 73 ± 12 | 75 ± 15 | 71 ± 14 | 73 ± 14 | 0.031 [2] |
| Systolic blood pressure (mmHg) | 130 ± 20 | 129 ± 20 | 132 ± 20 | 128 ± 21 | 132 ± 19 | 0.064 |
| Diastolic blood pressure (mmHg) | 76 ± 11 | 75 ± 11 | 78 ± 11 | 74 ± 11 | 76 ± 11 | 0.011 [3] |
| Time to first rehospitalization (days) | 1133 ± 683 | 1196 ± 667 | 1114 ± 697 | 1161 ± 707 | 1004 ± 646 | 0.012 [4] |
| NT-proBNP (pg/mL) | 5547 ± 7544 | 6109 ± 7213 | 6470 ± 8554 | 5275 ± 8313 | 4132 ± 5555 | 0.165 |
| Creatinine (mg/dL) | 1.2 ± 0.7 | 1.1 ± 0.6 | 1.2 ± 0.6 | 1.3 ± 0.9 | 1.3 ± 0.6 | 0.001 [5] |
| eGFR (mL/min/1.73 $m^2$) | 73 ± 25 | 77 ± 24 | 73 ± 25 | 71 ± 27 | 70 ± 24 | 0.001 [6] |
| Glucose (mg/dL) | 123 ± 55 | 110 ± 42 | 148 ± 66 | 107 ± 25 | 148 ± 74 | <0.001 [7] |
| Serum sodium (mmol/L) | 138 ± 6 | 138 ± 6 | 139 ± 9 | 138 ± 4 | 138 ± 4 | 0.470 |
| Interventricular septum width (mm) | 13 ± 2 | 12 ± 2 | 13 ± 2 | 13 ± 3 | 13 ± 2 | 0.001 [8] |
| Left ventricular end-diastolic diameter (mm) | 50 ± 7 | 47 ± 6 | 48 ± 6 | 52 ± 8 | 52 ± 7 | <0.001 [9] |
| Posterior wall width (mm) | 12 ± 2 | 12 ± 2 | 12 ± 2 | 12 ± 2 | 13 ± 2 | 0.034 [10] |
| Left ventricular mass (g) | 241 ± 78 | 216 ± 67 | 230 ± 64 | 257 ± 85 | 268 ± 80 | <0.001 [11] |
| Relative wall thickness | 0.49 ± 0.13 | 0.50 ± 0.14 | 0.52 ± 0.14 | 0.47 ± 0.13 | 0.48 ± 0.11 | <0.001 [12] |
| Left atrial diameter (mm) | 48 ± 10 | 48 ± 11 | 47 ± 7 | 48 ± 11 | 47 ± 8 | 0.133 |
| E/E′ | 14 ± 3 | 12 ± 3 | 13 ± 2 | 14 ± 3 | 14 ± 4 | 0.856 |
| Systolic pulmonary artery pressure (mmHg) | 51 ± 17 | 51 ± 16 | 55 ± 20 | 48 ± 17 | 50 ± 18 | 0.078 |

https://doi.org/10.3390/biomedicines14010173

**Table 1.** *Cont.*

| Variable | All Patients (N = 1018) | Non-Diabetic Women (G1, N = 334) | Diabetic Women (G2, N = 167) | Non-Diabetic Men (G3, N = 310) | Diabetic Men (G4, N = 207) | ANOVA *p*-Value |
|---|---|---|---|---|---|---|
| Tricuspid annulus plane systolic excursion (mm) | 19 ± 5 | 19 ± 5 | 19 ± 6 | 18 ± 4 | 19 ± 4 | 0.949 |
| Left ventricular ejection fraction (%) | 49 ± 6 | 51 ± 7 | 50 ± 6 | 49 ± 6 | 48 ± 6 | 0.001 [13] |

Superscript numbers correspond to significant post hoc pairwise comparisons listed below the table: [1] Age: G1 > G3 ($p$ = 0.005); G1 > G4 ($p$ < 0.001); G2 > G4 ($p$ = 0.001). [2] Heart rate: G2 > G3 ($p$ = 0.020). [3] Diastolic blood pressure: G2 > G3 ($p$ = 0.009). [4] Time to rehospitalization: G1 > G4 ($p$ = 0.006); G3 > G4 ($p$ = 0.046). [5] Creatinine: G1 < G3 ($p$ = 0.001); G1 < G4 ($p$ = 0.024). [6] eGFR: G1 > G3 ($p$ = 0.005); G1 > G4 ($p$ = 0.004). [7] Glucose: G1 < G2 ($p$ < 0.001); G1 < G4 ($p$ < 0.001); G2 > G3 ($p$ < 0.001); G3 < G4 ($p$ < 0.001). [8] Interventricular septum width: G1 < G2 ($p$ = 0.031); G1 < G4 ($p$ = 0.001). [9] Left ventricular end-diastolic diameter: G1 < G3 ($p$ < 0.001); G1 < G4 ($p$ < 0.001); G2 < G3 ($p$ < 0.001); G2 < G4 ($p$ < 0.001); G3 < G4 ($p$ < 0.001). [10] Posterior wall width: G1 < G4 ($p$ = 0.028); [11] Left ventricular mass: G1 < G3 ($p$ < 0.001); G1 < G4 ($p$ < 0.001); G2 < G3 ($p$ = 0.001); G2 < G4 ($p$ < 0.001). [12] Relative wall thickness: G1 > G3 ($p$ = 0.009); G2 > G3 ($p$ = 0.001); G2 > G4 ($p$ = 0.017). [13] Left ventricular ejection fraction: G1 > G3 ($p$ = 0.040); G1 > G4 ($p$ = 0.011); G2 > G3 ($p$ = 0.032); G2 > G4 ($p$ = 0.010); G3 > G4 ($p$ = 0.032).

Smoking was markedly more prevalent among non-diabetic and diabetic men, while women—especially those without diabetes—had lower smoking rates. The prevalence of hypercholesterolemia varied substantially, with diabetic men showing the highest rates. Atrial fibrillation was significantly more common among diabetic men, whereas non-diabetic women had lower-than-expected atrial fibrillation prevalence. Fatigue was more frequently reported by diabetic patients, particularly men. Conversely, chest pain and edema showed no significant distribution differences between groups. Severity grades of mitral and tricuspid regurgitation also displayed heterogeneous distributions, driven primarily by higher grades observed in male patients.

Use of beta-blockers, ACE inhibitors/angiotensin receptor–neprilysin inhibitors, mineralocorticoid receptor antagonists, loop diuretics, and amiodarone did not differ significantly across phenotypes. In contrast, SGLT2 inhibitor use differed significantly, being most frequent among diabetic men.

### 3.2. Cox Regression Analysis

In univariate Cox models (Table 2) evaluating the risk of HF rehospitalization, male sex (HR = 1.324, 95% CI 1.055–1.662, $p$ = 0.015) and type 2 diabetes mellitus (HR = 1.304, 95% CI 1.038–1.639, $p$ = 0.023) were both associated with a higher risk of rehospitalization. Active smoking also increased risk (HR = 1.206, 95% CI 1.016–1.432, $p$ = 0.032). Worse functional status reflected by a higher NYHA class strongly predicted adverse outcomes (HR = 1.289, 95% CI 1.081–1.537, $p$ = 0.005). Among echocardiographic measures, reduced TAPSE was a significant predictor of rehospitalization (HR = 0.937, 95% CI 0.894–0.982, $p$ = 0.007), and higher LVEF was modestly protective (HR = 0.982, 95% CI 0.964–0.999, $p$ = 0.039), consistent with the pathophysiological spectrum of HFmrEF/HFpEF.

Variables entered into the multivariate Cox proportional hazards model were selected based on clinical relevance rather than univariate statistical significance, in order to avoid omitted-variable bias. The final model included male sex, diabetes status, NYHA class, and LVEF.

After adjustment, male sex (HR 1.28, 95% CI 1.02–1.61, $p$ = 0.035) and diabetes (HR 1.28, 95% CI 1.02–1.61, $p$ = 0.036) remained independently associated with a higher risk of the composite outcome. NYHA class was also independently associated with the outcome (HR 1.24 per class increase, 95% CI 1.04–1.49, $p$ = 0.018). LVEF did not remain independently significant in the adjusted model (HR 0.99, $p$ = 0.234).

https://doi.org/10.3390/biomedicines14010173

To evaluate whether sex–diabetes phenotypes predicted HF rehospitalization beyond comorbidity burden, an additional Cox regression model was constructed including age, estimated glomerular filtration rate, and atrial fibrillation. After adjustment, the overall association between phenotype and outcome was attenuated ($p = 0.054$). Importantly, diabetic men continued to exhibit a significantly higher risk of HF rehospitalization compared with non-diabetic women (HR 1.58, 95% CI 1.14–2.18; $p = 0.006$), whereas other phenotypes were not independently associated with outcome.

**Table 2.** Univariate Cox regression analysis.

| Variable | *p*-Value | HR (95% CI) |
| --- | --- | --- |
| Age | <0.001 | 0.983 [0.974–0.992] |
| Sex (male) | 0.015 | 1.324 [1.055–1.662] |
| Type 2 diabetes mellitus | 0.023 | 1.304 [1.038–1.639] |
| Hypertension | 0.011 | 0.714 [0.552–0.925] |
| Hypercholesterolemia | 0.949 | 1.008 [0.795–1.277] |
| Smoking | 0.032 | 1.206 [1.016–1.432] |
| Chronic kidney disease | 0.759 | 1.038 [0.820–1.314] |
| NYHA class | 0.005 | 1.289 [1.081–1.537] |
| Sodium-glucose cotransporter-2 inhibitors use | 0.002 | 1.780 [1.233–2.571] |
| Betablockers use | 0.526 | 1.107 [0.809-1.515] |
| Loop diuretics use | 0.078 | 1.333 [0.969–1.835] |
| Left ventricular mass | 0.156 | 1.001 [1.000–1.002] |
| Relative wall thickness | 0.480 | 0.734 [0.311–1.733] |
| Systolic pulmonary artery pressure | 0.311 | 1.005 [0.995–1.016] |
| Tricuspid regurgitation severity grade | 0.681 | 1.021 [0.923–1.130] |
| Mitral regurgitation severity grade | 0.186 | 1.083 [0.962–1.218] |
| Tricuspid annulus plane systolic excursion | 0.007 | 0.937 [ 0.894–0.982] |
| Left ventricular ejection fraction | 0.039 | 0.982 [0.964–0.999] |

Abbreviations: CI, confidence interval; HR, hazard ratio.

### 3.3. *Kaplan–Meier Survival Analysis*

Kaplan–Meier survival curves (Figure 1) demonstrated significant differences in event-free survival across the four sex–diabetes groups (log-rank $p = 0.015$). Women without diabetes showed the most favorable prognosis, maintaining the highest survival probability throughout follow-up. Women with diabetes and men without diabetes displayed intermediate survival trajectories with comparable event-free survival curves. In contrast, men with diabetes experienced the poorest outcomes, exhibiting the steepest early decline and the lowest long-term event-free survival. This pattern highlights a clear interaction between sex and diabetic status, with the combination of male sex and diabetes conferring the greatest risk of adverse events.

https://doi.org/10.3390/biomedicines14010173

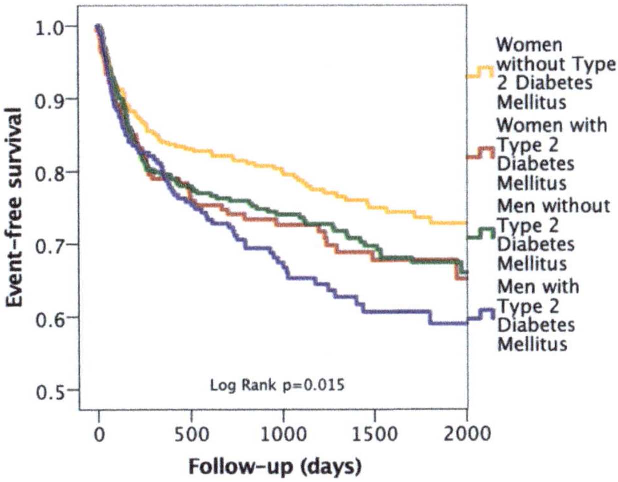


| **Number at risk** | | | | | |
|---|---|---|---|---|---|
| Women without T2DM | 331 | 274 | 204 | 135 | 31 |
| Women with T2DM | 166 | 127 | 189 | 62 | 19 |
| Men without T2DM | 308 | 240 | 180 | 128 | 32 |
| Men with T2DM | 206 | 157 | 96 | 55 | 10 |

**Figure 1.** Kaplan–Meier curves showing event-free survival across the four sex–diabetes phenotypes. Women without diabetes had the most favorable survival, whereas diabetic men showed the steepest early decline and worst long-term prognosis (log-rank $p$ = 0.015). Abbreviations: T2DM, type 2 Diabetes Mellitus.

## 4. Discussion

In this study of 1018 patients with HFmrEF/HFpEF, we investigated how sex and diabetes jointly influence clinical profiles and long-term outcomes. The main findings were that women without diabetes exhibited the most favorable prognosis, diabetic women and non-diabetic men showed intermediate risk, and diabetic men consistently demonstrated the highest likelihood of HF rehospitalization. This stepwise gradient appeared across baseline characteristics, structural and functional parameters, survival curves, and multivariable models. Importantly, our work provides novel insight by directly comparing combined sex–diabetes phenotypes within a single HFmrEF/HFpEF cohort—an approach not systematically explored in previous studies, which typically examined sex or diabetes effects in isolation. By evaluating these modifiers together, we highlight a biologically coherent and clinically meaningful interaction that helps explain the heterogeneous outcomes observed in patients with preserved-range EF.

### *4.1. Sex-Specific Differences in HFmrEF/HFpEF*

Previous epidemiologic and imaging studies have shown that women with HFpEF generally exhibit smaller LV cavities, greater concentric remodeling, higher ventricular–arterial elastance, and lower myocardial fibrosis [3]. These structural advantages correlate with the consistently lower mortality observed among women in trials such as CHARM-Preserved, I-Preserve, and TOPCAT [4]. In line with these findings, women without diabetes in our cohort showed the most favorable survival, suggesting that preserved diastolic reserve, superior microvascular function, and lower fibrosis burden contribute to a protective HF phenotype.

Men, on the other hand, demonstrated a phenotype characterized by larger LV volumes, increased fibrosis, greater arrhythmia burden, and worse right ventricular–

https://doi.org/10.3390/biomedicines14010173

pulmonary artery coupling [3,5]. These pathophysiological features likely underlie the poorer outcomes observed in both diabetic and non-diabetic men in our study. Notably, even non-diabetic men had worse prognosis than diabetic women, consistent with registry data showing male sex as an independent predictor of cardiac mortality [5].

### 4.2. Impact of Diabetes and Sex-Diabetes Interaction

Diabetes emerged as a key modifier of HF risk. Previous studies, including DIAMOND-CHF [11] and a meta-analysis [7], demonstrated a disproportionately higher relative HF risk in diabetic women. Mechanisms include enhanced endothelial dysfunction, microvascular rarefaction, inflammation, and loss of estrogen-mediated vascular protection [3,6]. In our cohort, diabetic women had significantly worse outcomes than non-diabetic women, and categorical findings—such as increased fatigue, atrial fibrillation prevalence, and valvular dysfunction—suggested that diabetes may accelerate structural deterioration in women.

Further support for the strong influence of diabetes on HF progression comes from metabolic and epidemiological studies [12]. Diabetes induces a shift from glucose oxidation toward fatty-acid oxidation, leading to impaired cardiac energetics, reduced contractile efficiency, and both systolic and diastolic dysfunction, even in the absence of coronary disease [13]. Hyperglycemia and insulin resistance additionally promote microvascular dysfunction, increased oxidative stress, local activation of the renin–angiotensin and sympathetic nervous systems, and enhanced myocardial fibrosis [13]. These metabolic disturbances explain why epidemiological data consistently demonstrate that diabetes increases the risk of developing HF more than twofold in men and more than fivefold in women [14,15], reinforcing the sex-specific vulnerability described in our cohort. Moreover, diabetes substantially increases recurrent HF hospitalizations and mortality across both reduced and preserved EF phenotypes, aligning with the markedly higher rehospitalization rates observed among diabetic patients in our study. The CHARM program similarly reported a twofold higher risk of cardiovascular death or HF hospitalization in insulin-treated diabetics [15]. Collectively, these data strengthen the biological plausibility of our findings, supporting diabetes as a major driver of adverse remodeling and HF morbidity, and highlighting the importance of integrating metabolic status into risk stratification frameworks.

However, despite this higher relative risk in women, men consistently experience a greater absolute burden of adverse HF events [8,16]. This pattern was reflected in our cohort: diabetic men had the highest comorbidity burden, most adverse remodeling, and the steepest decline in event-free survival. Thus, although diabetes confers a stronger relative risk in women, its absolute effect is most severe in men. These findings suggest that sex–diabetes phenotypes capture both biological differences and clustering of adverse comorbidities, with diabetic men representing a particularly high-risk subgroup even after adjustment for age, renal function, and atrial fibrillation. The observed association between atrial fibrillation and lower rehospitalization risk likely reflects differences in monitoring intensity and healthcare contact rather than a protective biological effect.

Although part of the observed risk gradient may reflect clustering of comorbidities across the four phenotypes, major HFpEF and HFmrEF trials and registries have consistently demonstrated that sex and diabetes are not merely correlates of comorbidity burden, but key modifiers of phenotype and prognosis. Sex-related differences in outcomes and clinical profiles have been reported across trials such as CHARM-Preserved [15], I-Preserve [17], and TOPCAT [18], with women demonstrating better survival despite a substantial comorbidity burden. Furthermore, diabetes has repeatedly been associated with higher rates of adverse outcomes in heart failure, including in the CHARM study. In

https://doi.org/10.3390/biomedicines14010173

this context, our findings are unlikely to be explained solely by comorbidity clustering and instead appear to reflect distinct risk phenotypes in which sex and diabetes act as meaningful modifiers within a broader comorbidity framework.

### 4.3. Structural, Functional, and Comorbidity Differences Across Phenotypes

Sex-specific anatomical and physiological differences further shape HF outcomes. Women generally have smaller LV size, lower stroke volume, and higher pulsatile load [19,20], while men more frequently develop ischemic scar and eccentric remodeling [21]. With diabetes, these differences intensify: women exhibit more pronounced concentric LV hypertrophy, while men demonstrate greater fibrosis and ischemic injury [21]. Additional evidence shows faster HF progression and more pronounced LV hypertrophy in diabetic patients, even at lower body-mass index, driven by oxidative stress, metabolic dysfunction, microvascular impairment, and extracellular matrix remodeling [22].

In our study, baseline continuous and categorical variables reinforced this gradient. Women without diabetes were the oldest yet had the most preserved renal function. Men—especially diabetic men—showed larger LV dimensions, greater LV mass, lower TAPSE, more atrial fibrillation, higher smoking prevalence, and more severe tricuspid regurgitation and mitral regurgitation grades. Diabetic patients exhibited higher glucose levels and shorter time to first rehospitalization, emphasizing the close link between metabolic dysfunction and HF progression.

### 4.4. Prognostic and Clinical Implications

Univariate Cox regression identified male sex, diabetes, smoking, NYHA class, TAPSE, and LVEF as significant predictors of rehospitalization. After adjustment for NYHA class and LVEF, male sex and diabetes remained independent predictors, underscoring their strong and consistent prognostic effect across HFmrEF/HFpEF. Kaplan–Meier curves showed a clear, stepwise separation of the four phenotypes, with women without diabetes showing the best outcomes and diabetic men the worst.

Taken together, these findings support a unified mechanistic framework. Women without diabetes appear to maintain a distinctly favorable cardiovascular profile, characterized by preserved microvascular function, less fibrosis, and more efficient ventricular–vascular coupling. The presence of diabetes alters this otherwise advantageous profile in women, shifting them into an intermediate-risk category by amplifying metabolic stress, microvascular dysfunction, and diastolic impairment. Men, even in the absence of diabetes, display a phenotype that is intrinsically more vulnerable, marked by greater myocardial fibrosis, more pronounced remodeling, and higher arrhythmic and ischemic burden. When diabetes is added to this already high-risk male substrate, the combined effect results in the most pronounced structural deterioration and the poorest clinical outcomes, producing the highest likelihood of recurrent HF decompensation observed in our cohort. This model aligns with the mechanistic and epidemiologic patterns previously described [3,8,16].

Clinically, our findings reinforce the need for sex-specific and diabetes-specific risk stratification in HFmrEF/HFpEF. Diabetic women require aggressive metabolic and cardiovascular risk control given their disproportionate relative risk, whereas diabetic men represent a uniquely high-risk endotype requiring intensified ischemic, arrhythmic, and RV-focused evaluation. Recognizing non-diabetic women as the lowest-risk group may refine follow-up intensity and therapeutic strategies. Incorporating sex–diabetes phenotyping into routine clinical assessment may therefore enhance personalized prognostication and improve management across the HFmrEF/HFpEF spectrum.

https://doi.org/10.3390/biomedicines14010173

### 4.5. Limitations

This study has several limitations. First, its retrospective, single-center design may limit generalizability and introduces the risk of residual confounding. Second, stratifying patients into four sex–diabetes phenotypes, although essential to the study's aim, reduced statistical power for some subgroup comparisons. Third, key variables such as glycemic control, diabetes duration, HbA1c values, diabetes-related complications, detailed antidiabetic treatment and detailed ischemic burden were not consistently available and may have influenced outcomes. Medication use was assessed at discharge only, without information on dose, duration, or adherence. The higher prevalence of SGLT2 inhibitor use among diabetic men likely reflects confounding by indication rather than treatment effect. In addition, anthropometric measurements such as body mass index (BMI) and body surface area (BSA) were not included because height was not systematically recorded, precluding accurate calculation for most patients; to avoid bias arising from incomplete or inconsistent data, BMI-, BSA-, and related indexed variables were therefore not analyzed. Consequently, the inability to account for glycemic control, diabetes severity and intensity of treatment may have led to residual confounding in the observed associations. Fourth, echocardiographic measurements were obtained from routine clinical practice rather than core-lab adjudication, allowing for potential measurement variability. Advanced echocardiographic parameters with demonstrated prognostic value in patients with HF were not included due to the retrospective design of the study, and the use of left atrial diameter instead of indexed left atrial volume may have underestimated atrial remodeling compared with guideline-recommended parameters. Finally, heart failure rehospitalization—while clinically relevant—may be affected by healthcare access and admission thresholds, and, given the retrospective design and reliance on discharge diagnoses, misclassification of events cannot be excluded, particularly for admissions occurring outside the study center.

## 5. Conclusions

Sex and diabetes strongly influence the risk of HF rehospitalization in HFmrEF/HFpEF, with women without diabetes experiencing the lowest risk and diabetic men the highest. Diabetes appears to attenuate the protective advantage seen in women, while male sex is associated with greater diabetes-related vulnerability, potentially reflecting underlying risk clustering. Our findings support integrating sex–diabetes profiling into routine risk assessment and management.

**Author Contributions:** Conceptualization, F.-M.S., D.-R.H. and C.F.; methodology, D.-R.H. and C.F.; formal analysis, F.-M.S., D.-R.H., C.-D.H., M.-L.I., G.-C.T., I.D., P.-A.C. and S.M.; investigation, F.-M.S., D.-R.H., C.-D.H., M.-L.I., G.-C.T., I.D., P.-A.C., S.M. and O.I.; resources, F.-M.S. and C.F.; data curation, F.-M.S. and D.-R.H.; writing—original draft preparation, F.-M.S., D.-R.H. and C.F.; writing—review and editing, F.-M.S., D.-R.H., C.-D.H., M.-L.I., G.-C.T., I.D., P.-A.C., S.M., O.I. and C.F.; supervision, D.-R.H., I.D., O.I. and C.F.; project administration, D.-R.H. and C.F.; funding acquisition, F.-M.S. and C.F. All authors have read and agreed to the published version of the manuscript.

**Funding:** This research was funded by The University of Medicine and Pharmacy of Craiova, Romania. The Article Processing Charges were funded by The University of Medicine and Pharmacy of Craiova, Romania.

**Institutional Review Board Statement:** The study was conducted in accordance with the Declaration of Helsinki, and approved by the Ethics Committee of The University of Medicine and Pharmacy of Craiova (protocol code 86, date of approval 19 February 2024).

**Informed Consent Statement:** Patient consent was waived due to the retrospective nature of the study.

https://doi.org/10.3390/biomedicines14010173

**Data Availability Statement:** The raw data supporting the conclusions of this article will be made available by the authors upon reasonable request and following approval by the University of Medicine and Pharmacy of Craiova, Romania. The data presented in this study are not publicly available due to privacy and ethical restrictions.

**Conflicts of Interest:** The authors declare no conflicts of interest. The funders had no role in the design of the study; in the collection, analyses, or interpretation of data; in the writing of the manuscript; or in the decision to publish the results.

## Abbreviations

The following abbreviations are used in this manuscript:

| | |
|---|---|
| BMI | Body mass index |
| BSA | Body surface area |
| EF | Ejection fraction |
| eGFR | Estimated glomerular filtration rate |
| ESC | European Society of Cardiology |
| HF | Heart failure |
| HFmrEF | Heart failure with mildly reduce ejection fraction |
| HFpEF | Heart failure with preserved ejection fraction |
| HR | Hazard ratio |
| LV | Left ventricle/ventricular |
| LVEF | Left ventricular ejection fraction |
| NT-proBNP | N-terminal pro-B-type natriuretic peptide |
| NYHA | New York Heart Association |
| TAPSE | Tricuspid annular plane systolic excursion |

## References

1. McDonagh, T.A.; Metra, M.; Adamo, M.; Baumbach, A.; Böhm, M.; Burri, H.; Čelutkiene, J.; Chioncel, O.; Cleland, J.G.F.; Coats, A.J.S.; et al. 2021 ESC Guidelines for the Diagnosis and Treatment of Acute and Chronic Heart Failure. *Eur. Heart J.* **2021**, *42*, 3599–3726. [CrossRef] [PubMed]
2. Marx, N.; Federici, M.; Schütt, K.; Müller-Wieland, D.; Ajjan, R.A.; Antunes, M.J.; Christodorescu, R.M.; Crawford, C.; Di Angelantonio, E.; Eliasson, B.; et al. 2023 ESC Guidelines for the Management of Cardiovascular Disease in Patients with Diabetes. *Eur. Heart J.* **2023**, *44*, 4043–4140. [CrossRef] [PubMed]
3. Sotomi, Y.; Hikoso, S.; Nakatani, D.; Mizuno, H.; Okada, K.; Dohi, T.; Kitamura, T.; Sunaga, A.; Kida, H.; Oeun, B.; et al. Sex Differences in Heart Failure with Preserved Ejection Fraction. *J. Am. Heart Assoc.* **2021**, *10*, e018574. [CrossRef] [PubMed]
4. Dewan, P.; Rørth, R.; Raparelli, V.; Campbell, R.T.; Shen, L.; Jhund, P.S.; Petrie, M.C.; Anand, I.S.; Carson, P.E.; Desai, A.S.; et al. Sex-Related Differences in Heart Failure with Preserved Ejection Fraction. *Circ Heart Fail.* **2019**, *12*, e006539. [CrossRef]
5. Duca, F.; Zotter-Tufaro, C.; Kammerlander, A.A.; Aschauer, S.; Binder, C.; Mascherbauer, J.; Bonderman, D. Gender-Related Differences in Heart Failure with Preserved Ejection Fraction. *Sci. Rep.* **2018**, *8*, 1080. [CrossRef]
6. Fourny, N.; Beauloye, C.; Bernard, M.; Horman, S.; Desrois, M.; Bertrand, L. Sex Differences of the Diabetic Heart. *Front. Physiol.* **2021**, *12*, 661297. [CrossRef]
7. Ohkuma, T.; Komorita, Y.; Peters, S.A.E.; Woodward, M. Diabetes as a Risk Factor for Heart Failure in Women and Men: A Systematic Review and Meta-Analysis of 47 Cohorts Including 12 Million Individuals. *Diabetologia* **2019**, *62*, 1550–1560. [CrossRef]
8. Chadalavada, S.; Jensen, M.T.; Aung, N.; Cooper, J.; Lekadir, K.; Munroe, P.B.; Petersen, S.E. Women With Diabetes Are at Increased Relative Risk of Heart Failure Compared to Men: Insights From UK Biobank. *Front. Cardiovasc. Med.* **2021**, *8*, 658726. [CrossRef]
9. Liu, Z.; Zhu, Y.; Chen, S.; Wu, M.; Huang, H.; Peng, K.; Zhang, L.; Zhao, W.; Peng, X.; Li, N.; et al. Sex Differences in Patients with Heart Failure and Mildly Reduced Left Ventricular Ejection Fraction. *Sci. Rep.* **2023**, *13*, 6832. [CrossRef]
10. Lang, R.M.; Badano, L.P.; Mor-Avi, V.; Afilalo, J.; Armstrong, A.; Ernande, L.; Flachskampf, F.A.; Foster, E.; Goldstein, S.A.; Kuznetsova, T.; et al. Recommendations for Cardiac Chamber Quantification by Echocardiography in Adults: An Update from the American Society of Echocardiography and the European Association of Cardiovascular Imaging. *Eur. Heart J. Cardiovasc. Imaging* **2015**, *16*, 233–271. [CrossRef]

https://doi.org/10.3390/biomedicines14010173

11. Gustafsson, I.; Brendorp, B.; Seibæk, M.; Burchardt, H.; Hildebrandt, P.; Køber, L.; Torp-Pedersen, C. Influence of Diabetes and Diabetes-Gender Interaction on the Risk of Death in Patients Hospitalized with Congestive Heart Failure. *J. Am. Coll. Cardiol.* **2004**, *43*, 771–777. [CrossRef] [PubMed]
12. Rosano, G.M.; Vitale, C.; Seferovic, P. Heart Failure in Patients with Diabetes Mellitus. *Card. Fail. Rev.* **2017**, *3*, 52. [CrossRef] [PubMed]
13. Rosano, G.; Fini, M.; Caminiti, G.; Barbaro, G. Cardiac Metabolism in Myocardial Ischemia. *Curr. Pharm. Des.* **2008**, *14*, 2551–2562. [CrossRef] [PubMed]
14. Metra, M.; Zacà, V.; Parati, G.; Agostoni, P.; Bonadies, M.; Ciccone, M.; Cas, A.D.; Iacoviello, M.; Lagioia, R.; Lombardi, C.; et al. Cardiovascular and Noncardiovascular Comorbidities in Patients with Chronic Heart Failure. *J. Cardiovasc. Med.* **2011**, *12*, 76–84. [CrossRef]
15. MacDonald, M.R.; Petrie, M.C.; Varyani, F.; Ostergren, J.; Michelson, E.L.; Young, J.B.; Solomon, S.D.; Granger, C.B.; Swedberg, K.; Yusuf, S.; et al. Impact of Diabetes on Outcomes in Patients with Low and Preserved Ejection Fraction Heart Failure: An Analysis of the Candesartan in Heart Failure: Assessment of Reduction in Mortality and Morbidity (CHARM) Programme. *Eur. Heart J.* **2008**, *29*, 1377–1385. [CrossRef]
16. Kim, H.-L.; Kim, M.-A.; Park, K.-T.; Choi, D.-J.; Han, S.; Jeon, E.-S.; Cho, M.-C.; Kim, J.-J.; Yoo, B.-S.; Shin, M.-S.; et al. Gender Difference in the Impact of Coexisting Diabetes Mellitus on Long-term Clinical Outcome in People with Heart Failure: A Report from the Korean Heart Failure Registry. *Diabet. Med.* **2019**, *36*, 1312–1318. [CrossRef]
17. Komajda, M.; Carson, P.E.; Hetzel, S.; McKelvie, R.; McMurray, J.; Ptaszynska, A.; Zile, M.R.; DeMets, D.; Massie, B.M. Factors Associated with Outcome in Heart Failure with Preserved Ejection Fraction: Findings From the Irbesartan in Heart Failure with Preserved Ejection Fraction Study (I-PRESERVE). *Circ. Heart Fail.* **2011**, *4*, 27–35. [CrossRef]
18. Pitt, B.; Pfeffer, M.A.; Assmann, S.F.; Boineau, R.; Anand, I.S.; Claggett, B.; Clausell, N.; Desai, A.S.; Diaz, R.; Fleg, J.L.; et al. Spironolactone for Heart Failure with Preserved Ejection Fraction. *N. Engl. J. Med.* **2014**, *370*, 1383–1392. [CrossRef]
19. Hartman, R.J.G.; Huisman, S.E.; Den Ruijter, H.M. Sex Differences in Cardiovascular Epigenetics—A Systematic Review. *Biol. Sex Differ.* **2018**, *9*, 19. [CrossRef]
20. Lejeune, S.; Roy, C.; Slimani, A.; Pasquet, A.; Vancraeynest, D.; Vanoverschelde, J.-L.; Gerber, B.L.; Beauloye, C.; Pouleur, A.-C. Diabetic Phenotype and Prognosis of Patients with Heart Failure and Preserved Ejection Fraction in a Real Life Cohort. *Cardiovasc. Diabetol.* **2021**, *20*, 48. [CrossRef]
21. Li, X.-M.; Jiang, L.; Guo, Y.-K.; Ren, Y.; Han, P.-L.; Peng, L.-Q.; Shi, R.; Yan, W.-F.; Yang, Z.-G. The Additive Effects of Type 2 Diabetes Mellitus on Left Ventricular Deformation and Myocardial Perfusion in Essential Hypertension: A 3.0 T Cardiac Magnetic Resonance Study. *Cardiovasc. Diabetol.* **2020**, *19*, 161. [CrossRef]
22. Hart, E.C.; Joyner, M.J.; Wallin, B.G.; Charkoudian, N. Sex, Ageing and Resting Blood Pressure: Gaining Insights from the Integrated Balance of Neural and Haemodynamic Factors. *J. Physiol.* **2012**, *590*, 2069–2079. [CrossRef]




https://doi.org/10.3390/biomedicines14010173

biomedicines

MDPI

*Article*

# Sex Differences in MASLD After Age 50: Presentation, Diagnosis, and Clinical Implications

**Ilaria Milani [1,*,†], Maria Eugenia Parrotta [2,3,†], Luca Colangeli [2,3], Marianna Chinucci [1], Simonetta Palleschi [4], Barbara Rossi [4], Paolo Sbraccia [2,3], Alessandro Mantovani [5,6], Frida Leonetti [1], Valeria Guglielmi [2,3,‡] and Danila Capoccia [1,‡]**

1 Department of Medico-Surgical Sciences and Biotechnologies, Faculty of Pharmacy and Medicine, University of Rome La Sapienza, 04100 Latina, Italy; chinuccimarianna@gmail.com (M.C.); frida.leonetti@uniroma1.it (F.L.); danila.capoccia@uniroma1.it (D.C.)

2 Department of Systems Medicine, University of Rome Tor Vergata, 00161 Rome, Italy; mariaeugenia.parrotta@ptvonline.it (M.E.P.); luca.colangeli.ptv@gmail.com (L.C.); sbraccia@med.uniroma2.it (P.S.); valeria.guglielmi@uniroma2.it (V.G.)

3 Obesity Medical Center, University Hospital Policlinico Tor Vergata, 00161 Rome, Italy

4 Department of Environment and Health, Istituto Superiore di Sanità, Viale Regina Elena 299, 00161 Rome, Italy; simonetta.palleschi@iss.it (S.P.); barbara.rossi@iss.it (B.R.)

5 Department of Medicine, University of Verona, 37129 Verona, Italy; alessandro.mantovani@univr.it

6 Metabolic Diseases Research Unit, IRCCS Sacro Cuore-Don Calabria Hospital, 37024 Negrar di Valpolicella, Italy

* Correspondence: ilaria.milani@uniroma1.it

† These Authors contributed equally to this work and served as co-first authors.

‡ These Authors contributed equally to this work and served as co-last authors.

## Abstract

**Background**: Age over 50, menopause, obesity and type 2 diabetes (T2D) are key risk factors for Metabolic dysfunction-associated steatotic liver disease (MASLD). This observational study aimed to assess sex differences in anthropometric and clinical profile, including non-invasive liver steatosis indices, in subjects with MASLD, obesity and/or T2D, aged $\geq$ 50 years. **Methods**: Anthropometric and clinical parameters, non-invasive indices for steatosis and fibrosis and FibroScan® data were collected. **Results**: Among 213 patients (65.7% women, median age 63.0 years and mean Body Mass Index (BMI 34.9 kg/m$^2$), men had higher body weight and waist circumference (WC), whereas women showed higher BMI and waist-to-height ratio (WHtR), and were more likely to exceed WC sex-specific and WHtR risk cut-offs. While transaminases values were higher in men, sex-specific cut-offs revealed that women more frequently exceeded these thresholds. No sex-differences were found for Fatty Liver Index (FLI), Fibrosis-4 (FIB-4) or FibroScan®, although higher rate of mild fibrosis in women. The diagnostic accuracy of FLI for detecting steatosis was significantly higher in men and unsatisfactory in women (Area Under the ROC Curve, AUC 0.863 vs. 0.655). **Conclusions**: While MASLD is more common in men, these results suggest that postmenopausal women with visceral obesity showed similar or worse liver and cardiometabolic profiles than men, despite appearing healthier based on standard clinical parameters. Notably, common markers like transaminases and the FLI were less accurate in detecting steatosis in women, underscoring the need for sex-specific diagnostic criteria and greater clinical attention to older women, particularly those with central obesity, to ensure early identification and management of MASLD.

**Keywords:** MASLD; obesity; sex-differences; type 2 diabetes; fibrosis; steatosis; FibroScan

## 1. Introduction

Metabolic dysfunction-associated steatotic liver disease (MASLD) is the most common chronic liver disease, affecting approximately 38.2% of adults worldwide [1–3]. It is diagnosed by the presence of liver steatosis with at least one cardiometabolic risk factor [4,5] and is associated with an increased risk of cardiovascular (CV) events and liver-related complications [4].

Early diagnosis of MASLD is essential, highlighting the need for the implementation of accessible, non-invasive and reliable diagnostic tools beyond standard reference imaging techniques [6,7]. Persistently elevated liver enzymes, such as alanine aminotransferase (ALT) and aspartate aminotransferase (AST), should prompt evaluation for metabolic dysfunction-associated steatohepatitis (MASH) and advanced fibrosis [4]. However, many patients with MASLD, MASH and advanced fibrosis may present with normal transaminase levels [4,8,9], underscoring the limited diagnostic accuracy of liver enzymes alone. To improve early detection, several non-invasive indices have been developed [10]. Among these, the Fatty Liver Index (FLI) is the only validated marker with strong diagnostic and prognostic value for identifying steatosis [6,11], while the Fibrosis-4 Index (FIB-4) is recommended for excluding advanced fibrosis [4,12]. Although clinical evaluation, laboratory tests, and ultrasound can help diagnose steatosis, they are not sufficient to assess its severity [7]. For this purpose, vibration-controlled transient elastography (VCTE) serves as a valuable second-line tool, providing a reliable estimate of both steatosis and fibrosis stage [4].

In the United States, the median age of people with MASLD was 50 years in 2015 [13], highlighting significant influence of aging on disease development [14,15]. Age-related changes in body composition are key contributors to the rising prevalence of MASLD and associated metabolic disorders [16]. Beyond age, sex and menopausal status also play a critical role [17–19], with notable sex-based differences in both the prevalence and severity of the disease. Men consistently exhibit a higher prevalence of MASLD [20–22] across all age groups, whereas in women, the risk increases with age [23], reaching or surpassing that of men after menopause [19,22,24–26].

While sex-differences in MASLD are increasingly acknowledged [17,19,24,27,28], sex-specific research, particularly studies using non-invasive diagnostic tools [29], in patients over 50 remains limited.

This observational study aims to evaluate sex differences in a population of MASLD over 50 years of age in anthropometric data, clinical profile (biochemical markers and non-invasive indicators of steatosis, fibrosis and cardiometabolic risk), and FibroScan® assessments. We will also evaluate the influence of sex on the FLI prediction of liver steatosis (assessed by FibroScan®).

## 2. Materials and Methods

### 2.1. Study Design and Participants

A cross-sectional observational study was conducted in all consecutive patients aged $\geq$ 50 years and diagnosed with MASLD [4,30,31] at the Diabetes and Obesity Clinic of the Santa Maria Goretti Hospital in Latina, Italy, and the Obesity Medical Center of the Policlinico Tor Vergata Hospital in Rome, Italy from September 2024 to April 2025.

This is a secondary analysis of cohort data collected for the purpose of evaluating the efficacy of serum biomarkers in monitoring MASLD progression and metabolic complications over time. Considering that menopause typically occurs between the ages of 45–55 [32], with 51 being the generally accepted as the average age [33,34], the use of age > 50 as a proxy for menopausal status in women may help address this gap in the literature.

Exclusion criteria were as follows: (1) age younger than 50 years; (2) concomitant chronic liver diseases (e.g., viral, drug-induced liver injury, autoimmune hepatitis, hemochromatosis, primary sclerosing cholangitis, primary biliary cholangitis); (3) the presence of hepatocellular carcinoma or other malignancies; (4) excessive alcohol intake, defined as intake of >140 g/week for women, >210 g/week for men, detected with Alcohol Use Disorders Identification Test (AUDIT) questionnaire [4]; (5) clinical features suggestive of endocrinological disorders (Cushing syndrome, hypothyroidism, acromegaly, or Polycystic ovary syndrome); (6) pregnancy.

### 2.2. Anthropometric Assessment

Anthropometric measurements, including height (Ht), body weight (BW), and waist circumference (WC), were obtained during the clinical visit. Height and weight were measured with participants barefoot and wearing light clothing using a calibrated scale equipped with a standard stadiometer. BMI was calculated as weight in kilograms divided by the square of height in meters ($kg/m^2$).

WC was measured at the level of the umbilicus with the patient standing upright, using a non-elastic anthropometric tape. Abdominal obesity was defined as a waist circumference $\geq$80 cm in women and $\geq$94 cm in men as a marker of visceral obesity and high risk of metabolic complications [4]. waist-to-height ratio (WHtR) was calculated by dividing WC (cm) by Ht (cm), with a fixed threshold of $\geq$0.50 used to indicate increased cardiometabolic risk [35].

### 2.3. Laboratory Analysis

After an overnight fast of at least 10 h, venous blood samples were obtained from all participants for biochemical analysis. Laboratory assessments included measurements of triglycerides (TG, mg/dL), total cholesterol (TC, mg/dL), high-density lipoprotein cholesterol (HDL-C, mg/dL), fasting plasma glucose (FPG, mg/dL), alanine aminotransferase (ALT, IU/L), aspartate aminotransferase (AST, IU/L), gamma-glutamyl transferase (GGT, IU/L), serum creatinine (Cr, mg/dL), and C-reactive protein (CRP, mg/L).

Quantitative estimation of biochemical parameters was performed in the central laboratory of the institute hospital participating to this study according to standard laboratory procedures using proprietary reagents on the fully automated biochemical analyzer Abbott Architect c16000 (Abbott Diagnostics, Abbott Park, IL, USA). Hemoglobin A1c (HbA1c) levels were determined using a high-performance liquid chromatography analyzer (Lifotronic Technology Co., Ltd., Shenzhen, China). Low-density lipoprotein cholesterol (LDL-C, mg/dL) was estimated using the Friedewald formula.

### 2.4. Non-Invasive Liver Disease Assessment (NILDAs)

MASLD was diagnosed in patients with evidence of hepatic steatosis detected by abdominal ultrasound or a FLI score $\geq$ 60 [30], in the presence of at least one cardiometabolic risk factor and in the absence of daily alcohol consumption (defined as <210 g/week for women and <140 g/week for men) or other liver diseases, in accordance with recent clinical recommendations [4,31].

The FLI was calculated using the established formula that includes BMI, WC, TG, and GGT levels [30]: [e^(0.953 × ln(TG) + 0.139 × BMI + 0.718 × ln(GGT) + 0.053 × WC − 15.745)]/[1 + e^(0.953 × ln(TG) + 0.139 × BMI + 0.718 × ln(GGT) + 0.053 × WC − 15.745)] × 100.

Although the FLI was originally validated to identify NAFLD, recent studies have shown that MASLD overlaps almost completely with the NAFLD population [36–39].

FIB-4 was determined by using the following formula: age [years] × AST [U/L]/(platelet count [$10^9$/L] × $\surd$(ALT [U/L])) (where $\surd$ denotes the square root of ALT, measured in units per liter) [4].

For each patients the AST/ALT ratio was calculated and a value of 0.8 was considered to exclude advanced fibrosis [4].

All participants underwent VCTE with FibroScan® Compact 530 after an overnight fast. With the patient in the supine position and the right arm abducted, the ultrasound probe was performed by professionally trained operators on the right lobe of the liver through intercostal spaces. The median value of the successful liver stiffness measurement (LSM) was expressed in kilopascals (kPa), while the median value of the successful Controlled Attenuation Parameter (CAP) score was expressed in decibels per meter (dB/m). Only LSM and CAP score data, acquired from at least 10 valid measurements, and IQR/median for both LSM and CAP score of <30% were considered reliable. Hepatic steatosis was simultaneously assessed by CAP and reported in decibels per meter (dB/m). Grades of steatosis were defined as follows:

- S0 (no steatosis): <248 dB/m
- S1 (mild): 248–267 dB/m
- S2 (moderate): 268–279 dB/m
- S3 (severe): ≥280 dB/m [4,40].

Fibrosis staging was based on the following LSM thresholds:

- F0: <7.9 kPa
- F1 (significant): ≥7.9–9.4 kPa
- F2 (advanced): 9.5–12.4 kPa
- F3–4 (cirrhosis): ≥12.5 kPa

These cut-offs are consistent with established guidelines for non-invasive assessment of liver fibrosis [4,41,42].

### 2.5. Cardiometabolic and Adipose Tissue Dysfunction Indices

Visceral adiposity index (VAI) was defined using WC, BMI, TG, and HDL-C according to the published sex-specific formula:

$$\text{Males: VAI} = (\text{WC [cm]}/(39.68 + (1.88 \times \text{BMI [kg/m}^2\text{]}))) \times (\text{TG [mmol/L]}/1.03) \times (1.31/\text{HDL [mmol/L]})$$

$$\text{Females: VAI} = (\text{WC [cm]}/(36.58 + (1.89 \times \text{BMI [kg/m}^2\text{]}))) \times (\text{TG [mmol/L]}/0.81) \times (1.52/\text{HDL [mmol/L]})$$

A VAI cut-off of 1.93 was used to identify adipose tissue dysfunction [43,44].

The TyG Index (TyG) was calculated as: Ln [TG (mg/dL) × fasting glucose (mg/dL)/2].

TyG-WHtR [45,46] was calculated as TyG index x WHtR.

### 2.6. Lifestyle and Other Variables

Data on age (years), sex, T2D and hypertension diagnosis, use of antihypertensive or antihyperlipidemic medication, cigarette smoking status were collected through oral interviews. Smoking status was classified into three categories: current smoker, former smoker, and never smoker. T2D was defined as fasting plasma glucose ≥ 126 mg/dL, 2-h plasma glucose ≥ 200 mg/dL after a 75-g oral glucose tolerance test, HbA1c ≥ 6.5%, clinical diagnosis by a clinician, and/or use of antidiabetic medication [47]. Hypertension was defined as systolic blood pressure (SBP) ≥ 140 mmHg and/or diastolic blood pressure (DBP) ≥ 90 mmHg or current use of antihypertensive medication.

Alcohol consumption was assessed using the validated Alcohol Use Disorders Identification Test (AUDIT) questionnaire. A score below 8 was considered indicative of low-risk drinking or abstinence [48]. In addition, alcohol consumption was quantified as the average number of standard drinks consumed per week.

Adherence to the Mediterranean diet was assessed using the adapted 14-item Mediterranean Diet Adherence Screener (MEDAS). A score of ≥9 was used as the threshold to categorize participants into high or low adherence groups [49].

### 2.7. Ethics Statement

The study was conducted in accordance with the Declaration of Helsinki [50], and approved by the "Comitato Etico Lazio 1", the Ethics Committee of the Lazio Region (reference no. 7421, protocol code 0038/2024 and date of approval 10 January 2024). Written informed consent was obtained from all participants.

### 2.8. Statistical Analysis

Calculation of the sample size required to detect a medium effect size (Cohen's d = 0.5) at a significance level of $\alpha = 0.05$ and a desired power of 0.90 indicated a total sample size of 192 participants, including 64 in group 1 and 128 in group 2 (Power estimation was performed using G-Power version 3.1.9.4 software; Heinrich Heine University Düsseldorf, Düsseldorf, Germany).

The Kolmogorov-Smirnov test was used to assess the normality of the data distribution. Group comparisons were performed using Student's *t*-test, Mann-Whitney U test, or chi-squared test, as appropriate. Continuous variables were expressed as mean ± standard deviation (SD) for normally distributed data or as median and interquartile range (IQR) for non-normally distributed data. Categorical variables were expressed as percentages.

A multivariate analysis of variance (MANOVA) was performed to examine the significant effect of sex on two dependent variables (CAP and LSM), adjusted for potential confounders.

Logistic regression model was used to evaluate the predictive ability of FLI for liver steatosis defined by CAP ≥ 248 dB/m, with adjustment for confounders and Receiver Operating Characteristic (ROC) curves were generated separately for men and women. To test the statistical difference between the two AUC curves, the DeLong test statistic was applied (using the free statistical software R 4.5.1).

Sex-stratified cross-tab analyses were performed to assess the association between FLI ≥ 60 and liver steatosis (CAP ≥ 248 dB/m), and sensitivity, specificity, positive and negative predictive values and accuracy of this cut-off were calculated for each sex.

In all analyses, a *p*-value < 0.05 was considered statistically significant. Statistical analyses were performed with SPSS software (version 21.0, IBM Corp, Armonk, NY, USA).

## 3. Results

A total of 213 patients were included in the analysis (Participants flow-diagram Figure S1 Supplementary Materials). General and anthropometric characteristics of the study population are shown in Table 1. In the overall population, women accounted for 65.7% (140/213). The median age was 63.0 years (62.7–64.6), with a mean BMI of 34.9 kg/m$^2$ (±5.5), median BW of 92.5 kg (92.9–97.7), mean WC of 114.2 (±12.7), and mean WHtR of 0.68 (±0.07). High prevalence of comorbidities such as T2D, hypertension and dyslipidemia have been observed in the overall population, without significant sex-differences. Higher significant percentage of men were active smokers (16.4% vs. 8.6%, $p < 0.001$) and former smokers (39.7% vs. 17.1%, $p < 0.001$), while no significant sex differences were observed for adherence to the Mediterranean diet assessed by MEDAS scores.

With regard to anthropometric parameters, body weight and WC were significantly higher in men compared to women (102.0 (99.5–107.5) vs. 88.5 (88.1–93.7) kg $p < 0.001$; 117.1 ± 12.4 vs. 112.7 ± 12.6 cm, $p < 0.001$). In contrast, WHtR was significantly higher in women than in men (0.69 ± 0.07 vs. 0.64 ± 0.12, $p < 0.001$), with the same trend for BMI (35.4 ± 5.6 vs. 33.8 ± 5.1 kg/m$^2$, $p = 0.051$). Furthermore, the proportion of patients exceeding the sex-specific WC risk thresholds and those with a WHtR > 0.5 was significantly higher in women than in men (100% vs. 91.8%; 100% vs. 95.9%, $p = 0.02$, respectively).

With regard to biochemical and clinical parameters (Table 2), women had significantly higher levels of total cholesterol (181.3 ± 41.1 vs. 162.5 ± 38.8, mg/dL, $p = 0.001$) and

LDL-C (105.6 ± 36.9 vs. 88.6 ± 34.3, mg/dL, $p$ = 0.02) than men. HDL-C levels were higher in women (52.0 (51.9–55.9) mg/dL) vs. 47.0 (45.5–56.7) mg/dL), although lower than the normal sex-specific HDL-cholesterol cut-offs. With regard to chronic low-grade inflammation, CRP concentrations were significantly higher in women than in men (0.35 (0.4–0.7) vs. 0.30 (0.2–0.6) mg/dL, $p$ = 0.001).

**Table 1.** General and anthropometric characteristics of the study population.

| | Overall (n = 213) | Men (n = 73) | Women (n = 140) | *p*-Value |
|---|---|---|---|---|
| Age (years) | 63.0 (62.7–64.6) | 63.0 (62.8–66.2) | 63.0 (61.9–64.2) | 0.10 |
| Diabetes (%) | 48.3 | 57.5 | 43.6 | 0.06 |
| Hypertension (%) | 76.5 | 76.7 | 76.4 | 0.97 |
| Dyslipidemia (%) | 63.4 | 63.0 | 63.6 | 0.88 |
| Antihyperlipidemic drugs (%) | 56.8 | 61.6 | 55.0 | 0.46 |
| Current smokers (%) | 11.3 | 16.4 | 8.6 | 0.001 |
| Ex-smokers (%) | 24.8 | 39.7 | 17.1 | 0.001 |
| Medas ≥ 9 (%) | 41.8 | 43.8 | 40.7 | 0.88 |
| Weight (kg) | 92.5 (92.9–97.7) | 102.0 (99.5–107.5) | 88.5 (88.1–93.7) | 0.001 |
| BMI (kg/m$^2$) | 34.9 ± 5.5 | 33.8 ± 5.1 | 35.4 ± 5.6 | 0.051 |
| WC (cm) | 114.2 ± 12.7 | 117.1 ± 12.4 | 112.7 ± 12.6 | 0.001 |
| WC > 80 cm for women and WC > 94 cm for men (%) | | 91.8 | 100 | 0.02 |
| WHtR | 0.68 ± 0.07 | 0.64 ± 0.12 | 0.69 ± 0.07 | 0.001 |
| WHtR > 0.5 (%) | 98.6 | 95.9 | 100 | 0.02 |

Data are expressed as mean ± standard deviation (SD), number [n (%)] or median (interquartile range [IQR]). Medas: Mediterranean Diet Adherence Screener; BMI: Body Mass Index; WC: Waist circumference; WHtR: Waist-to-height ratio.

**Table 2.** Biochemical and clinical parameters.

| | Overall | Men | Women | *p*-Value |
|---|---|---|---|---|
| SBP (mmHg) | 130.0 (129.1–133.5) | 130.0 (126.3–134.4) | 131.2 (129.2–134.4) | 0.67 |
| DBP (mmHg) | 80.0 (78.5–81.4) | 80.0 (78.2–82.8) | 80.0 (77.7–81.6) | 0.87 |
| FBG (mg/dL) | 99.0 (100.6–109.0) | 104.0 (104.9–118.8) | 96.0 (95.8–106.5) | 0.001 |
| HbA1c (%) | 6.0 (5.9–6.8) | 6.2 (6.0–6.7) | 5.9 (5.8–7.0) | 0.17 |
| HbA1c (mmol/mol) | 42.0 (42.9–46.3) | 44.0 (42.9–52.2) | 41.0 (41.1–44.8) | 0.15 |
| Total cholesterol (mg/dL) | 175.1 ± 41.3 | 162.5 ± 38.8 | 181.3 ± 41.1 | 0.001 |
| LDL cholesterol (mg/dL) | 98.5 ± 36.8 | 88.6 ± 34.3 | 105.6 ± 36.9 | 0.02 |
| HDL cholesterol (mg/dL) | 50.0 (50.7–55.1) | 47.0 (45.5–56.7) | 52.0 (51.9–55.9) | 0.001 |
| TG (mg/dL) | 122.5 ± 49.5 | 124.9 ± 60.4 | 119.7 ± 42.9 | 0.52 |
| CRP (mg/dL) | 0.30 (0.4–0.6) | 0.30 (0.2–0.6) | 0.35 (0.4–0.7) | 0.001 |
| Creatinine (mg/dL) | 0.85 ± 0.2 | 1.01 ± 0.19 | 0.77 ± 0.18 | 0.001 |
| ALT (U/L) | 21.0 (22.7–27.1) | 25.5 (24.4–32.4) | 19.0 (20.6–25.7) | 0.001 |
| AST (U/L) | 21.0 (21.7–24.7) | 22.0 (22.2–27.9) | 21.0 (20.6–23.6) | 0.29 |
| GGT (U/L) | 24.0 (21.8–32.2) | 28.5 (28.9–45.5) | 22.0 (21.8–32.2) | 0.02 |

Data are expressed as mean ± standard deviation (SD), number [n (%)] or median (interquartile range [IQR]). SBP: Systolic Blood Pressure; DBP: Diastolic Blood Pressure; FBG: Fasting Blood Glucose; HbA1c: Glycated Hemoglobin; LDL: Low-Density Lipoprotein; HDL: High-Density Lipoprotein; TG: Triglycerides; CRP: C-Reactive Protein; ALT: Alanine Aminotransferase; AST: Aspartate Aminotransferase; GGT: Gamma-Glutamyl Transferase.

The median values of ALT, AST and GGT were within the normal range in the whole population (21.0 (22.7–27.1); 21.0 (21.7–24.7); 24.0 (21.8–32.2)). However, ALT and GGT were significantly higher in men than in women (25.5 (24.4–32.4) vs. 19.0 (20.6–25.7) U/L, $p$ = 0.001; 28.5 (28.9–45.5) vs. 22.0 (21.8–32.2) U/L, $p$ = 0.02).

As regards NILDAs, FLI and FIB-4 did not show statistically significant differences between men and women (Table 3). However, the median AST/ALT ratio was significantly higher in women (1.1 (1.0–1.1) vs. 0.9 (0.8–1.0), $p < 0.001$), as was the proportion of women exceeding the clinical threshold of 0.8 (71.4% vs. 57.5%, $p < 0.001$). Furthermore, using the updated sex-specific cut-offs (≥33 U/L in men and ≥ 25 U/L in women for ALT and ≥30 U/L in men and ≥26 U/L in women for AST) [4,51], the proportions of women with above-normal levels was higher than that of men for ALT and AST respectively (30.7% vs. 24.6%, $p = 0.06$; 24.3% vs. 19.2%, $p = 0.09$), although not significant. Among indicators of cardiometabolic risk and visceral adiposity, VAI was significantly higher in women than in men (1.97 (1.9–2.3) vs. 1.38 (1.4–2.1), $p < 0.001$). Similarly, the TyG-WHtR index was significantly higher in women (6.0 ± 0.8 vs. 5.6 ± 1.0, $p = 0.02$). A higher proportion of women than men had VAI values above the critical threshold of 1.93 (47.8% vs. 30.1%, $p = 0.046$).

**Table 3.** NILDAs, cardiometabolic and adipose tissue dysfunction indices.

| | Overall | Men | Women | *p*-Value |
|---|---|---|---|---|
| FLI | 88.1 (77.7–83.5) | 92.5 (75.9–86.9) | 86.0 (76.7–83.6) | 0.31 |
| FIB-4 | 1.45 (1.4–1.6) | 1.43 (1.43–1.71) | 1.45 (1.40–1.52) | 0.75 |
| AST/ALT | 1.0 (1.0–1.1) | 0.9 (0.8–1.0) | 1.1 (1.0–1.1) | 0.001 |
| AST/ALT > 0.8 (%) | 66.7 | 57.5 | *71.4* | 0.001 |
| ALT ≥ 33 U/L for men and ≥25 U/L n (%) | | 24.6 | 30.7 | 0.06 |
| AST ≥ 30 U/L for men and ≥26 U/L n (%) | | 19.2 | 24.3 | 0.09 |
| TyG-WHtR | 5.9 ± 0.9 | 5.6 ± 1.0 | 6.0 ± 0.8 | 0.02 |
| VAI | 1.82 (1.8–2.1) | 1.38 (1.4–2.1) | 1.97 (1.9–2.3) | 0.001 |
| VAI > 1.93 (%) | 41.8 | 30.1 | 47.8 | 0.046 |

Data are expressed as mean ± standard deviation (SD), number [n (%)] or median (interquartile range [IQR]). FLI: Fatty Liver Index; FIB-4: Fibrosis-4 Index; AST/ALT: Alanine Aminotransferase to Aspartate Aminotransferase ratio; ALT: Alanine Aminotransferase; AST: Aspartate Aminotransferase; TyG-WHtR: TyG-waist-to-height ratio; VAI: Visceral Adiposity Index (cut-off > 1.93 indicates increased cardiometabolic risk).

The results obtained by VCTE (Table 4) showed that the median CAP value was significantly higher in men than in women (294.5 ± 50.9 vs. 277.4 ± 46.9 dB/m, $p = 0.02$), while no sex-differences were observed for LSM values. When analyzing the distribution of the degree of steatosis and fibrosis based on CAP and LSM thresholds, no significant sex-differences were found. However, a slight trend towards a higher prevalence of mild fibrosis (F1) was observed in women compared to men (16.5% vs. 8.2%, $p = 0.07$).

**Table 4.** Sex-based analysis of VCTE parameters.

| | Overall | Men | Women | *p*-Value |
|---|---|---|---|---|
| CAP (dB/m) | 283.0 ± 48.9 | 294.5 ± 50.9 | 277.4 ± 46.9 | 0.02 |
| CAP S0 (%) | 22.5 (48/213) | 16.4 (12/73) | 25.7 (36/140) | 0.29 |
| CAP S1 (%) | 13.6 (29/213) | 17.8 (13/73) | 11.4 (16/140) | 0.29 |
| CAP S2 (%) | 11.7 (25/213) | 10.9 (8/73) | 12.1 (17/140) | 0.82 |
| CAP S3 (%) | 53.0 (113/213) | 57.5 (42/73) | 50.7 (71/140) | 0.32 |
| LSM (kPa) | 5.8 (5.9–6.5) | 5.9 (5.9–6.9) | 5.8 (5.7–6.4) | 0.38 |
| LSM F0 (%) | 77.5 (165/213) | 79.4 (58/73) | 76.4 (107/140) | 0.60 |
| LSM F1 (%) | 13.6 (29/213) | 8.2 (6/73) | 16.5 (23/140) | 0.07 |
| LSM F2 (%) | 7.5 (16/213) | 9.6 (7/73) | 6.5 (9/140) | 0.58 |
| LSM F3–F4 (%) | 1.4 (3/213) | 2.7 (2/73) | 0.7 (1/140) | 0.55 |

Data are expressed as mean ± standard deviation (SD), number [n (%)] or median (interquartile range [IQR]). CAP: Controlled Attenuation Parameter; LSM: Liver Stiffness Measurement.

Multivariate analysis (Table S2 in the Supplementary Materials) confirmed the significant effect of sex and diabetes on CAP and LSM. Univariate analysis showed a significant effect of sex on CAP and diabetes on LSM. Other factors (age, smoking, dyslipidemia, hypertension) had no significant effect on CAP and/or LSM.

Logistic Regression Model used to evaluate the ability of FLI to predict liver steatosis as defined by CAP ≥ 248 dB/m (Table 5) reports the estimated coefficients and their confidence intervals, which were significantly higher in men compared to women. Furthermore, the model showed lower explanatory power in women, accounting for less variability in CAP values than in men, even after adjustment for confounders (diabetes, dyslipidemia, hypertension, smoking, age) (Table S3 Supplementary Materials).

**Table 5.** Logistic regression model with FLI predicting CAP ≥ 248 dB/m.

| | | Beta Coefficient | Sig. | Exp(B) | 95% CI per EXP(B) | |
|---|---|---|---|---|---|---|
| | | | | | Lower | Upper |
| Women | *FLI* | *0.026* | *0.018* | 1.026 | 1.004 | 1.049 |
| | Constant | −0.845 | 0.331 | 0.430 | | |
| Men | *FLI* | *0.064* | *0.001* | 1.066 | 1.026 | 1.107 |
| | Constant | −3.220 | 0.022 | 0.040 | | |

FLI: Fatty Liver Index; Sig.: significance level; EXP(B): exponentiated coefficient (odds ratio); CI: confidence interval. For women: −2 Log Likelihood = 124.764, Cox & Snell $R^2 = 0.046$, Nagelkerke $R^2 = 0.069$. For men: −2 Log Likelihood = 38.039, Cox & Snell $R^2 = 0.228$, Nagelkerke $R^2 = 0.375$.

The ROC analysis (Figure 1) showed that the diagnostic accuracy of the FLI for the detection of liver steatosis (CAP ≥ 248 dB/m) was significantly higher in men than in women. In women the Area Under the ROC Curve (AUC) was 0.655 (95% CI: 0.542–0.767, $p = 0.014$), whereas in men the AUC was 0.863 (95% CI: 0.731–0.995, $p < 0.001$). The difference between the two AUC was significant ($p = 0.015$).

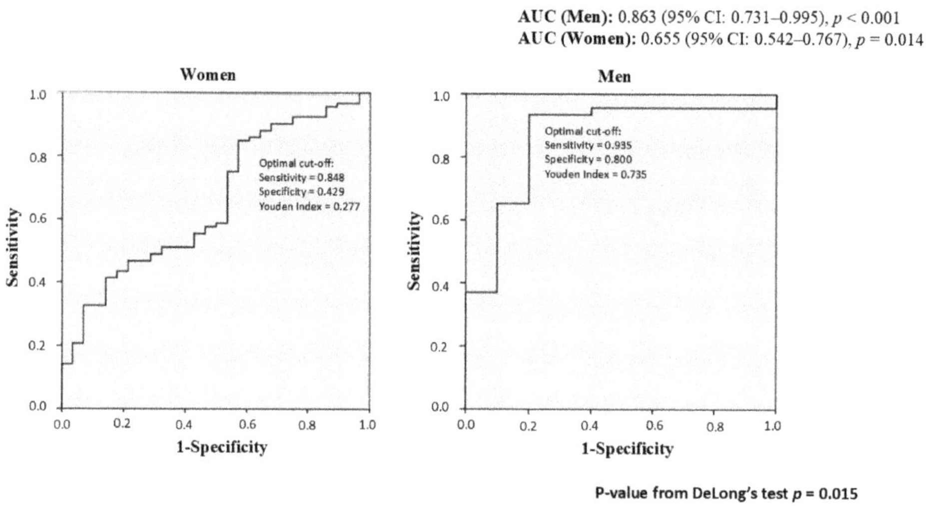


**Figure 1.** Sex-based analysis of Receiver Operating Characteristic (ROC) curves for FLI in predicting liver steatosis (CAP ≥ 248 dB/m). AUC: Area Under the ROC Curve.

In men, FLI ≥ 60 showed a sensitivity of 93.5%, a specificity of 70.0% with an overall diagnostic accuracy of 89.3% in predicting hepatic steatosis (i.e., CAP ≥ 248 dB/m). In women, while sensitivity remained high (90.2%), specificity dropped to 28.6% and negative predictive value (NPV) was only 47.1%, indicating a significant rate of false negatives (Table 6). The detailed contingency tables and chi-square test statistics are provided in Supplementary Materials Table S1.

**Table 6.** Sex-based analysis of diagnostic performance of the FLI ≥ 60 in predicting liver steatosis (CAP ≥ 248 dB/m).

| | Men | Women |
|---|---|---|
| Sensitivity (%) | 93.5 (CI 95%: 86.4–99.6%) | 90.2 (CI 95%: 82.6–95.3%) |
| Specificity (%) | 70.0 (CI 95%: 41.6–98.4%) | 28.6 (CI 95%: 13.2–48.7%) |
| Positive Predictive Value (%) | 93.5 (CI 95%: 86.4–99.6%) | 80.6 (CI 95%: 71.9–87.6%) |
| Negative Predictive Value (%) | 70.0 (CI 95%: 41.6–98.4%) | 47.1 (CI 95%: 23.0–72.2%) |
| Overall Accuracy (%) | 89.3 (CI 95%: 81.2–97.4%) | 75.8 (CI 95%: 67.1–83.1%) |

Data are expressed as percentage (%). CI (Confidence Interval).

## 4. Discussion

Although MASLD is less prevalent in women [52,53], particularly during their reproductive years [24,54], its prevalence rises significantly after menopause [27], reaching levels comparable to those in men by around age 50, especially among individuals with obesity [17,55–58].

However, despite this convergence in prevalence, sex-related differences in clinical presentation, disease progression, and outcomes beyond midlife remain poorly defined.

Our results suggest that, after menopause, although women may appear metabolically healthier based on conventional clinical and biochemical parameters, a sex-specific analysis reveals a more unfavorable liver profile: despite mean transaminase levels were higher in men, applying sex-specific cut-offs (i.e., ALT ≥ 25 U/L and AST ≥ 26 U/L) [4,51] revealed a higher proportion of abnormal ALT and AST in women. Anyway, the proportion of individuals with elevated enzyme levels remained low in both sexes, consistent with existing evidence that plasma aminotransferases are weak indicators of MASLD [4,59], highlighting the need for additional diagnostic assessment.

NILDAs analysis also revealed similar mean FLI and FIB-4 levels in both sexes, and men had significantly higher mean CAP than women. On the other hand, when the population was stratified based on standard CAP and LSM cut-offs for steatosis and fibrosis, we found that the rate of early fibrosis (F1) was almost doubled in women. In addition, women had a higher AST/ALT ratio than men, which is recognized as a marker of liver inflammation [4]. This is supported by several studies reporting a higher AST/ALT ratio in women despite lower liver enzyme levels [60–62] or similar [63] to those in men. Moreover, a greater proportion of women also exceeded the AST/ALT ratio threshold associated with long-term liver complications, such as fibrosis [64], which is the strongest predictor of poor outcomes in MASLD [4].

Adipose tissue distribution changes with age and differs significantly between sexes [15,57], approximately doubling in men and quadrupling in women over time [65]. The age-related increase in VAT is reflected in higher WC, WHtR and BMI [66,67], particularly after menopause [68,69]. Notably, women tend to develop MASLD at higher total body fat percentages than men, with abdominal obesity [70], especially visceral obesity being an independent risk factor [69]. A large meta-analysis also reported consistently higher WHtR values in women with MASLD compared to men [71]. However, WC may underestimate visceral fat in postmenopausal women [69], as it primarily reflects SAT [72,73], or a mix of SAT and VAT [74], and appears to be a reliable marker of liver fat only in men [62,69]. Despite this limitation, the use of sex-specific WC thresholds improves risk stratification in women [75], as seen in our study.

In men, lower subcutaneous adipose tissue (SAT) levels contribute to greater ectopic fat deposition throughout life [25,70]. In contrast, women generally have higher SAT and benefit from a protective premenopausal hormonal environment [56,76]. During the menopausal transition, fat initially accumulates in SAT, followed by a gradual increase in total body fat and VAT [77], which helps delay the accumulation of fat in the liver [25].

The difference in SAT and VAT before and after menopause in women may also explain the differences in ALT and CRP levels observed in our study. Elevated liver enzymes, reflecting fat accumulation in the liver [78], are generally higher in men. However, both SAT and VAT are associated with markers of chronic inflammation, such as CRP [79]. In addition, the correlation between visceral fat and CRP appears to be stronger in women than in men [80–82]. Supporting this, studies on normal-weight subjects with obesity (NOW) have reported higher ALT in men, but higher CRP in women, compared to lean controls [83].

In our study, although higher absolute body weight and WC in men suggest a greater metabolic burden, sex-specific cut-offs show that women have increased total and visceral adiposity combined with more frequent early-stage liver fibrosis (F1). In addition, elevated cardiometabolic indices (e.g., VAI, TyG-WHtR) and more frequent exceeding of thresholds for these indices suggest greater metabolic dysregulation after menopause, likely driven by visceral adipose dysfunction and increased systemic inflammation, which may contribute to the observed sex differences after menopause [25].

In our population, a FLI was a good indicator of steatosis in men, but lacked specificity and negative predictive value in women. ROC analysis, also, highlights a low ability of FLI showed to detect hepatic steatosis in women, even after adjustment for several risk factors. On the other hand, the AASLD has reported limited accuracy of the CAP in quantifying hepatic steatosis [84], with measurement failures more common in female sex, probably due to differences in fat distribution [85]. These findings suggest that cut-off values for both CAP and LSM should be sex-specific, especially for postmenopausal women. Similarly, since sex-specific thresholds have been proposed for FLI components such as WC [86], the same consideration applies to the FLI score, which incorporates WC, and to the FIB-4 index, which has reduced accuracy in older adults [4]. However, neither index currently includes sex as a variable.

Although a FLI $\geq$ 60 in postmenopausal women shows limited accuracy in predicting steatosis as assessed by CAP, it still indicates an underlying metabolic risk especially when combined with other parameters associated with increased risk, such as WHtR, BMI, and WC. Some studies recommend substantially lower FLI cut-offs for women [11,86–88] up to 50% lower in those over 50, to improve detection accuracy in this group [11].

Overall, this supports the hypothesis that women with a positive FLI but no steatosis detected by CAP may be in the early stages of MASLD (S0-S1), that CAP may underestimate steatosis in women, and that fibrosis may develop even in the absence of CAP-detected liver steatosis.

This may explain the observed "sex paradox", characterized by a non-linear trajectory of MASLD prevalence and progression across different ages and metabolic risk factors [19,25]. In fact, in our study, women showed more frequent early-stage fibrosis (F1) despite lower CAP values, whereas men had higher absolute CAP. This aligns with a large meta-analysis showing that although women have a lower overall prevalence of MASLD, they face a similar risk of MASH and an even higher risk of advanced fibrosis than men, especially after menopause [25]. Severe metabolic and inflammatory profiles were particularly evident in women at risk for MASH [89], and metabolic syndrome increased the risk of liver fibrosis by 2.7-fold, even without clear hepatic steatosis [90]. These findings are consistent with less favorable metabolic and inflammatory profiles observed in women, which may drive fibrosis progression despite lower CAP values.

The latest EASL-EASD-EASO guidelines identify men over 50, postmenopausal women, and individuals with multiple cardiometabolic risk factors as being at increased risk for progressive fibrosis and cirrhosis [4]. Therefore, this results highlight that women with MASLD, especially those over 50, should be evaluated for MASLD, MASH and fibrosis with equal or greater clinical vigilance compared to men [57].

These results must be interpreted taking into account the objectives and limitations. First, the identification of MASLD relied on non-invasive methods such as FLI and abdominal ultrasound rather than liver biopsy, which remains the gold standard for diagnosing liver fibrosis. However, FLI has shown predictive value for steatosis in various populations [30,91,92]. Second, the study was conducted exclusively in individuals with obesity, limiting the generalizability of the findings to MASLD patients without obesity. Third, the study is based on self-reported dietary and alcohol intake data obtained through questionnaires, which may be subject to recall bias. Fourth, because of its cross-sectional design, the study can detect associations but not causal relationships.

Key strengths include a well-characterized study population and the use of multiple non-invasive indices to assess steatosis, fibrosis, and cardiometabolic risk. Notably, the inclusion of VCTE provides a reliable assessment of hepatic steatosis and fibrosis. VCTE is a validated, cost-effective and non-invasive alternative to magnetic resonance elastography and liver biopsy, with strong correlation to histological findings in patients with NAFLD [93]. Importantly, this study specifically focused on individuals over 50 years of age, enabling a more accurate examination of sex-differences in a population where MASLD prevalence and progression are more comparable between men and women, thus addressing a critical gap in the existing literature.

## 5. Conclusions

While MASLD has historically been considered more prevalent in men, projections suggest a faster rate of increase among women by 2040 [94]. Similar to CV disease [95,96], MASLD in women often appears later, typically after menopause, and is associated with a greater burden of comorbidities and complications [25], higher mortality [97], increased cirrhosis prevalence [98], and a higher likelihood of being listed for liver transplantation [99]. Despite these risks, research focusing on middle-aged women remains limited, revealing a diagnostic gap similar to that in cardiovascular care, where the lack of sex-specific analyses has contributed to underdiagnosis and undertreatment in women [100].

The results of this study suggest that a more tailored approach is critical in postmenopausal women, especially those over 50, where increases in visceral fat may accelerate the progression of MASLD in women, even though healthier clinical appearances may delay diagnosis [25]. In particular, transaminase levels and FLI do not reliably identify liver steatosis in women compared to men, whereas the use of anthropometric height-adjusted measures and non-invasive markers of fibrosis and steatosis, together with the application of sex-specific cut-offs, may help to reveal a comparable or slightly higher burden of disease associated with visceral fat accumulation in women. Therefore, to prevent disease progression, especially after age 50, women may require closer clinical monitoring [25] using multiple non-invasive tools in addition to sex-specific cut-offs to enable early detection of these higher-risk female patients.

**Supplementary Materials:** The following supporting information can be downloaded at: https://www.mdpi.com/article/10.3390/biomedicines13092292/s1, Figure S1: Participants flow diagram; Table S1: Contingency tables (FLI vs. CAP $\geq$ 248) and chi-square test results, stratified by sex; Table S2: Multivariate analysis of variance (MANOVA) on the dependent variables (CAP and LSM); Table S3: Logistic Regression Model for the Association Between FLI and CAP $\geq$ 248 adjusted for confounders; Strobe Checklist S4: Strobe Checklist.

**Author Contributions:** Conceptualization, D.C., V.G. and A.M.; Methodology, L.C., M.E.P., M.C., S.P. and B.R.; Data curation: I.M., M.E.P. and M.C.; writing—original draft, and formal analysis, I.M.; writing—review and editing, I.M., M.E.P. and L.C.; Supervision, D.C. and V.G.; Validation, D.C., V.G., A.M., P.S. and F.L. All authors have read and agreed to the published version of the manuscript.

**Funding:** D.C., V.G., A.M. are supported by PRIN PNRR 2022, P2022C5Y3H.

**Institutional Review Board Statement:** The study was conducted in accordance with the Declaration of Helsinki [50], and approved by the "Comitato Etico Lazio 1", the Ethics Committee of the Lazio Region (reference no. 7421, protocol code 0038/2024 and date of approval 10 January 2024).

**Informed Consent Statement:** Informed consent was obtained from all subjects involved in the study.

**Data Availability Statement:** The original contributions presented in this study are included in the article. Further inquiries can be directed to the corresponding author.

**Acknowledgments:** During the preparation of this manuscript, the authors used ChatGPT (version GPT-4, OpenAI, San Francisco, CA, USA) for the text summary and revision. The authors have reviewed and edited the output and take full responsibility for the content of this publication.

**Conflicts of Interest:** The authors declare no conflicts of interest.

## References

1. Younossi, Z.M.; Golabi, P.; Paik, J.M.; Henry, A.; Van Dongen, C.; Henry, L. The global epidemiology of nonalcoholic fatty liver disease (NAFLD) and nonalcoholic steatohepatitis (NASH): A systematic review. *Hepatology* **2023**, *77*, 1335–1347. [CrossRef]
2. Miao, L.; Targher, G.; Byrne, C.D.; Cao, Y.-Y.; Zheng, M.-H. Current status and future trends of the global burden of MASLD. *Trends Endocrinol. Metab.* **2024**, *35*, 697–707. [CrossRef]
3. Paik, J.M.; Golabi, P.; Younossi, Y.; Srishord, M.; Mishra, A.; Younossi, Z.M. The Growing Burden of Disability Related to Nonalcoholic Fatty Liver Disease: Data from the Global Burden of Disease 2007–2017. *Hepatol. Commun.* **2020**, *4*, 1769–1780. [CrossRef]
4. Tacke, F.; Horn, P.; Wong, V.W.-S.; Ratziu, V.; Bugianesi, E.; Francque, S.; Zelber-Sagi, S.; Valenti, L.; Roden, M.; Schick, F.; et al. EASL–EASD–EASO Clinical Practice Guidelines on the management of metabolic dysfunction-associated steatotic liver disease (MASLD). *J. Hepatol.* **2024**, *81*, 492–542. [CrossRef]
5. Gastaldelli, A.; Newsome, P.N. NAFLD vs. MASLD (Metabolic Dysfunction–Associated Steatotic Liver Disease)—Why the Need for a Change of Nomenclature? *J. Clin. Endocrinol. Metab.* **2025**, *110*, e2407–e2410. [CrossRef]
6. Eslam, M.; Newsome, P.N.; Sarin, S.K.; Anstee, Q.M.; Targher, G.; Romero-Gomez, M.; Zelber-Sagi, S.; Wong, V.W.-S.; Dufour, J.-F.; Schattenberg, J.M.; et al. A new definition for metabolic dysfunction-associated fatty liver disease: An international expert consensus statement. *J. Hepatol.* **2020**, *73*, 202–209. [CrossRef]
7. Bozic, D.; Podrug, K.; Mikolasevic, I.; Grgurevic, I. Ultrasound Methods for the Assessment of Liver Steatosis: A Critical Appraisal. *Diagnostics* **2022**, *12*, 2287. [CrossRef]
8. Sun, D.-Q.; Zheng, K.I.; Xu, G.; Ma, H.-L.; Zhang, H.-Y.; Pan, X.-Y.; Zhu, P.-W.; Wang, X.-D.; Targher, G.; Byrne, C.D.; et al. PNPLA3 rs738409 is associated with renal glomerular and tubular injury in NAFLD patients with persistently normal ALT levels. *Liver Int.* **2020**, *40*, 107–119. [CrossRef]
9. Umehara, T. Nonalcoholic fatty liver disease with elevated alanine aminotransferase levels is negatively associated with bone mineral density: Cross-sectional study in U.S. adults. *PLoS ONE* **2018**, *13*, e0197900. [CrossRef]
10. Noureddin, M.; Loomba, R. Nonalcoholic fatty liver disease: Indications for liver biopsy and noninvasive biomarkers. *Clin. Liver Dis.* **2012**, *1*, 104–107. [CrossRef]
11. Crudele, L.; De Matteis, C.; Novielli, F.; Di Buduo, E.; Petruzzelli, S.; De Giorgi, A.; Antonica, G.; Berardi, E.; Moschetta, A. Fatty Liver Index (FLI) is the best score to predict MASLD with 50% lower cut-off value in women than in men. *Biol. Sex. Differ.* **2024**, *15*, 43. [CrossRef]
12. Hernandez Roman, J.; Siddiqui, M.S. The role of noninvasive biomarkers in diagnosis and risk stratification in nonalcoholic fatty liver disease. *Endocrinol. Diabetes Metab.* **2020**, *3*, e00127. [CrossRef]
13. Younossi, Z.M.; Koenig, A.B.; Abdelatif, D.; Fazel, Y.; Henry, L.; Wymer, M. Global epidemiology of nonalcoholic fatty liver disease—Meta-analytic assessment of prevalence, incidence, and outcomes. *Hepatology* **2016**, *64*, 73. [CrossRef]
14. He, Q.-J.; Li, Y.-F.; Zhao, L.-T.; Lin, C.-T.; Yu, C.-Y.; Wang, D. Recent advances in age-related metabolic dysfunction-associated steatotic liver disease. *World J. Gastroenterol.* **2024**, *30*, 652–662. [CrossRef]
15. Yadav, A.K.; MacNeill, J.J.; Krylov, A.; Ashrafi, N.; Mimi, R.A.; Saxena, R.; Liu, S.; Graham, S.F.; Wan, J.; Morral, N. Sex- and age-associated factors drive the pathophysiology of MASLD. *Hepatol. Commun.* **2024**, *8*, e0523. [CrossRef]

16. Chen, H.; Liu, Y.; Liu, D.; Liang, Y.; Zhu, Z.; Dong, K.; Li, H.; Bao, Y.; Wu, J.; Hou, X.; et al. Sex- and age-specific associations between abdominal fat and non-alcoholic fatty liver disease: A prospective cohort study. *J. Mol. Cell Biol.* **2023**, *15*, mjad069. [CrossRef]
17. Lonardo, A.; Suzuki, A. Sexual Dimorphism of NAFLD in Adults. Focus on Clinical Aspects and Implications for Practice and Translational Research. *J. Clin. Med.* **2020**, *9*, 1278. [CrossRef]
18. Booijink, R.; Ramachandran, P.; Bansal, R. Implications of innate immune sexual dimorphism for MASLD pathogenesis and treatment. *Trends Pharmacol. Sci.* **2024**, *45*, 614–627. [CrossRef]
19. Milani, I.; Chinucci, M.; Leonetti, F.; Capoccia, D. MASLD: Prevalence, Mechanisms, and Sex-Based Therapies in Postmenopausal Women. *Biomedicines* **2025**, *13*, 855. [CrossRef]
20. Le, M.H.; Le, D.M.; Baez, T.C.; Wu, Y.; Ito, T.; Lee, E.Y.; Lee, K.; Stave, C.D.; Henry, L.; Barnett, S.D.; et al. Global incidence of non-alcoholic fatty liver disease: A systematic review and meta-analysis of 63 studies and 1,201,807 persons. *J. Hepatol.* **2023**, *79*, 287–295. [CrossRef]
21. Liu, J.; Ayada, I.; Zhang, X.; Wang, L.; Li, Y.; Wen, T.; Ma, Z.; Bruno, M.J.; de Knegt, R.J.; Cao, W.; et al. Estimating Global Prevalence of Metabolic Dysfunction-Associated Fatty Liver Disease in Overweight or Obese Adults. *Clin. Gastroenterol. Hepatol.* **2022**, *20*, e573–e582. [CrossRef]
22. Eng, P.C.; Forlano, R.; Tan, T.; Manousou, P.; Dhillo, W.S.; Izzi-Engbeaya, C. Non-alcoholic fatty liver disease in women—Current knowledge and emerging concepts. *JHEP Rep.* **2023**, *5*, 100835. [CrossRef]
23. Nagral, A.; Bangar, M.; Menezes, S.; Bhatia, S.; Butt, N.; Ghosh, J.; Manchanayake, J.H.; Mahtab, M.A.; Singh, S.P. Gender Differences in Nonalcoholic Fatty Liver Disease. *Euroasian J. Hepatogastroenterol.* **2022**, *12*, S19–S25. [CrossRef]
24. Burra, P.; Bizzaro, D.; Gonta, A.; Shalaby, S.; Gambato, M.; Morelli, M.C.; Trapani, S.; Floreani, A.; Marra, F.; Brunetto, M.R.; et al. Clinical impact of sexual dimorphism in non-alcoholic fatty liver disease (NAFLD) and non-alcoholic steatohepatitis (NASH). *Liver Int.* **2021**, *41*, 1713–1733. [CrossRef]
25. Balakrishnan, M.; Patel, P.; Dunn-Valadez, S.; Dao, C.; Khan, V.; Ali, H.; El-Serag, L.; Hernaez, R.; Sisson, A.; Thrift, A.P.; et al. Women have Lower Risk of Nonalcoholic Fatty Liver Disease but Higher Risk of Progression vs. Men: A Systematic Review and Meta-analysis. *Clin. Gastroenterol. Hepatol.* **2021**, *19*, 61–71.e15. [CrossRef]
26. Tang, Z.; Pham, M.; Hao, Y.; Wang, F.; Patel, D.; Jean-Baptiste, L.; Fan, L.; Wang, W.; Wang, Y.; Cheng, F. Sex, Age, and BMI Modulate the Association of Physical Examinations and Blood Biochemistry Parameters and NAFLD: A Retrospective Study on 1994 Cases Observed at Shuguang Hospital, China. *Biomed. Res. Int.* **2019**, *2019*, 1246518. [CrossRef]
27. Burnside, J.; Cinque, F.; Sebastiani, G.; Ramji, A.; Patel, K.; Swain, M.; Saeed, S. Sex differences in the prevalence and cardiometabolic risk profiles of steatotic liver disease: A Canadian Longitudinal Study on Aging analysis. *Can. J. Public Health* **2025**. [CrossRef]
28. Duan, H.; Gong, M.; Yuan, G.; Wang, Z. Sex Hormone: A Potential Target at Treating Female Metabolic Dysfunction-Associated Steatotic Liver Disease? *J. Clin. Exp. Hepatol.* **2025**, *15*, 102459. [CrossRef]
29. Lonardo, A.; Nascimbeni, F.; Ballestri, S.; Fairweather, D.; Win, S.; Than, T.A.; Abdelmalek, M.F.; Suzuki, A. Sex Differences in Nonalcoholic Fatty Liver Disease: State of the Art and Identification of Research Gaps. *Hepatology* **2019**, *70*, 1457–1469. [CrossRef]
30. Bedogni, G.; Bellentani, S.; Miglioli, L.; Masutti, F.; Passalacqua, M.; Castiglione, A.; Tiribelli, C. The Fatty Liver Index: A simple and accurate predictor of hepatic steatosis in the general population. *BMC Gastroenterol.* **2006**, *6*, 33. [CrossRef]
31. Rinella, M.E.; Lazarus, J.V.; Ratziu, V.; Francque, S.M.; Sanyal, A.J.; Kanwal, F.; Romero, D.; Abdelmalek, M.F.; Anstee, Q.M.; Arab, J.P.; et al. A multisociety Delphi consensus statement on new fatty liver disease nomenclature. *Hepatology* **2023**, *78*, 1966–1986. [CrossRef]
32. Heer, E.; Harper, A.; Escandor, N.; Sung, H.; McCormack, V.; Fidler-Benaoudia, M.M. Global burden and trends in premenopausal and postmenopausal breast cancer: A population-based study. *Lancet Glob. Health* **2020**, *8*, e1027–e1037. [CrossRef]
33. Koukouliata, A.; Nena, E.; Koutlaki, N.; Liberis, V.; Constantinidis, T. Correlation of age at natural menopause with occupational status and other epidemiologic factors in women from Prefecture of Kavala, Greece. *Hippokratia* **2017**, *21*, 32–37.
34. Namazi, M.; Sadeghi, R.; Behboodi Moghadam, Z. Social Determinants of Health in Menopause: An Integrative Review. *Int. J. Womens Health* **2019**, *11*, 637–647. [CrossRef]
35. Gibson, S.; Ashwell, M. A simple cut-off for waist-to-height ratio (0·5) can act as an indicator for cardiometabolic risk: Recent data from adults in the Health Survey for England. *Br. J. Nutr.* **2020**, *123*, 681–690. [CrossRef]
36. Hagström, H.; Vessby, J.; Ekstedt, M.; Shang, Y. 99% of patients with NAFLD meet MASLD criteria and natural history is therefore identical. *J. Hepatol.* **2024**, *80*, e76–e77. [CrossRef]
37. Ratziu, V.; Boursier, J.; AFEFGroup for the Study of Liver Fibrosis. Confirmatory biomarker diagnostic studies are not needed when transitioning from NAFLD to MASLD. *J. Hepatol.* **2024**, *80*, e51–e52. [CrossRef]
38. Song, S.J.; Lai, J.C.-T.; Wong, G.L.-H.; Wong, V.W.-S.; Yip, T.C.-F. Can we use old NAFLD data under the new MASLD definition? *J. Hepatol.* **2024**, *80*, e54–e56. [CrossRef]

39. Elsabaawy, M.; Naguib, M.; Abuamer, A.; Shaban, A. Comparative application of MAFLD and MASLD diagnostic criteria on NAFLD patients: Insights from a single-center cohort. *Clin. Exp. Med.* **2025**, *25*, 36. [CrossRef]
40. Karlas, T.; Petroff, D.; Sasso, M.; Fan, J.-G.; Mi, Y.-Q.; de Lédinghen, V.; Kumar, M.; Lupsor-Platon, M.; Han, K.-H.; Cardoso, A.C.; et al. Individual patient data meta-analysis of controlled attenuation parameter (CAP) technology for assessing steatosis. *J. Hepatol.* **2017**, *66*, 1022–1030. [CrossRef]
41. Ozercan, A.M.; Ozkan, H. Vibration-controlled Transient Elastography in NAFLD: Review Study. *Euroasian J. Hepatogastroenterol.* **2022**, *12*, S41–S45. [CrossRef]
42. Singh, S.; Muir, A.J.; Dieterich, D.T.; Falck-Ytter, Y.T. American Gastroenterological Association Institute Technical Review on the Role of Elastography in Chronic Liver Diseases. *Gastroenterology* **2017**, *152*, 1544–1577. [CrossRef]
43. Amato, M.C.; Giordano, C.; Pitrone, M.; Galluzzo, A. Cut-off points of the visceral adiposity index (VAI) identifying a visceral adipose dysfunction associated with cardiometabolic risk in a Caucasian Sicilian population. *Lipids Health Dis.* **2011**, *10*, 183. [CrossRef]
44. Amato, M.C.; Giordano, C. Visceral Adiposity Index: An Indicator of Adipose Tissue Dysfunction. *Int. J. Endocrinol.* **2014**, *2014*, 730827. [CrossRef]
45. Xue, Y.; Xu, J.; Li, M.; Gao, Y. Potential screening indicators for early diagnosis of NAFLD/MAFLD and liver fibrosis: Triglyceride glucose index–related parameters. *Front. Endocrinol.* **2022**, *13*, 951689. [CrossRef]
46. Malek, M.; Khamseh, M.E.; Chehrehgosha, H.; Nobarani, S.; Alaei-Shahmiri, F. Triglyceride glucose-waist to height ratio: A novel and effective marker for identifying hepatic steatosis in individuals with type 2 diabetes mellitus. *Endocrine* **2021**, *74*, 538–545. [CrossRef]
47. American Diabetes Association Professional Practice Committee: 2. Diagnosis and Classification of Diabetes: Standards of Care in Diabetes—2025. *Diabetes Care* **2024**, *48*, S27–S49. [CrossRef]
48. Saunders, J.B.; Aasland, O.G.; Babor, T.F.; De La Fuente, J.R.; Grant, M. Development of the Alcohol Use Disorders Identification Test (AUDIT): WHO Collaborative Project on Early Detection of Persons with Harmful Alcohol Consumption-II. *Addiction* **1993**, *88*, 791–804. [CrossRef]
49. García-Conesa, M.-T.; Philippou, E.; Pafilas, C.; Massaro, M.; Quarta, S.; Andrade, V.; Jorge, R.; Chervenkov, M.; Ivanova, T.; Dimitrova, D.; et al. Exploring the Validity of the 14-Item Mediterranean Diet Adherence Screener (MEDAS): A Cross-National Study in Seven European Countries around the Mediterranean Region. *Nutrients* **2020**, *12*, 2960. [CrossRef]
50. Shephard, D.A. The 1975 Declaration of Helsinki and consent. *Can. Med. Assoc. J.* **1976**, *115*, 1191–1192.
51. Loo, S.-Y.; Chan, W.-K. Emerging new standard for non-invasive assessment of liver disease mortality in non-alcoholic fatty liver disease. *Hepatobiliary Surg. Nutr.* **2017**, *6*, 135–137. [CrossRef]
52. Feng, G.; Targher, G.; Byrne, C.D.; Yilmaz, Y.; Wai-Sun Wong, V.; Adithya Lesmana, C.R.; Adams, L.A.; Boursier, J.; Papatheodoridis, G.; El-Kassas, M.; et al. Global burden of metabolic dysfunction-associated steatotic liver disease, 2010 to 2021. *JHEP Rep.* **2024**, *7*, 101271. [CrossRef]
53. Sheka, A.C.; Adeyi, O.; Thompson, J.; Hameed, B.; Crawford, P.A.; Ikramuddin, S. Nonalcoholic Steatohepatitis: A Review. *JAMA* **2020**, *323*, 1175–1183. [CrossRef]
54. Riazi, K.; Azhari, H.; Charette, J.H.; Underwood, F.E.; King, J.A.; Afshar, E.E.; Swain, M.G.; Congly, S.E.; Kaplan, G.G.; Shaheen, A.-A. The prevalence and incidence of NAFLD worldwide: A systematic review and meta-analysis. *Lancet Gastroenterol. Hepatol.* **2022**, *7*, 851–861. [CrossRef]
55. Polyzos, S.A.; Goulis, D.G. Menopause and metabolic dysfunction-associated steatotic liver disease. *Maturitas* **2024**, *186*, 108024. [CrossRef]
56. Ntikoudi, A.; Spyrou, A.; Evangelou, E.; Dokoutsidou, E.; Mastorakos, G. The Effect of Menopausal Status, Insulin Resistance and Body Mass Index on the Prevalence of Non-Alcoholic Fatty Liver Disease. *Healthcare* **2024**, *12*, 1081. [CrossRef]
57. Cherubini, A.; Torre, S.D.; Pelusi, S.; Valenti, L. Sexual dimorphism of metabolic dysfunction-associated steatotic liver disease. *Trends Mol. Med.* **2024**, *30*, 1126–1136. [CrossRef]
58. Zuo, Q.; Park, N.H.; Lee, J.K.; Santaliz-Casiano, A.; Madak-Erdogan, Z. Navigating nonalcoholic fatty liver disease (NAFLD): Exploring the roles of estrogens, pharmacological and medical interventions, and life style. *Steroids* **2024**, *203*, 109330. [CrossRef]
59. Klisic, A.; Kavaric, N.; Ninic, A. Predictive Values of Serum Uric Acid and Alanine-aminotransferase for Fatty Liver Index in Montenegrin Population. *J. Med. Biochem.* **2019**, *38*, 407–417. [CrossRef]
60. Kim, H.-J.; Lim, C.-W.; Lee, J.H.; Park, H.-B.; Suh, Y.; Cho, Y.-H.; Choi, T.-Y.; Hwang, E.-S.; Cho, D.-K.; Kim, H.-J. Gender-based differences in the relationship between fatty liver disease and atherosclerosis. *Cardiovasc. J. Afr.* **2016**, *27*, 281–286. [CrossRef]
61. Mera, J.R.; Dickson, B.; Feldman, M. Influence of Gender on the Ratio of Serum Aspartate Aminotransferase (AST) to Alanine Aminotransferase (ALT) in Patients with and Without Hyperbilirubinemia. *Dig. Dis. Sci.* **2008**, *53*, 799–802. [CrossRef]
62. Ishibashi, E.; Eguchi, Y.; Eguchi, T.; Matsunobu, A.; Oza, N.; Nakashita, S.; Kitajima, Y.; Kuroki, S.; Ozaki, I.; Kawaguchi, Y.; et al. Waist circumference correlates with hepatic fat accumulation in male Japanese patients with non-alcoholic fatty liver disease, but not in females. *J. Gastroenterol. Hepatol.* **2008**, *23*, 908–913. [CrossRef]

63. Oye-Somefun, A.; Blyuss, E.; Ardern, C.I. Trends in Serum AST-to-ALT Ratio Among U.S. Adults: Analysis of the U.S. National Health and Nutrition Examination Survey. *Metab. Syndr. Relat. Disord.* **2021**, *19*, 498–506. [CrossRef]
64. Botros, M.; Sikaris, K.A. The De Ritis Ratio: The Test of Time. *Clin. Biochem. Rev.* **2013**, *34*, 117–130.
65. Chen, J.; Li, J.; Qu, H.; Ning, T.; Xie, H.; Lu, G. A Mendelian randomization study: Years of education and nonalcoholic fatty liver disease. *Medicine* **2024**, *103*, e38761. [CrossRef] [PubMed]
66. Pernoud, L.E.; Gardiner, P.A.; Fraser, S.D.; Dillon-Rossiter, K.; Dean, M.M.; Schaumberg, M.A. A systematic review and meta-analysis investigating differences in chronic inflammation and adiposity before and after menopause. *Maturitas* **2024**, *190*, 108119. [CrossRef] [PubMed]
67. Tang, R.; Fan, Y.; Luo, M.; Zhang, D.; Xie, Z.; Huang, F.; Wang, Y.; Liu, G.; Wang, Y.; Lin, S.; et al. General and Central Obesity Are Associated with Increased Severity of the VMS and Sexual Symptoms of Menopause Among Chinese Women: A Longitudinal Study. *Front. Endocrinol.* **2022**, *13*, 814872. [CrossRef] [PubMed]
68. Lee, B.J.; Kim, J.U.; Lee, S. Association of menopausal status with body composition and anthropometric indices in Korean women. *PLoS ONE* **2024**, *19*, e0298212. [CrossRef]
69. Cao, Y.; Zhang, W.; Lou, Y.; Yan, Q.; Zhang, Y.; Qi, F.; Xiang, L.; Lv, T.; Fang, Z.; Yu, J.; et al. Sex- and reproductive status-specific relationships between body composition and non-alcoholic fatty liver disease. *BMC Gastroenterol.* **2023**, *23*, 364. [CrossRef]
70. Lee, J.-H.; Jeon, S.; Lee, H.S.; Kwon, Y.-J. Gender Differences in the Risk for Incident Non-Alcoholic Fatty Liver Disease According to the Transition of Abdominal Obesity Status: A 16-Year Cohort Study. *Nutrients* **2023**, *15*, 2880. [CrossRef]
71. Ismaiel, A.; Hosiny, B.E.; Ismaiel, M.; Leucuta, D.-C.; Popa, S.-L.; Catana, C.S.; Dumitrascu, D.L. Waist to height ratio in nonalcoholic fatty liver disease—Systematic review and meta-analysis. *Clin. Res. Hepatol. Gastroenterol.* **2023**, *47*, 102160. [CrossRef]
72. Camhi, S.M.; Bray, G.A.; Bouchard, C.; Greenway, F.L.; Johnson, W.D.; Newton, R.L.; Ravussin, E.; Ryan, D.H.; Smith, S.R.; Katzmarzyk, P.T. The Relationship of Waist Circumference and BMI to Visceral, Subcutaneous, and Total Body Fat: Sex and Race Differences. *Obesity* **2011**, *19*, 402–408. [CrossRef] [PubMed]
73. Shuster, A.; Patlas, M.; Pinthus, J.H.; Mourtzakis, M. The clinical importance of visceral adiposity: A critical review of methods for visceral adipose tissue analysis. *Br. J. Radiol.* **2012**, *85*, 1–10. [CrossRef]
74. Kullberg, J.; Von Below, C.; Lönn, L.; Lind, L.; Ahlström, H.; Johansson, L. Practical approach for estimation of subcutaneous and visceral adipose tissue. *Clin. Physiol. Funct. Imaging* **2007**, *27*, 148–153. [CrossRef]
75. De Vincentis, A.; Tavaglione, F.; Spagnuolo, R.; Pujia, R.; Tuccinardi, D.; Mascianà, G.; Picardi, A.; Antonelli Incalzi, R.; Valenti, L.; Romeo, S.; et al. Metabolic and genetic determinants for progression to severe liver disease in subjects with obesity from the UK Biobank. *Int. J. Obes.* **2022**, *46*, 486–493. [CrossRef] [PubMed]
76. Rehman, A.; Lathief, S.; Charoenngam, N.; Pal, L. Aging and Adiposity—Focus on Biological Females at Midlife and Beyond. *Int. J. Mol. Sci.* **2024**, *25*, 2972. [CrossRef]
77. Lovejoy, J.; Champagne, C.; de Jonge, L.; Xie, H.; Smith, S. Increased visceral fat and decreased energy expenditure during the menopausal transition. *Int. J. Obes.* **2008**, *32*, 949–958. [CrossRef]
78. Buday, B.; Pach, P.F.; Literati-Nagy, B.; Vitai, M.; Kovacs, G.; Vecsei, Z.; Koranyi, L.; Lengyel, C. Sex influenced association of directly measured insulin sensitivity and serum transaminase levels: Why alanine aminotransferase only predicts cardiovascular risk in men? *Cardiovasc. Diabetol.* **2015**, *14*, 55. [CrossRef]
79. Pou, K.M.; Massaro, J.M.; Hoffmann, U.; Vasan, R.S.; Maurovich-Horvat, P.; Larson, M.G.; Keaney, J.F.; Meigs, J.B.; Lipinska, I.; Kathiresan, S.; et al. Visceral and Subcutaneous Adipose Tissue Volumes Are Cross-Sectionally Related to Markers of Inflammation and Oxidative Stress. *Circulation* **2007**, *116*, 1234–1241. [CrossRef]
80. Khera, A.; Vega, G.L.; Das, S.R.; Ayers, C.; McGuire, D.K.; Grundy, S.M.; de Lemos, J.A. Sex Differences in the Relationship between C-Reactive Protein and Body Fat. *J. Clin. Endocrinol. Metab.* **2009**, *94*, 3251–3258. [CrossRef] [PubMed]
81. Thorand, B.; Baumert, J.; Döring, A.; Herder, C.; Kolb, H.; Rathmann, W.; Giani, G.; Koenig, W. Sex differences in the relation of body composition to markers of inflammation. *Atherosclerosis* **2006**, *184*, 216–224. [CrossRef] [PubMed]
82. Carbone, F.; Lattanzio, M.S.; Minetti, S.; Ansaldo, A.M.; Ferrara, D.; Molina-Molina, E.; Belfiore, A.; Elia, E.; Pugliese, S.; Palmieri, V.O.; et al. Circulating CRP Levels Are Associated with Epicardial and Visceral Fat Depots in Women with Metabolic Syndrome Criteria. *Int. J. Mol. Sci.* **2019**, *20*, 5981. [CrossRef]
83. Kang, S.; Kyung, C.; Park, J.S.; Kim, S.; Lee, S.-P.; Kim, M.K.; Kim, H.K.; Kim, K.R.; Jeon, T.J.; Ahn, C.W. Subclinical vascular inflammation in subjects with normal weight obesity and its association with body Fat: An 18 F-FDG-PET/CT study. *Cardiovasc. Diabetol.* **2014**, *13*, 70. [CrossRef]
84. Rinella, M.E.; Neuschwander-Tetri, B.A.; Siddiqui, M.S.; Abdelmalek, M.F.; Caldwell, S.; Barb, D.; Kleiner, D.E.; Loomba, R. AASLD Practice Guidance on the clinical assessment and management of nonalcoholic fatty liver disease. *Hepatology* **2023**, *77*, 1797. [CrossRef] [PubMed]
85. Nishimura, T.; Tada, T.; Akita, T.; Kondo, R.; Suzuki, Y.; Imajo, K.; Kokubu, S.; Abe, T.; Kuroda, H.; Hirooka, M.; et al. Diagnostic performance of attenuation imaging versus controlled attenuation parameter for hepatic steatosis with MRI-based proton density fat fraction as the reference standard: A prospective multicenter study. *J. Gastroenterol.* **2025**, *60*, 727–737. [CrossRef]

86. Wu, J.; Tian, S.; Li, H.; Xu, Z.; Li, S.; Chen, Y.; Liang, X.; Xiao, J.; Song, J.; She, R.; et al. Population-specific cut-off points of fatty liver index: A study based on the National Health and Nutrition Examination Survey data. *BMC Gastroenterol.* **2022**, *22*, 265. [CrossRef]
87. Chen, L.-W.; Huang, P.-R.; Chien, C.-H.; Lin, C.-L.; Chien, R.-N. A community-based study on the application of fatty liver index in screening subjects with nonalcoholic fatty liver disease. *J. Formos. Med. Assoc.* **2020**, *119*, 173–181. [CrossRef]
88. Dehnavi, Z.; Razmpour, F.; Belghaisi Naseri, M.; Nematy, M.; Alamdaran, S.A.; Vatanparast, H.A.; Azimi Nezhad, M.; Abbasi, B.; Ganji, A. Fatty Liver Index (FLI) in Predicting Non-alcoholic Fatty Liver Disease (NAFLD). *Hepat. Mon.* **2018**, *18*, e63227. [CrossRef]
89. Noureddin, M.; Truong, E.; Mayo, R.; Martínez-Arranz, I.; Mincholé, I.; Banales, J.M.; Arrese, M.; Cusi, K.; Arias-Loste, M.T.; Bruha, R.; et al. Serum identification of at-risk MASH: The metabolomics-advanced steatohepatitis fibrosis score (MASEF). *Hepatology* **2024**, *79*, 135. [CrossRef] [PubMed]
90. Gangireddy, V.G.R.; Pilkerton, C.; Xiang, J.; Tinajero, R.; Ashcraft, A.M. Hepatic Fibrosis and Steatosis in Metabolic Syndrome. *J. Obes. Metab. Syndr.* **2022**, *31*, 61–69. [CrossRef]
91. Zelber-Sagi, S.; Webb, M.; Assy, N.; Blendis, L.; Yeshua, H.; Leshno, M.; Ratziu, V.; Halpern, Z.; Oren, R.; Santo, E. Comparison of fatty liver index with noninvasive methods for steatosis detection and quantification. *World J. Gastroenterol.* **2013**, *19*, 57–64. [CrossRef]
92. Koehler, E.M.; Schouten, J.N.L.; Hansen, B.E.; Hofman, A.; Stricker, B.H.; Janssen, H.L.A. External validation of the fatty liver index for identifying nonalcoholic fatty liver disease in a population-based study. *Clin. Gastroenterol. Hepatol.* **2013**, *11*, 1201–1204. [CrossRef]
93. Vuppalanchi, R.; Siddiqui, M.S.; Van Natta, M.L.; Hallinan, E.; Brandman, D.; Kowdley, K.; Neuschwander-Tetri, B.A.; Loomba, R.; Dasarathy, S.; Abdelmalek, M.; et al. Performance characteristics of vibration-controlled transient elastography for evaluation of nonalcoholic fatty liver disease. *Hepatology* **2018**, *67*, 134–144. [CrossRef] [PubMed]
94. Le, M.H.; Yeo, Y.H.; Zou, B.; Barnet, S.; Henry, L.; Cheung, R.; Nguyen, M.H. Forecasted 2040 global prevalence of nonalcoholic fatty liver disease using hierarchical bayesian approach. *Clin. Mol. Hepatol.* **2022**, *28*, 841–850. [CrossRef] [PubMed]
95. Maas, A.H.E.M.; Rosano, G.; Cifkova, R.; Chieffo, A.; van Dijken, D.; Hamoda, H.; Kunadian, V.; Laan, E.; Lambrinoudaki, I.; Maclaran, K.; et al. Cardiovascular health after menopause transition, pregnancy disorders, and other gynaecologic conditions: A consensus document from European cardiologists, gynaecologists, and endocrinologists. *Eur. Heart J.* **2021**, *42*, 967–984. [CrossRef]
96. Earle, N.J.; Doughty, R.N.; Devlin, G.; White, H.; Riddell, C.; Choi, Y.; Kerr, A.J.; Poppe, K.K. Sex differences in outcomes after acute coronary syndrome vary with age: A New Zealand national study. European Heart Journal. *Acute Cardiovasc. Care* **2024**, *13*, 284–292. [CrossRef]
97. DiStefano, J.K. NAFLD and NASH in Postmenopausal Women: Implications for Diagnosis and Treatment. *Endocrinology* **2020**, *161*, bqaa134. [CrossRef]
98. Jutras, G.; Flemming, J.A. Global Epidemiology of Cirrhosis in Women. *Off. J. Am. Coll. Gastroenterol. ACG* **2025**, *120*, 518. [CrossRef]
99. Paklar, N.; Mijic, M.; Filipec-Kanizaj, T. The Outcomes of Liver Transplantation in Severe Metabolic Dysfunction-Associated Steatotic Liver Disease Patients. *Biomedicines* **2023**, *11*, 3096. [CrossRef]
100. Gauci, S.; Cartledge, S.; Redfern, J.; Gallagher, R.; Huxley, R.; Lee, C.M.Y.; Vassallo, A.; O'Neil, A. Biology, Bias, or Both? The Contribution of Sex and Gender to the Disparity in Cardiovascular Outcomes Between Women and Men. *Curr. Atheroscler. Rep.* **2022**, *24*, 701–708. [CrossRef]

*biomedicines*

MDPI

*Article*

# Intersecting Endocrine Pathways in Cardiomyopathy: The Role of Metabolic Burden in Structural Heart Disease

**Ovidiu Țica [1,2,†], Mircea Ioan Șandor [3,*,†], Anca Huniadi [3], Cristian Daina [3], Sanda Monica Filip [4], Ilarie Brihan [5], Monica Sabău [5], Ioana Zaha [3] and Otilia Țica [6,*,†]**

1. Department of Morphological Disciplines, Faculty of Medicine and Pharmacy, University of Oradea, 410073 Oradea, Romania; tica.ovidiu@didactic.uoradea.ro
2. Department of Pathology, Emergency County Clinical Hospital of Bihor, 410165 Oradea, Romania
3. Department of Surgical Disciplines, Faculty of Medicine and Pharmacy, University of Oradea, 410073 Oradea, Romania; ahuniadi@uoradea.ro (A.H.); cdaina@uoradea.ro (C.D.); izaha@uoradea.ro (I.Z.)
4. Department of Physics, Faculty of Informatics and Sciences, University of Oradea, 410087 Oradea, Romania; sfilip@uoradea.ro
5. Department of Psycho-Neurosciences, and Recovery, Faculty of Medicine and Pharmacy, University of Oradea, 410073 Oradea, Romania; ibrihan@uoradea.ro (I.B.); msabau@uoradea.ro (M.S.)
6. Cardiology Clinic, Emergency County Clinical Hospital of Bihor, 410165 Oradea, Romania

* Correspondence: msandor@uoradea.ro (M.I.Ș.); otilia_cristea01@yahoo.com (O.Ț.)

† These authors have contributed equally to this work.

## Abstract

**Background:** Dilated cardiomyopathy (DCM) is a major contributor to heart failure-related morbidity and mortality. While type 2 diabetes mellitus (T2DM), obesity, and thyroid dysfunction are individually linked to cardiovascular disease, their combined effects on DCM remain poorly understood. **Objective:** To evaluate the independent and synergistic associations of diabetes (stratified by treatment), thyroid dysfunction, and obesity with the prevalence of DCM and 30-day hospital readmission. We further examined the utility of a composite Metabolic Burden Score for risk stratification. **Methods:** In this retrospective cohort study, electronic health record data from 1079 adult patients at a tertiary care center were analyzed. Multivariable logistic regression, including ridge regularization, was used to identify predictors of DCM. Endocrine phenotypes were stratified by diabetes and thyroid status. A Metabolic Burden Score (range: 0–3) based on diabetes, obesity, and thyroid dysfunction was developed and correlated with clinical outcomes. **Results:** DCM was diagnosed in 46% of the cohort. Non-insulin-treated diabetes (OR: 6.93; 95% CI: 3.78–12.73), hypothyroidism (OR: 1.78; 95% CI: 1.02–3.11), and male sex (OR: 2.33; 95% CI: 1.36–4.00) were independently associated with increased DCM risk. Obesity was not independently predictive but contributed to DCM prevalence when assessed within the Metabolic Burden Score. DCM prevalence increased across burden strata, reaching 50% in the high-risk group. Notably, the moderate-risk group had the highest 30-day readmission rate (42.8%). **Conclusions:** Non-insulin-treated diabetes and hypothyroidism are key metabolic drivers of DCM. A simple composite burden score offers a clinically useful tool for stratifying risk of DCM and early readmission. These findings support integrated endocrine–cardiac screening strategies to improve early identification and prevention of structural heart disease.

**Keywords:** dilated cardiomyopathy; diabetes mellitus; thyroid dysfunction; obesity; metabolic burden; cardiovascular risk; endocrine comorbidity; hospital readmission

## 1. Introduction

Cardiovascular disease (CVD) remains the leading cause of morbidity and mortality globally, and its bidirectional relationship with metabolic dysfunction is increasingly recognized. Among metabolic exposures, type 2 diabetes mellitus (T2DM) significantly raises the risk of heart failure across phenotypes, including dilated cardiomyopathy (DCM), due to complex mechanisms involving microvascular injury, inflammation, lipotoxicity, and autonomic dysfunction [1]. Precision prognostic efforts are underway to better predict which individuals with T2DM will develop cardiovascular complications, yet current models remain only modestly effective [2,3].

Obesity independently contributes to myocardial dysfunction through mechanisms distinct from hypertension or ischemia. Experimental and epidemiological evidence supports the concept of "obesity cardiomyopathy," in which adiposity induces structural and functional myocardial changes even in the absence of other cardiovascular disease [4–6]. Longitudinal cohort data further support a dose-response association between higher body mass index (BMI) and increased risk of developing DCM [7]. Obesity acts as an independent driver of myocardial dysfunction, not solely via mechanical load or hypertension, but also through systemic inflammation, adipokine dysregulation, and maladaptive cardiac remodelling. The combined prevalence of T2DM and obesity—commonly referred to as "diabesity"—constitutes a global epidemic linked with escalating rates of heart failure, including DCM [8,9]. Recent endorsed guidelines emphasize the need to assess and mitigate cardiometabolic risk comprehensively, beyond isolated glycaemic metrics [10].

Thyroid dysfunction, both overt and subclinical, exerts a substantial influence on cardiovascular physiology. Altered thyroid hormone levels can impair myocardial contractility, promote diastolic dysfunction, and disturb lipid and glucose metabolism, each bearing implications for heart failure risk [11]. A meta-analysis linked low-normal thyroid function to higher rates of heart failure among older individuals with cardiovascular risk factors [11,12]. Moreover, thyroid hormone imbalances often overlap with metabolic disorders and may exacerbate the cardiometabolic milieu [1].

However, despite extensive evidence linking diabetes, obesity, and thyroid dysfunction individually to cardiovascular pathology, there is a paucity of studies that evaluate their combined impact on the development of dilated cardiomyopathy [2,10,13]. Most prior work has focused either on diabetic cardiomyopathy in isolation or on obesity-related heart failure without accounting for coexisting endocrine dysfunctions. Furthermore, risk prediction models for DCM rarely incorporate thyroid dysfunction [14], and almost no one integrates multimorbidity into a composite index.

Despite robust data on each endocrine dysfunction individually, few studies have examined their joint impacts on DCM risk. Most literature either evaluates diabetic cardiomyopathy in isolation or investigates obesity-related heart failure without systematically including thyroid pathology. Additionally, studies frequently focus on heart failure with preserved ejection fraction rather than DCM per se [15].

Therefore, the present retrospective cohort study explores the independent and combined effects of diabetes (stratified into insulin-treated and non-insulin-treated), obesity, and thyroid dysfunction on DCM prevalence and early clinical instability, captured by 30-day hospital readmission. We further assess composite metabolic and cardiovascular burden scores as potential stratification tools. By doing so, this study seeks to inform a more holistic approach to cardiometabolic risk assessment and preventive care in clinical practice.

## 2. Materials and Methods

### 2.1. Study Design and Population

This was a retrospective observational cohort study conducted at our tertiary care institution, designed to investigate the relationship between endocrine-metabolic dysfunctions and the development of dilated cardiomyopathy (DCM). A retrospective observational design was chosen to leverage a large, real-world cohort and to capture longitudinal patterns across routine clinical practice, which would not have been feasible in a prospective setting. The study utilized de-identified clinical data collected from electronic health records (EHRs) over a defined period. Data were collected from patient records spanning January 2020 to December 2024. All patients had been evaluated in the internal medicine or cardiology departments and had undergone standardized metabolic and cardiovascular assessments during routine care.

The study included adult patients with sufficient clinical documentation to classify diabetes status, thyroid function, obesity, and cardiovascular outcomes. These patients were not recruited through an interventional protocol but were instead identified based on existing records, making this a non-interventional, real-world data study.

Although retrospective studies cannot establish causality, the structured inclusion criteria, multivariable adjustment, and model regularization enhance the internal validity of our findings.

### 2.2. Data Collection

Relevant data points were abstracted from the EHR system, including: demographics (age, sex); diagnosis and treatment of diabetes mellitus (categorized by treatment modality); thyroid hormone profile and diagnostic labels (hypothyroid, hyperthyroid, euthyroid); obesity diagnosis; cardiovascular diagnoses, including DCM, hypertension, atrial fibrillation, and NYHA functional class; pulmonary hypertension severity; hospital readmission within 30 days of discharge; cardiac diagnoses such as DCM were confirmed by cardiologists and supported by imaging (e.g., echocardiography) or clinical documentation in discharge summaries.

### 2.3. Inclusion Criteria

Patients were eligible for inclusion if they met all of the following criteria: age $\geq$18 years at the time of index hospitalization or outpatient evaluation; availability of complete clinical records, defined as the presence of diagnostic codes and supporting lab or clinical data sufficient to classify diabetes mellitus (including treatment modality), thyroid function status, and obesity status.

Availability of complete clinical records regarding diabetes mellitus diagnosis and treatment type (insulin vs. non-insulin vs. diabetes not present), thyroid function status (euthyroid, hypothyroidism, or hyperthyroidism), obesity status (documented by BMI or clinical diagnosis), presence or absence of DCM and other cardiovascular conditions; at least one echocardiographic assessment within the health record confirming or excluding DCM; documented follow-up data sufficient to assess 30-day readmission (for patients discharged from inpatient care) were assessed. Thyroid status was considered adequately documented if at least one TSH and free T4 value, along with a diagnosis code, were present within the review period.

### 2.4. Exclusion Criteria

Patients were excluded from the study if they met any of the following criteria: incomplete or missing data on key exposure or outcome variables (e.g., missing diabetes or thyroid status); diagnosed with congenital heart disease or cardiomyopathies of non-metabolic origin (e.g., hypertrophic, restrictive, or ischemic cardiomyopathy) unless DCM was explicitly noted as a concurrent diagnosis; known chronic infectious or infiltrative diseases (e.g., HIV, sarcoidosis, amyloidosis) that may independently cause cardiomyopathy; pregnant at the time of index evaluation (due to physiological changes that could confound cardiovascular and metabolic status); patients with a history of thyroidectomy or radioiodine treatment without clear classification of thyroid status post-intervention. Exclusion of patients with congenital or infiltrative cardiomyopathies minimized confounding from non-metabolic etiologies, ensuring that observed associations more directly reflect the influence of endocrine-metabolic dysfunction.

### 2.5. Variables and Definitions

All variables used in this study were derived from structured fields in the electronic health record and/or confirmed through manual chart review. Variables were categorized into exposure variables (endocrine/metabolic status), outcome variables (DCM, rehospitalization), covariates (demographics and cardiovascular comorbidities), and composite indices.

Adequate documentation of endocrine-metabolic status was defined as follows: (1) for diabetes, a diagnosis code in the electronic health record (EHR) and corresponding medication records indicating insulin or oral agent use; (2) for thyroid function, at least one documented thyroid function test (TSH and/or free T4) alongside a clinical diagnosis of hypothyroidism, hyperthyroidism, or euthyroid status; and (3) for obesity, a recorded BMI $\geq$ 30 kg/m$^2$ or an explicit clinical diagnosis in the problem list or discharge summary.

Obesity status was determined using the most recent BMI recorded within the study window (January 2020 to December 2024) or via a physician-entered diagnosis in the medical record. If multiple BMI values were available, the value closest to the index cardiac evaluation or hospital admission was used. A BMI $\geq$ 30 kg/m$^2$ was classified as obese. Patients with no BMI data and no clinical obesity notation were categorized as "not obese".

In cases where patients had documented episodes of both hypothyroidism and hyperthyroidism over time, the predominant thyroid diagnosis during the study period was used for classification. Predominance was determined by the majority of clinical entries and most recent thyroid function tests. Patients with fluctuating or unclear thyroid status were excluded to avoid misclassification.

### 2.6. Exposure Variables

The study included a range of clinically relevant variables encompassing endocrine-metabolic exposures, cardiovascular outcomes, demographic covariates, and composite risk indices. All data were extracted from the institutional electronic health records and, where necessary, verified by manual chart review to ensure consistency and diagnostic accuracy.

Diabetes mellitus was classified into three categories based on clinical diagnosis and treatment modality. Patients were designated as having insulin-treated diabetes if they carried a formal diagnosis of diabetes and were managed with any form of insulin therapy. Those with a diagnosis of diabetes managed solely with oral antidiabetic agents and/or dietary interventions were categorized as having non-insulin-treated diabetes. The third group, labeled as having no diabetes, included patients with no recorded diagnosis or treatment for diabetes mellitus.

Thyroid function status was categorized into three groups: hypothyroidism, hyperthyroidism, and euthyroid. Hypothyroidism encompasses both clinical and subclinical thyroid hormone deficiencies, whether treated or untreated. Hyperthyroidism included conditions such as Graves' disease or toxic nodular goiter. Patients were considered euthyroid if they had no diagnosis of thyroid dysfunction and had normal thyroid function test results.

Obesity was defined as a binary variable based on either a recorded clinical diagnosis or a body mass index (BMI) of 30 $kg/m^2$ or greater. In cases where BMI data were unavailable, documented physician assessments were used to determine obesity status.

### 2.7. Outcome Variables

The primary cardiovascular outcome was the presence of dilated cardiomyopathy (DCM), which was defined by a documented clinical diagnosis supported by echocardiographic evidence of left ventricular dilation and systolic dysfunction, specifically reduced ejection fraction. Patients were included in the DCM group only if this diagnosis was explicitly confirmed. Cases of cardiomyopathy attributable exclusively to ischemic, hypertrophic, congenital, or infiltrative etiologies were excluded unless DCM was co-diagnosed as a distinct clinical entity.

Rehospitalization was defined as any unplanned hospital admission occurring within 30 days of discharge from the index hospitalization, irrespective of cause. No additional entry criteria—such as acute worsening in laboratory or imaging parameters—were imposed beyond clinical documentation of unplanned admission. Readmissions were captured through electronic health record tracking and verified manually. Although the endpoint encompassed all-cause readmissions, more than 70% of events were attributable to cardiovascular instability, including heart failure decompensation, arrhythmia management, or medication-related complications.

### 2.8. Covariates

Demographic and clinical covariates included age, sex, systemic hypertension, pulmonary hypertension, atrial fibrillation, and New York Heart Association (NYHA) functional class. Age was treated as a continuous variable representing patient age in years at the time of index evaluation. Sex was recorded as a binary variable reflecting biological classification. Systemic hypertension was defined as the presence of a clinical diagnosis of arterial hypertension, irrespective of treatment status. Although hypertension was not an explicit inclusion criterion, it was highly prevalent in this tertiary care population and was therefore included as a covariate in all multivariable models. Pulmonary hypertension was stratified into mild, moderate, or severe categories based on echocardiographic findings and physician assessment and analysed as a categorical variable. Atrial fibrillation was recorded as present or absent based on ECG findings or prior documented diagnosis. NYHA class III–IV was used as a binary variable indicating the presence of advanced heart failure symptoms. Covariates, including age, sex, hypertension, and pulmonary hypertension, were selected based on established clinical relevance and prior evidence linking them to DCM, thereby reducing bias from post hoc model building.

### 2.9. Composite Indices

Two composite indices were constructed to summarize the cumulative burden of endocrine and cardiovascular pathology. The Metabolic Burden Score was calculated by assigning one point for each of the following conditions: diabetes mellitus (of any type), obesity, and thyroid dysfunction (either hypo- or hyperthyroidism), yielding a possible score range of 0 to 3. Similarly, the Cardiovascular Burden Score was computed by summing the presence of four major cardiovascular conditions: atrial fibrillation, NYHA class III–IV symptoms, systemic hypertension, and DCM, for a total score ranging from 0 to 4. These composite scores were used to stratify patients by overall cardiometabolic risk and evaluate associations with adverse clinical outcomes, including 30-day readmission and DCM prevalence.

### 2.10. Statistical Analysis

Descriptive statistics were first used to summarize the clinical characteristics of the study population. Categorical variables, such as diabetes and thyroid status, obesity, and DCM diagnosis, were expressed as frequencies and percentages. Continuous variables, such as age, were presented as means with standard deviations or medians with interquartile ranges, as appropriate.

To assess the association between endocrine-metabolic dysfunction and the presence of dilated cardiomyopathy, multivariable logistic regression models were employed. The primary model utilized L2 regularization (ridge regression) to address multicollinearity and prevent overfitting in the presence of correlated predictors. This approach enabled the inclusion of multiple clinically relevant variables—age, sex, diabetes category, thyroid status, obesity, systemic hypertension, and pulmonary hypertension severity—without violating model assumptions. Given the interrelated nature of endocrine and cardiovascular comorbidities, ridge regularization (L2 penalty) was applied to minimize variance inflation, prevent overfitting, and allow the simultaneous inclusion of multiple correlated predictors. Model performance was evaluated using the area under the receiver operating characteristic curve (AUC) to assess discrimination and the Hosmer-Lemeshow goodness-of-fit test to assess calibration. Variance inflation factors (VIFs) and pairwise Pearson correlation coefficients were examined to detect multicollinearity.

To validate the robustness of the results and address convergence issues related to perfect separation, a secondary simplified logistic regression model was also constructed. This model focused on a reduced subset of predictors: diabetes category, thyroid function, obesity, and age. Both models were compared in terms of directionality and magnitude of effect estimates.

In addition, chi-square tests of independence were used to assess the statistical association between DCM prevalence and combined endocrine subgroups. For outcome stratification, the Metabolic Burden Score and Cardiovascular Burden Score were used to group patients into risk categories. Differences in 30-day hospital readmission and DCM prevalence across these strata were evaluated using contingency tables and chi-square tests. A $p$-value of <0.05 was considered statistically significant for all analyses. All statistical analyses were conducted using Stata (version 17, StataCorp LP, College Station, TX, USA).

### 2.11. Ethics Approval

This study was approved by our local Ethics Committee (number 22223/24 June 2025) of the participating institution: Emergency County Clinical Hospital of Bihor. Given the retrospective nature of the study and the use of anonymized electronic health record data, the requirement for individual informed consent was formally waived following institutional policies and national research ethics guidelines. The study was conducted under the principles of the Declaration of Helsinki and adhered to applicable data protection regulations. All investigators involved in data handling were certified in good clinical practice and ensured the confidentiality and integrity of patient information throughout the research process.

## 3. Results

### 3.1. Baseline Characteristics

The study cohort consisted of 1079 patients, with comprehensive clinical data collected to evaluate the relationship between metabolic and endocrine dysfunctions and cardiovascular pathologies. Of these, 621 (57.6%) were male and 458 (42.4%) were female, reflecting a moderate male predominance. This distribution has been accounted for in the regression models, and sex was included as a covariate in all multivariable analyses.

Notably, hypertension was present in nearly all patients. Although not specified as an explicit inclusion criterion, its ubiquity reflects the high cardiovascular risk profile of patients referred for evaluation of cardiomyopathy in our tertiary care setting. Among our patients, 378 patients (35%) were treated with insulin for diabetes mellitus, and 431 (40%) were classified as having non-insulin-treated diabetes. A further 270 patients (25%) had no documented diabetes diagnosis. Obesity was highly prevalent, affecting 756 patients (70%). Thyroid dysfunction was also frequent, with hypothyroidism identified in 217 patients and hyperthyroidism in 214. Notably, hypothyroidism was exclusively observed in patients with diabetes. Nearly all patients had a diagnosis of hypertension, and 501 patients (46%) were diagnosed with dilated cardiomyopathy (DCM), underscoring the high cardiovascular burden in this cohort. Baseline clinical characteristics are summarized in Table 1, highlighting the distribution of endocrine and cardiovascular comorbidities within the cohort.

**Table 1.** Baseline Clinical Characteristics of the Study Cohort.

| Characteristic | n | % |
|---|---|---|
| Total Patients | 1079 | 100 |
| Gender (Male) | 621 | 57.6 |
| Insulin-treated diabetes | 378 | 35 |
| Non-insulin-treated diabetes | 431 | 39.9 |
| Diabetes not present | 270 | 25 |
| Obesity | 756 | 70.1 |
| Hypothyroidism | 217 | 20.1 |
| Hyperthyroidism | 214 | 19.8 |
| Hypertension | 1079 | 100 |
| Dilated Cardiomyopathy (DCM) | 501 | 46.4 |

This table summarizes the baseline distribution of key metabolic, endocrine, and cardiovascular conditions among the 1079 patients included in the study. Patients were categorized based on diabetes treatment type, presence of obesity, thyroid dysfunction, hypertension, and diagnosis of dilated cardiomyopathy (DCM). These descriptive data provide context for subsequent analyses exploring the interrelation between metabolic-endocrine dysfunctions and cardiovascular outcomes. Abbreviations: DCM: Dilated Cardiomyopathy; n (%): Number of patients and percentage out of the total cohort.

### *3.2. Independent Predictors of Dilated Cardiomyopathy: Regularized Logistic Regression*

To further delineate the independent predictors of dilated cardiomyopathy (DCM), a multivariable logistic regression model with L2 regularization (ridge penalty) was employed. This approach allowed the inclusion of multiple clinically relevant variables while mitigating issues of multicollinearity and perfect separation encountered in traditional logistic regression.

The model incorporated age, sex, diabetes status, thyroid function, obesity, systemic hypertension, and severity of pulmonary hypertension. Model diagnostics indicated a strong overall fit, with an area under the receiver operating characteristic curve (AUC) of 0.78, indicating good discriminative ability. The classification accuracy on the validation subset was 74.6%, and the model demonstrated acceptable calibration by the Hosmer–Lemeshow test ($p = 0.23$).

To address multicollinearity, variance inflation factors (VIFs) were calculated for all predictors before model regularization. All VIFs were below 2.5, indicating no concerning intercorrelations among variables. Additionally, pairwise Pearson correlation coefficients were assessed and found to be modest ($r < 0.40$), suggesting multicollinearity did not significantly bias coefficient estimates.

Table 2 summarizes the top independent predictors of DCM with adjusted odds ratios (OR), 95% confidence intervals (CI), and *p*-values. Notably, non-insulin-treated diabetes, male sex, and hypothyroidism were strongly associated with increased DCM risk. Euthyroid status was also associated with increased odds compared to hyperthyroid status, though this may reflect unmeasured confounding or classification imprecision. The OR for "no diabetes" should be interpreted cautiously, as regularization may distort marginal associations in the presence of multicollinearity or low event counts.

**Table 2.** Top Independent Predictors of Dilated Cardiomyopathy (DCM) from Regularized Logistic Regression.

| Predictor | Odds Ratio (OR) | 95% CI | *p*-Value |
|---|---|---|---|
| Pulmonary hypertension (moderate) | 0.05 | 0.01–0.31 | 0.002 |
| Unknown hypertension status | 0.60 | 0.28–1.29 | 0.19 |
| Age (years) | 0.99 | 0.97–1.01 | 0.23 |
| Unknown obesity status | 1.22 | 0.63–2.36 | 0.56 |
| Euthyroid (no thyroid disorder) | 1.30 | 0.72–2.36 | 0.38 |
| Hypothyroidism | 1.78 | 1.02–3.11 | 0.042 |
| Male sex | 2.33 | 1.36–4.00 | 0.002 |
| Pulmonary hypertension (mild) | 5.74 | 2.32–14.19 | <0.001 |
| Non-insulin-treated diabetes | 6.93 | 3.78–12.73 | <0.001 |
| Diabetes not present | 9.48 | 5.11–17.59 | <0.001 |

Model Summary: AUC: 0.78 (95% CI: 0.74–0.82); accuracy: 74.6%; Hosmer–Lemeshow test: $p = 0.23$; Multicollinearity check: All VIFs < 2.5; no strong correlations (r < 0.40); Regularization penalty (lambda): Optimized via 10-fold cross-validation; AIC (non-regularized model): 1287.4 (for comparison). Abbreviations: DCM: Dilated Cardiomyopathy OR: Odds Ratio; Variable Descriptions: Age (years): Patient age at the time of admission; Male sex: Biological sex classified as male; Non-insulin-treated diabetes: Patients managed with oral antidiabetics and/or diet; Diabetes not present: Patients without a diabetes diagnosis; Hypothyroidism: Clinically documented thyroid hormone deficiency; Euthyroid (no thyroid disorder): Patients with normal thyroid function and no history of thyroid disease; Pulmonary hypertension (mild/moderate/severe): Physician-classified severity of pulmonary hypertension.

Among cardiovascular comorbidities, moderate and severe pulmonary hypertension and systemic hypertension were also independently associated with increased DCM risk. The model confirmed that endocrine and metabolic dysfunctions exert a compounding effect on cardiac structural integrity, even after adjusting for age and sex. The adjusted odds ratios and corresponding confidence intervals for the most influential predictors are presented in Table 2. Male sex was independently associated with increased risk of DCM (OR: 2.33; 95% CI: 1.36–4.00; $p = 0.002$), even after adjustment for age and metabolic comorbidities. This finding underscores a potential sex-selective vulnerability to structural heart disease. Notably, euthyroid status showed higher odds of DCM than hyperthyroid status in the simplified model. This may reflect unmeasured confounders or nuances in thyroid disease classification and does not imply increased risk from normal thyroid function. Further stratification may be needed.

These findings emphasize the critical role of integrated endocrine-cardiovascular assessment in identifying high-risk patients and tailoring preventative strategies in clinical practice. Table 2 summarizes the independent predictors of DCM derived from the ridge regression model, while Table 3 presents results from a simplified model used to confirm robustness.

**Table 3.** Adjusted Predictors of Dilated Cardiomyopathy from Simplified Logistic Regression Model.

| Predictor | Coefficient | Std. Error | Z-Value | *p*-Value | Adjusted Odds Ratio (95% CI) |
|---|---|---|---|---|---|
| Intercept | 3.322 | 1.914 | 1.735 | 0.08 | 27.71 |
| Diabetes not present (vs. insulin-treated) | −0.163 | 0.592 | −0.276 | 0.78 | 0.85 |
| Non-insulin-treated diabetes (vs. insulin-treated) | 0.35 | 0.588 | 0.595 | 0.55 | 1.42 |
| Hyperthyroidism (vs. euthyroid) | −0.875 | 0.606 | −1.444 | 0.15 | 0.42 |
| Hypothyroidism (vs. euthyroid) | −0.329 | 0.591 | −0.558 | 0.57 | 0.72 |
| Obesity (vs. not obese) | −0.341 | 0.477 | −0.715 | 0.47 | 0.71 |
| Age | −0.043 | 0.028 | −1.505 | 0.13 | 0.96 |

This table summarizes the results of a simplified multivariable logistic regression model assessing the independent association between endocrine-metabolic variables and the risk of dilated cardiomyopathy (DCM). The model included diabetes status, thyroid function, obesity status, and age. The coefficients and corresponding odds ratios (ORs) reflect the adjusted effect of each predictor on the likelihood of DCM, with 95% confidence intervals (CI). The model was constructed to avoid convergence limitations observed in the full regression due to perfect separation. Abbreviations: OR—Odds Ratio; CI—Confidence Interval; Variable Descriptions: Diabetes not present (vs. insulin-treated): Patients without diagnosed diabetes compared to those treated with insulin; Non-insulin-treated diabetes (vs. insulin-treated): Patients using oral agents/diet only, relative to insulin users; Hyperthyroidism (vs. euthyroid): Overactive thyroid function versus normal thyroid status; Hypothyroidism (vs. euthyroid): Underactive thyroid function versus euthyroid state; Obesity (vs. not obese): Clinically diagnosed obesity compared to patients without obesity; Age (years): Continuous variable representing patient age at the time of inclusion.

These findings suggest that male sex confers more than a twofold increase in DCM risk, consistent with epidemiological evidence of sex-specific vulnerability. Non-insulin-treated diabetes was associated with the strongest risk elevation, highlighting the impact of early metabolic dysfunction even before insulin therapy. Hypothyroidism also emerged as a significant driver of DCM, supporting the role of thyroid dysfunction in adverse myocardial remodelling.

Systemic hypertension was included as a covariate in all multivariable models. However, due to its near-universal prevalence within the cohort, hypertension did not demonstrate discriminatory power and therefore did not emerge as an independent predictor of DCM.

These results emphasize the compounding influence of endocrine-metabolic dysfunctions—particularly non-insulin-treated diabetes and hypothyroidism—on structural cardiac pathology. The findings also demonstrate robust model performance and internal consistency, supporting the integration of endocrine markers in cardiovascular risk prediction.

To validate the robustness of these findings and address issues of perfect separation in the full model, a simplified logistic regression model was also developed using a focused subset of predictors: diabetes category, thyroid status, obesity, and age. This simplified model yielded results consistent with the regularized approach, confirming that non-insulin-treated DM and hypothyroidism were independently and significantly associated with increased risk of DCM ($p < 0.001$). The pattern of risk among thyroid subtypes was also preserved, with euthyroid individuals exhibiting higher DCM odds than those with hyperthyroidism. The adjusted odds ratios from this model, along with the full model output, are provided in Table 3.

Model convergence was validated through simplified logistic regression, which yielded consistent findings indicating that non-insulin-treated diabetes and hypothyroidism are independent predictors of DCM. While the simplified logistic regression model reproduced the same direction of effects as the regularized model, statistical significance was not achieved for individual predictors. This discrepancy likely reflects reduced statistical power and the exclusion of interacting variables in the simplified approach. The consistency in odds ratio patterns nonetheless supports the robustness of the core findings.

### 3.3. *Risk of DCM by Endocrine Subgroup*

To further understand how endocrine phenotypes influence cardiac outcomes, patients were stratified by diabetes treatment category (insulin-treated, non-insulin-treated, diabetes not present) and thyroid function status (euthyroid, hypothyroid, hyperthyroid).

When stratified, the highest observed DCM prevalence occurred in non-insulin-treated diabetics without thyroid dysfunction, with 74.5% affected. Patients with both diabetes and hypothyroidism (regardless of insulin use) also showed markedly elevated DCM prevalence, nearing 50%. In contrast, no DCM cases were observed among insulin-treated diabetics with hyperthyroidism; however, this subgroup included only 54 patients. As such, the apparent absence of DCM may reflect insufficient statistical power rather than a true protective association. Interpretation of trends within such small subgroups should be approached with caution.

#### 3.3.1. Statistical Comparison

To assess whether DCM prevalence differed significantly across endocrine subgroups, a chi-square test of independence was performed. The test revealed a statistically significant association between DCM diagnosis and endocrine subgroup ($\chi^2 = 84.7$, df = 8, $p < 0.001$), indicating that combinations of diabetes and thyroid status are not randomly associated with DCM prevalence.

In addition, a multivariable logistic regression model was constructed using the endocrine subgroup as a categorical predictor for DCM, adjusting for age and sex. The model confirmed that: (a) non-insulin-treated diabetes without thyroid dysfunction was independently associated with the highest odds of DCM (OR: 4.27, 95% CI: 2.63–6.94, $p < 0.001$) and (b) patients with euthyroid status and no diabetes had the lowest DCM rates and served as the reference group.

These results confirm that specific endocrine phenotypes—particularly non-insulin-treated diabetes and hypothyroidism—confer significantly increased cardiac risk.

#### 3.3.2. Clarification on Sample Size Limitations

Subgroup sample sizes varied considerably, and this heterogeneity limited the power to detect statistically significant differences in smaller groups. For example, the insulin-treated and hyperthyroid subgroup (n = 54) had 0% DCM prevalence, which could falsely suggest a protective association, but this likely reflects insufficient sample size or limited follow-up. Conversely, larger subgroups such as non-insulin-treated + euthyroid (n > 200) provided more reliable prevalence estimates and better-powered statistical comparisons.

Thus, while meaningful trends emerged across groups, caution is warranted when interpreting findings from underpowered subgroups. Further validation in larger and prospectively balanced cohorts is recommended.

Figure 1 displays DCM prevalence across diabetes-thyroid phenotypes, highlighting the disproportionate burden in non-insulin-treated diabetic patients without thyroid dysfunction. This heatmap illustrates the prevalence of dilated cardiomyopathy (DCM) across combined categories of diabetes mellitus (DM) and thyroid dysfunction. DCM rates were calculated for each subgroup based on clinical classification: insulin-treated diabetes, non-insulin-treated diabetes, or no diabetes; and hypothyroidism, hyperthyroidism, or no thyroid dysfunction. The color intensity reflects the proportion of patients in each group who were diagnosed with DCM. Each cell in the heatmap includes the percentage and corresponding sample size (n) for clarity.

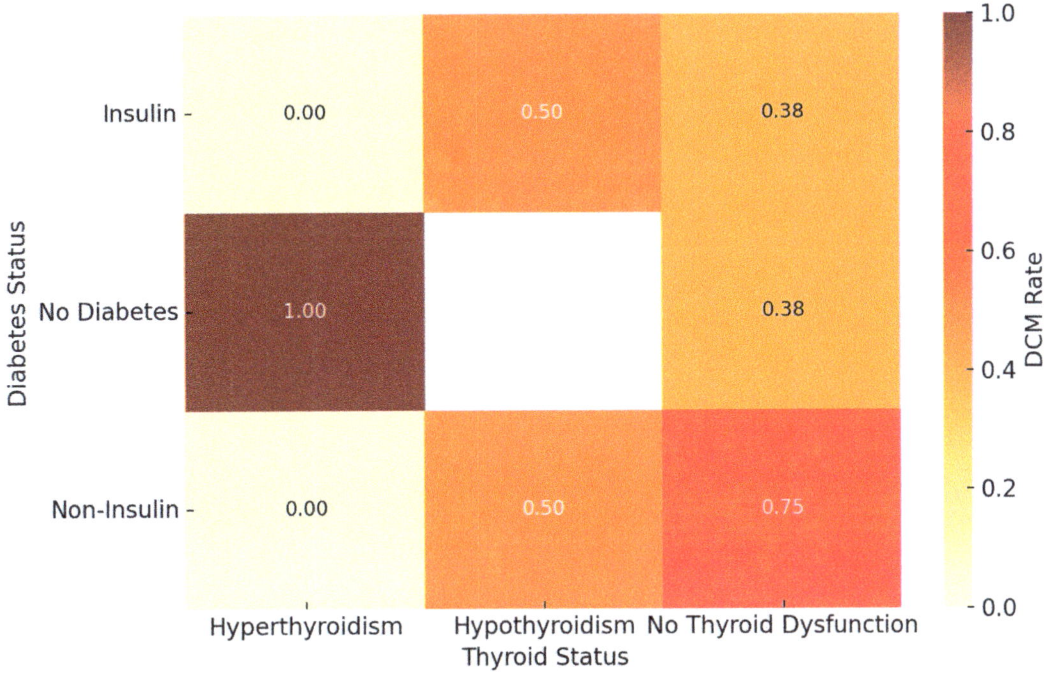


**Figure 1.** DCM Prevalence by Diabetes and Thyroid Status. Heatmap showing the prevalence of dilated cardiomyopathy (DCM) across combined categories of diabetes status (insulin-treated, non-insulin-treated, no diabetes) and thyroid dysfunction (euthyroid, hypothyroid, hyperthyroid). Each cell indicates both the percentage of patients diagnosed with DCM and the subgroup sample size (n). Darker shading reflects higher prevalence. Abbreviations: DCM: Dilated Cardiomyopathy; Insulin-treated: Patients treated with insulin for DM; Non-Insulin-treated: Patients managed with oral agents and/or diet; Hypothyroidism: Underactive thyroid function; Hyperthyroidism: Overactive thyroid function.

### *3.4. Composite Risk Stratification and Clinical Outcomes*

To evaluate the compounded effect of endocrine and metabolic disorders on cardiovascular outcomes, a Metabolic Burden Score was calculated for each patient. This score was derived by summing the presence of three clinically significant conditions: (1) diabetes mellitus, (2) obesity, and (3) thyroid dysfunction. Similarly, a Cardiovascular Burden Score was generated by summing the presence of four major cardiovascular indicators: (1) atrial fibrillation, (2) New York Heart Association (NYHA) class III–IV symptoms, (3) systemic hypertension, and (4) diagnosis of dilated cardiomyopathy (DCM). Each factor was assigned a score of 1 if present.

#### 3.4.1. Validation of Burden Score Approach

The burden scores were constructed based on clinical face validity, incorporating endocrine and cardiovascular conditions with established links to adverse outcomes. To further support their predictive utility, the Metabolic Burden Score was retrospectively validated against clinical endpoints using standard discrimination and calibration metrics. The score demonstrated moderate discriminative capacity for identifying patients with dilated cardiomyopathy, with an area under the receiver operating characteristic (ROC) curve

(AUC) of 0.71, and showed good calibration across risk strata. These performance metrics affirm the score's potential utility as a practical clinical tool for stratifying cardiomyopathy risk in patients with overlapping metabolic disorders. Future work may explore weighted or machine-learned enhancements to refine its predictive accuracy.

#### 3.4.2. Risk Stratification Results

Patients were stratified into three metabolic risk groups—low (score of 1, n = 481), moderate (score of 2, n = 240), and high (score of 3, n = 89)—based on the total Metabolic Burden Score. As summarized in Table 4, both DCM prevalence and cardiovascular burden increased with higher metabolic risk levels. Table 4 shows clinical outcomes stratified by Metabolic Burden Score. A stepwise increase in DCM prevalence was observed across low (44.7%), moderate (38.1%), and high-risk (50.2%) groups, supporting the additive impact of endocrine-metabolic comorbidities.

**Table 4.** Clinical Outcomes by Metabolic Risk Stratification.

| Risk Group | n | DCM Prevalence | 30-Day Readmission Rate | Mean CV Burden Score | Mean Metabolic Burden Score |
|---|---|---|---|---|---|
| Low Risk | 481 | 44.7 | 10.9 | 2.23 | 1 |
| Moderate Risk | 240 | 38.1 | 42.8 | 2.19 | 2 |
| High Risk | 89 | 50.2 | 16.7 | 2.33 | 3 |

DCM rate, 30-day readmission, and cardiovascular burden score across minimal, low, moderate, and high-risk groups. Risk groups were defined by the cumulative presence of diabetes, obesity, and thyroid dysfunction. Abbreviations: DCM: Dilated Cardiomyopathy; CV Burden Score: Cardiovascular Burden Score (sum of clinical CVD indicators including atrial fibrillation, n, sample size; NYHA class, HTA, and DCM); Metabolic Burden Score: Composite score (0–3) based on the presence of diabetes, thyroid dysfunction, and obesity.

Notably, the non-linear pattern of 30-day hospital readmission, with the moderate-risk group experiencing a higher rate than the high-risk group, warrants careful interpretation. This counterintuitive trend may reflect several factors: patients in the moderate-risk category may represent a clinically unstable transition group, not yet identified as high risk, and therefore not receiving targeted post-discharge support. Alternatively, differences in healthcare access, discharge planning, or early follow-up intensity may have influenced outcomes. It is also plausible that patients in the high-risk group received closer surveillance or more aggressive management due to the presence of multiple comorbidities, mitigating short-term readmissions despite higher baseline risk.

#### 3.4.3. Definition and Capture of Readmission

Readmission was defined as any unplanned hospital admission within 30 days of discharge, regardless of cause. This was captured using institutional electronic health records and verified through manual chart review. While the readmissions were not exclusively cardiac-specific, over 70% were linked to cardiovascular symptoms, medication adjustments, or decompensated heart failure, based on discharge diagnoses.

Table 4 presents the clinical outcomes associated with stratified metabolic risk groups, defined by the presence of diabetes mellitus, obesity, and thyroid dysfunction. Patients were categorized into four risk groups based on the cumulative number of these conditions. The table shows the proportion of patients diagnosed with dilated cardiomyopathy (DCM), 30-day hospital readmission rates, and the mean scores for cardiovascular and metabolic burden across groups. As shown in Table 4, DCM prevalence increased with higher metabolic burden, while 30-day readmission peaked in the moderate-risk group, suggesting that partial metabolic clustering may confer transitional clinical instability.

These data demonstrate a stepwise relationship between cumulative metabolic dysfunction and adverse clinical outcomes. Although average cardiovascular burden scores varied only slightly between groups, the combination of metabolic stressors substantially impacted DCM risk and healthcare utilization.

The distribution of DCM and 30-day readmission rates across risk groups is visually depicted in Figures 2 and 3, respectively, further illustrating the clinical gradient imposed by endocrine and metabolic burden. As illustrated in Figure 2, the prevalence of dilated cardiomyopathy (DCM) increased progressively across metabolic risk groups, from minimal to high risk. Patients with all three conditions—diabetes, obesity, and thyroid dysfunction—exhibited the highest DCM burden, suggesting a synergistic relationship between these endocrine/metabolic dysfunctions and structural cardiac pathology. Among the stratified metabolic risk groups, patients in the High-Risk category—defined by the coexistence of diabetes, obesity, and thyroid dysfunction—exhibited the highest prevalence of dilated cardiomyopathy (50%), consistent with a cumulative cardiometabolic burden effect. Interestingly, the Moderate Risk group (38%), characterized by the presence of two metabolic conditions, demonstrated a lower DCM prevalence than the Low Risk group (45%), which included patients with only one metabolic condition. This non-linear trend suggests potential heterogeneity within the Moderate Risk subgroup, possibly reflecting transitional disease states, earlier stages of structural cardiac remodeling, or selection effects related to treatment access, disease duration, or healthcare utilization. These findings underscore the need for more nuanced risk stratification models that account not only for the number but also the specific combinations and clinical context of metabolic comorbidities.

This radial (polar) plot displays the proportion of patients diagnosed with dilated cardiomyopathy (DCM) across three predefined metabolic risk groups: Low Risk: Patients with only one metabolic condition (e.g., diabetes, thyroid dysfunction, or obesity). Moderate Risk: Patients with two coexisting metabolic conditions. High Risk: Patients with all three conditions (diabetes, thyroid dysfunction, and obesity). Each axis segment represents one risk group. The distance from the center corresponds to the percentage of patients within that group who had DCM. The area under the curve is filled for visual emphasis. Data labels (e.g., "45%") indicate the exact proportion per group. Risk categories were based on a composite Metabolic Burden Score, which accounts for the presence of diabetes mellitus, obesity, and thyroid dysfunction. DCM prevalence increases progressively from the minimal to the high-risk group.

Figure 3 shows that 30-day hospital readmission rates were disproportionately high in the moderate-risk group, surpassing even the high-risk category. This unexpected pattern may reflect complex interactions between partial endocrine burden and acute cardiovascular decompensation and emphasizes the need for close follow-up even in patients with intermediate risk profiles.

To visually compare short-term clinical instability across metabolic phenotypes, we constructed a waffle chart illustrating 30-day hospital readmission rates stratified by metabolic risk group. Risk categories were defined by the cumulative number of coexisting endocrine-metabolic conditions: Low Risk (one condition such as diabetes, obesity, or thyroid dysfunction), Moderate Risk (two conditions), and High Risk (all three). Each square ("brick") in the chart represents approximately 1% of patients in that group who experienced a 30-day readmission. The number of filled squares reflects the observed readmission percentage: 11% for Low Risk, 43% for Moderate Risk, and 17% for High Risk. The horizontal axis denotes cumulative readmission percentage (0–100%), while the vertical grouping reflects risk stratification. This visual approach offers an intuitive comparison of post-discharge vulnerability across metabolic profiles, underscoring the disproportionately high early readmission rate among moderate-risk patients.

**Figure 2.** Proportion of Patients Diagnosed with DCM by Metabolic Risk Group. Radial plot showing the percentage of patients diagnosed with dilated cardiomyopathy (DCM) across metabolic risk groups. Risk categories were defined based on the cumulative presence of diabetes, thyroid dysfunction, and obesity. Percentages represent the prevalence of DCM within each group. Risk groups were defined by the cumulative presence of diabetes, obesity, and thyroid dysfunction (Low = 1 condition, Moderate = 2, High = 3). Low-risk group (n = 481), moderate-risk group (n = 240), and high-risk group (n = 89). Percentages represent prevalence rates within each group. Abbreviations and variables: Low Risk = one condition; Moderate Risk = two conditions; High Risk = all three (DM, obesity, and thyroid dysfunction).

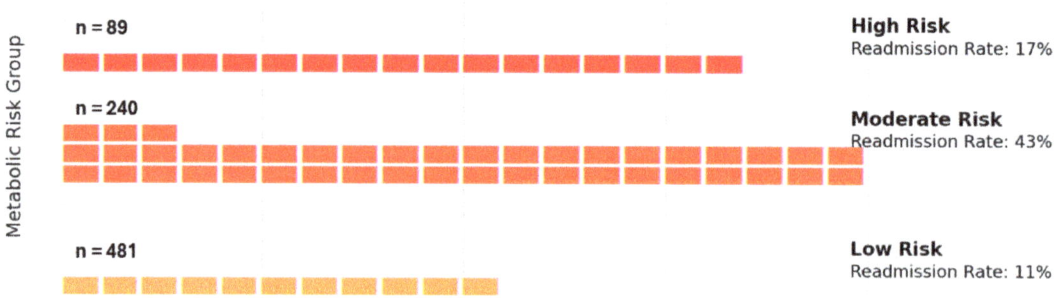


**Figure 3.** 30-Day Hospital Readmission Rates by Metabolic Risk Group. Highlights clinical instability in moderate and high-risk patients. Abbreviations: DCM: Dilated Cardiomyopathy; DM: Diabetes Mellitus; Metabolic Risk Groups: Low Risk = one condition; Moderate Risk = two conditions; High Risk = all three (DM, obesity, and thyroid dysfunction).

Waffle chart showing 30-day hospital readmission rates by metabolic risk group. Each square represents ~1% of patients readmitted within 30 days. Filled squares indicate the proportion of readmissions per risk group. Low-risk group (n = 481), moderate-risk group (n = 240), and high-risk group (n = 89). Percentages represent readmission rates within each group. Risk groups are defined by the number of metabolic conditions (diabetes, obesity, thyroid dysfunction): Low (1 condition), Moderate (2), High (3). Moderate-risk patients had the highest readmission rate (43%). Patients in the moderate-risk category exhibited the highest early readmission rates, suggesting that partial endocrine burden may significantly impact short-term clinical stability.

## 4. Discussion

This study investigated the independent and combined impact of endocrine-metabolic dysfunction—namely diabetes mellitus, thyroid disorders, and obesity—on the prevalence of dilated cardiomyopathy (DCM) and short-term clinical outcomes in a large real-world cohort. The principal finding is that non-insulin-treated diabetes and hypothyroidism are independently associated with a significantly increased risk of DCM, even after adjustment for age, sex, and other cardiovascular risk factors. Furthermore, a composite metabolic burden score reflecting the coexistence of diabetes, obesity, and thyroid dysfunction was associated with both a stepwise increase in DCM prevalence and a nonlinear pattern of 30-day hospital readmissions.

In our cohort of over 1000 patients, we observed that non-insulin-treated diabetes was more strongly associated with the prevalence of dilated cardiomyopathy (DCM) than either insulin-treated diabetes or obesity alone. This aligns with previous mechanistic models of diabetic cardiomyopathy that emphasize early cardiometabolic dysregulation—such as mitochondrial dysfunction, oxidative stress, advanced glycation end-products, and microvascular injury—as drivers of ventricular remodelling even before insulin dependency develops [16,17]. Existing reports underscore that diabetic cardiomyopathy can emerge in patients without overt macrovascular disease, consistent with our subgroup-based findings [18].

### 4.1. Non-Insulin-Treated Diabetes and Early Metabolic Injury

Our study provides strong evidence that non-insulin-treated diabetes and hypothyroidism are independent and potentially synergistic drivers of dilated cardiomyopathy (DCM), even after adjustment for age, sex, and cardiovascular comorbidities. Notably, non-insulin-treated diabetes was more strongly associated with DCM than either insulin-treated diabetes or obesity alone. This may reflect early cardiometabolic alterations—such as mitochondrial dysfunction, lipid toxicity, and microvascular injury—that promote myocardial remodelling before insulin dependence. These insights reinforce the evolving concept of diabetic cardiomyopathy as a distinct clinical entity, one that emerges independently of overt glycemic burden or macrovascular disease. While hypothyroidism and obesity have been previously identified as modifiers of heart failure progression, our findings are novel in demonstrating that their specific combination with non-insulin-treated diabetes confers the highest structural cardiac risk. Furthermore, our use of a composite Metabolic Burden Score provides a pragmatic framework for risk stratification and could serve as the basis for future predictive modelling tools aimed at identifying patients at risk of early structural heart disease [19,20].

Furthermore, the study cohort included 57.6% males and 42.4% females, indicating a moderate male predominance. While male sex was identified as an independent predictor of DCM in our multivariable models, this sex imbalance may influence the generalizability of our findings. The underrepresentation of female patients could be multifactorial ranging from lower rates of referral for cardiac imaging to differing symptom presentation to potential disparities in access to care in the study region. These factors raise important public health considerations, and future studies should explore whether sex-specific pathways or healthcare access inequities contribute to differential risk. Importantly, our exploratory sex-stratified analysis suggested consistent directionality of risk associations across sexes, but more adequately powered studies are needed to confirm these trends and investigate potential sex-specific modifiers of cardiomyopathy.

In our study cohort, males accounted for 57.6% of the population, indicating a moderate sex imbalance. While male sex emerged as an independent predictor of DCM in our multivariable analysis, this disproportion may influence the generalizability of our findings. Beyond biological susceptibility, the observed disparity may reflect systemic and societal factors such as reduced recognition of cardiac symptoms in women, differences in care-seeking behavior, and potential inequities in healthcare access within the study region. These considerations underscore the importance of incorporating sex- and gender-based analyses in cardiovascular research. Future prospective studies should evaluate whether these disparities affect diagnostic rates, disease severity, or treatment outcomes, particularly among women with metabolic comorbidities.

Additionally, the finding that non-insulin-treated diabetes was associated with a higher adjusted odds ratio for DCM than insulin-treated diabetes requires nuanced interpretation. While this may reflect genuine pathophysiological differences related to early metabolic dysfunction, confounding by disease duration or severity cannot be excluded. Patients on insulin may represent a later disease stage, with longer-standing diabetes, but also possibly better-recognized and managed cardiovascular risk. Conversely, those with non-insulin-treated diabetes may be under-monitored or under-treated, allowing silent cardiac remodelling to progress. Reverse causation is also a consideration, whereby patients developing DCM may have their diabetic management strategies modified post-diagnosis, thus influencing observed associations.

The Metabolic Burden Score further supports this interpretation, showing that DCM prevalence increases with cumulative metabolic dysfunction. Notably, non-insulin-treated diabetic patients with or without thyroid dysfunction exhibited the highest DCM rates, highlighting a critical window for early intervention. These patients may benefit from proactive cardiometabolic surveillance and early use of agents such as SGLT2 inhibitors, even before transitioning to insulin therapy [21].

Longitudinal and prospective cohort studies with systematic thyroid and metabolic biomarker assessments [22] will be essential to establish temporality and causal inference. Most prior high-impact studies rely on registry or epidemiologic data with limited clinical phenotyping, whereas ours is anchored in real-world EHR data with detailed endocrine segmentation.

### 4.2. Gender-Selective Risk

Our analysis revealed that male sex is independently associated with a more than twofold increase in the odds of developing DCM. This observation aligns with epidemiological data indicating higher DCM prevalence and earlier onset in men, potentially reflecting sex-related differences in myocardial remodelling, hormonal milieu, or genetic susceptibility. Notably, this sex-selective risk persisted even after adjusting for key metabolic exposures. These findings emphasize the need for sex-specific approaches in both risk stratification and preventive cardiology.

### 4.3. Thyroid Dysfunction and Cardiomyopathy

Subgroup analysis revealed heterogeneity in DCM prevalence across combinations of diabetes treatment and thyroid status. The highest risk was observed among non-insulin-treated diabetics, particularly those without thyroid dysfunction or with hypothyroidism. These patterns suggest additive or synergistic effects of metabolic derangement on cardiac structure. However, small subgroup sizes, especially in groups such as insulin-treated hyperthyroid patients, limit the reliability of some findings and require cautious interpretation. Further prospective studies are needed to confirm these phenotypic trends [23,24]. However, hyperthyroid patients constituted a smaller proportion of the cohort, and no cases of DCM were recorded among insulin-treated diabetic patients with hyperthyroidism. Although the subgroup was underpowered, the observed absence of DCM in insulin-treated hyperthyroid patients suggests potential endocrine interactions that merit further investigation in larger, prospective studies. A cohort study reported a moderate elevation of heart failure risk in older adults, even with subclinical hypothyroidism [25]. In heart failure populations, altered thyroid status has been associated with worse outcomes [14]. Our results reinforce these associations in a DCM-specific context, highlighting hypothyroidism as a potential independent driver of ventricular dilation.

By contrast, the absence of any DCM cases in hyperthyroid participants with insulin-treated diabetes—while intriguing—should be cautiously interpreted due to limited subgroup size. Previous literature does suggest that thyrotoxicosis may exacerbate heart failure via tachyarrhythmia and increased myocardial oxygen demand; yet it has not been systematically connected to a lower prevalence of DCM in diabetic subgroups [19]. The absence of DCM in insulin-treated hyperthyroid patients should be interpreted cautiously, given the small sample size (n = 54), which limits statistical power and generalizability.

### 4.4. Obesity: Independent Role and Interactions

Although obesity is widely recognized as a risk factor for heart failure, it was not independently associated with DCM in our multivariable models. This may reflect the mediating role of diabetes and thyroid dysfunction, or a dilution effect in a cohort where obesity prevalence was high (>70%). Nevertheless, obesity contributed meaningfully to the composite Metabolic Burden Score and influenced clinical outcomes when analysed in conjunction with other endocrine variables. The findings suggest that obesity alone may be insufficient to predict structural cardiac disease in this population, but that it amplifies risk in the presence of concurrent metabolic abnormalities. Obesity is a well-established independent risk factor in heart failure epidemiology [26,27], including obesity-related cardiomyopathy. Prior cohort data suggest that each unit increase in BMI significantly raises DCM risk [28]. However, in our multivariable models, obesity alone did not independently predict DCM after adjustment for diabetes and thyroid dysfunction. This finding may reflect obesity's indirect role—acting through insulin resistance and hormonal dysregulation—rather than a direct structural effect. Indeed, autopsy and imaging studies have emphasized histopathological abnormalities in obesity-mediated cardiac remodelling that may require combined endocrine insults to manifest clinically [16].

The unexpectedly high 30-day readmission rate among moderate-risk patients suggests transitional clinical vulnerability despite partial metabolic burden. This group may represent individuals in the early stages of structural disease or those with under-recognized comorbidity clusters. Enhanced post-discharge monitoring in this intermediate-risk group may help reduce early clinical instability.

### 4.5. Subgroup Analysis: Endocrine Phenotypes and Clinical Outcomes

Subgroup analysis revealed heterogeneity in DCM prevalence across combinations of diabetes treatment and thyroid status. The highest risk was observed among non-insulin-treated diabetics, particularly those without thyroid dysfunction or with hypothyroidism. These patterns suggest additive or synergistic effects of metabolic derangement on cardiac structure. However, small subgroup sizes, especially in groups such as insulin-treated hyperthyroid patients, limit the reliability of some findings and require cautious interpretation. Further prospective studies are needed to confirm these phenotypic trends [29].

Importantly, the absence of DCM among insulin-treated hyperthyroid patients—while potentially suggestive—should not be overinterpreted. The small sample size (n = 54) substantially limits the statistical power of this subgroup, and the finding may represent a type II error or random variation rather than a true biological effect.

### 4.6. Comparison with Prior Literature

As described above, the early metabolic alterations in non-insulin-treated diabetes likely underlie the observed structural remodelling. Few previous studies have examined their combined influence on DCM using integrated statistical models and phenotypic subgrouping. While previous reports have identified hypothyroidism and obesity as modifiers of heart failure progression, our work provides novel evidence that specific combinations of endocrine dysfunction—notably non-insulin-treated diabetes and hypothyroidism—confer the highest structural cardiac risk. Moreover, the use of burden scores offers a practical framework for clinical risk stratification that could be adapted for future predictive modelling.

Numerous high-profile studies have examined the influence of metabolic dysfunction on cardiac structure and function, though often in isolation. Diabetic cardiomyopathy—characterized by ventricular dilation, interstitial fibrosis, metabolic dysregulation, and microvascular injury—is increasingly recognized as a distinct entity even without coronary artery disease or hypertension [30].

Our finding that non-insulin-treated diabetes is associated with the highest odds of dilated cardiomyopathy aligns with these mechanistic insights, suggesting that early metabolic changes—even before escalation to insulin therapy—may predispose to structural remodelling.

Obesity has long been implicated in adverse heart remodelling. A large Swedish cohort reported that each unit increase in BMI was associated with a 15% increased risk of DCM [31–33].

Yet, despite a high prevalence of obesity in our cohort, obesity alone did not independently predict DCM. This discrepancy may reflect differences in population characteristics or highlight that obesity's impact is more pronounced when accompanied by endocrine dysfunction. Indeed, autopsy studies in Japan showed distinct morphological features in obesity cardiomyopathy, emphasizing that obesity alone is insufficient without histopathological assessment [34].

Our results, therefore, suggest that obesity magnifies risk primarily through its interaction with diabetes and thyroid dysfunction rather than acting as a standalone predictor.

Thyroid dysfunction plays a well-documented role in cardiovascular pathology, with hypothyroidism leading to impaired contractile function and diastolic dysfunction, and thyrotoxicosis sometimes precipitating cardiomyopathy in susceptible individuals [35].

Consistent with these data, hypothyroidism in our study was independently associated with increased DCM risk. Findings regarding hyperthyroidism were more nuanced: although no DCM cases were observed in the hyperthyroid subgroup with insulin-treated diabetes, this could reflect the small subgroup size. Other studies have demonstrated that thyrotoxicosis may accelerate heart failure progression and increase arrhythmic risk [36].

Recent reviews emphasize the concept of cardio-renal-metabolic multimorbidity in DCM, noting the high prevalence of metabolic syndrome, diabetes, and renal disease in patients with DCM and their compounded adverse impact on prognosis [26,37].

Our study supports this paradigm by demonstrating that composite endocrine burden scores correlate more closely with DCM and early clinical instability than isolated variables. This reinforces the notion that metabolic clustering, rather than single risk factor presence, drives disease progression.

### 4.7. Differences in Stratification and Modelling

Prior literature often treats diabetes, obesity, or thyroid disease as independent exposures, without systematically evaluating their joint effects. There is a dearth of clinical models that incorporate phenotypic subtyping, such as insulin treatment status or specific thyroid dysfunction categories. Our analytic approach, which used both ridge regression to reduce multicollinearity and subgroup phenotyping, diverges from traditional models and may explain why non-insulin-treated diabetes emerged with a stronger association than insulin-treated or obese phenotypes. This finding may reflect unique pathophysiologic features in early-stage metabolic disease, including lipid accumulation and inflammation, which precede insulin dependence [36].

### 4.8. Strengths over Prior Studies

In contrast to small single-center case series or autopsy studies of obesity-associated cardiomyopathy [38]. Our cohort includes over 1000 real-world patients with well-characterized endocrine and cardiac phenotypes. Importantly, the use of composite burden scores enabled us to stratify risk in a way that reflects clinical multimorbidity—a contrast to prior registry- or genetics-focused work that prioritized molecular aetiology or staging of DCM [39,40].

Compared with smaller studies or histopathologic investigations, our relatively large, phenotype cohort is a strength. The use of regularized regression to address multicollinearity among interrelated metabolic exposures also improves robustness.

However, limitations remain: our retrospective design and reliance on diagnosed thyroid and diabetes status—instead of standardized laboratory or biomarker thresholds—may underestimate subclinical disease prevalence. Prior prospective studies that used serial TSH, HbA1c, and imaging assessments have offered stronger causal insights but lacked subgroup-level analysis linking metabolic profiles to DCM.

### 4.9. Implications for Clinical Practice

These findings support a more nuanced and interdisciplinary approach to cardiovascular risk assessment. Routine cardiology evaluation of patients with diabetes and thyroid dysfunction may be warranted even in the absence of classic heart failure symptoms. Endocrinologists, cardiologists, and primary care providers should be aware of the compounded cardiac risk posed by coexisting metabolic disorders and consider joint screening strategies to identify patients at the highest risk of DCM and clinical decompensation. Incorporating composite metabolic indices into electronic health record alerts or care pathways could improve early detection and management of cardiomyopathy in high-risk populations. Incorporating the Metabolic Burden Score into EHR-based risk alerts may support earlier identification of DCM in high-risk endocrine profiles.

### 4.10. Limitations

This study is subject to several limitations that must be considered when interpreting the findings. First, the study cohort included a moderate predominance of male patients (57.6%), which may limit the generalizability of findings to the broader female population. Although sex was adjusted for in our multivariable models, underrepresentation of women may obscure sex-specific disease associations or risk modifiers. Second, the retrospective design inherently limits the ability to establish causal relationships. Although multivariable adjustment was performed to account for confounders, residual confounding from unmeasured clinical variables, such as medication adherence, socioeconomic status, or duration of endocrine disease, may still have influenced the results. Additionally, the reliance on existing electronic health records introduces potential documentation bias. Clinical diagnoses such as hypothyroidism or obesity may have been underreported or inconsistently coded, leading to potential misclassification of exposure variables. Moreover, detailed information on pharmacological therapy (e.g., antihypertensive drugs, oral antidiabetic agents, thyroid hormone replacement) and patient adherence was inconsistently documented in the retrospective dataset. As such, these factors could not be systematically included in the analysis. This limitation introduces the possibility of residual confounding, as variations in medication type, intensity, and compliance may have influenced the risk of DCM or readmission. Future prospective studies should incorporate standardized medication data and adherence measures to better account for these important modifiers of cardiometabolic risk.

Third, the heterogeneity in subgroup sample sizes, particularly in some endocrine phenotype combinations, reduced the statistical power to detect differences across all strata. For example, no cases of dilated cardiomyopathy were observed in the subgroup of insulin-treated patients with hyperthyroidism, but the small size of this group (n = 54) limits the generalizability and interpretability of this finding. Furthermore, the study lacked detailed data on the duration of endocrine conditions, particularly diabetes and thyroid dysfunction, which may significantly influence cardiovascular outcomes. Laboratory measures such as HbA1c for glycemic control and information on thyroid hormone replacement therapy were not systematically available and therefore could not be adjusted for. The absence of these variables may have introduced residual confounding and limited the ability to assess disease severity or treatment adequacy as modifiers of DCM risk. Similarly, stratified comparisons involving multiple combinations of thyroid and diabetes status may have been underpowered to detect nuanced interactions.

Fourth, the presence of perfect separation in some variable categories precluded the use of traditional logistic regression models, necessitating the use of regularized regression methods. While ridge regression allowed for robust model convergence and mitigated overfitting, it may introduce bias in the estimation of individual coefficients and complicate direct interpretation of marginal effects.

Fifth, the definitions of thyroid dysfunction and diabetes were based on documented clinical diagnoses rather than systematic laboratory measurements. This introduces the possibility that subclinical or undiagnosed cases may have been overlooked, particularly among euthyroid individuals or patients with impaired glucose tolerance who did not meet diagnostic thresholds for diabetes. Furthermore, the classification of thyroid status did not distinguish between treated and untreated disease, nor did it account for biochemical severity or duration of dysfunction, which may have important implications for cardiovascular risk. In particular, misclassification of thyroid status is a notable concern, as some patients had only limited or inconsistent thyroid function testing (e.g., missing free T4 or TSH outside of the index period), potentially leading to incorrect categorization as euthyroid or hypothyroid. This variability in testing may have attenuated or exaggerated the observed associations between thyroid dysfunction and DCM.

An important observation is that virtually all patients in our cohort had systemic hypertension, consistent with its high background prevalence in populations at risk of cardiomyopathy. While hypertension is well recognized as a contributor to adverse cardiac remodeling, its lack of statistical signal as an independent predictor in our models likely reflects this limited variability, rather than a true absence of association. Future prospective studies with more heterogenous cohorts are needed to disentangle the independent effects of hypertension from overlapping metabolic comorbidities.

Some subgroup analyses were underpowered due to small sample sizes. In particular, combinations such as insulin-treated diabetes with hyperthyroidism included fewer than 60 patients, limiting the ability to detect associations or generalize findings. The absence of observed events in these groups may reflect insufficient power rather than true clinical differences.

Finally, although the outcome of 30-day hospital readmission was clearly defined and validated, it encompassed all-cause admissions and was not restricted to cardiovascular events. While most readmissions were related to cardiac symptoms, the inclusion of non-cardiac causes may have diluted associations between metabolic risk and clinical instability. Moreover, the single-centre setting of the study may limit the external validity of the findings, as practice patterns and patient demographics may differ in other institutions or healthcare systems.

Subgroup sample sizes were sometimes small, reducing statistical power in stratified analyses. Additionally, perfect separation in some variables may have limited convergence in traditional regression models, necessitating regularization. Underpowered subgroup analyses may have produced misleading trends, particularly where no events were observed. Finally, causality cannot be inferred due to the observational nature of the data. Medication use (e.g., statins, thyroid hormone therapy) was not systematically captured or adjusted for, which may confound associations.

Taken together, these limitations underscore the need for future prospective studies using standardized diagnostic criteria, longer follow-up periods, and broader populations to confirm and extend the current findings. Despite these constraints, the study provides meaningful insights into the interdependence of endocrine and cardiovascular health and highlights key targets for risk stratification and preventive care in patients with metabolic disorders.

## 5. Conclusions

In this retrospective cohort study of over 1000 patients, endocrine-metabolic dysfunctions, particularly non-insulin-treated diabetes and hypothyroidism and male sex were independently associated with increased odds of DCM in this cohort. While the associations with non-insulin-treated diabetes, hypothyroidism, and male sex were consistent and statistically robust, subgroup trends (e.g., absence of DCM in insulin-treated hyperthyroid patients, nonlinear readmission rates across risk strata) should be interpreted as exploratory and hypothesis-generating due to limited sample size. These findings support the implementation of interdisciplinary screening and management approaches that integrate endocrine and cardiovascular care to slow the progression of heart failure. As a retrospective study, causality cannot be established, and prospective validation is required, as residual confounding cannot be excluded.

By critically contextualizing our results within existing high-impact literature, this study strengthens the evidence for non-insulin-treated diabetes and hypothyroidism as independent—and potentially synergistic—drivers of DCM. Moreover, it highlights obesity's indirect role through its contribution to broader metabolic clustering. Collectively, our findings emphasize the importance of integrated cardiometabolic risk assessment and timely, targeted interventions aimed at preventing structural heart disease in metabolically vulnerable populations. Our results underscore the importance of integrating endocrine and sex-specific factors into DCM risk assessment. Prospective studies are warranted to validate these findings and to refine risk-stratification strategies that account for combined metabolic and demographic influences.

**Author Contributions:** Conceptualization, O.Ţ. (Otilia Ţica) and O.Ţ. (Ovidiu Ţica); methodology, O.Ţ. (Ovidiu Ţica) and M.I.Ş.; software, A.H. and S.M.F.; validation, M.I.Ş., A.H., and I.Z.; formal analysis, O.Ţ. (Otilia Ţica) and I.B.; investigation, O.Ţ. (Ovidiu Ţica), A.H., and I.Z.; resources, C.D. and M.S.; data curation, S.M.F. and I.B.; writing—original draft preparation, O.Ţ. (Ovidiu Ţica) and M.I.Ş.; writing—review and editing, O.Ţ. (Otilia Ţica), I.Z., and C.D.; visualization, A.H. and S.M.F.; supervision, M.I.Ş. and O.Ţ. (Otilia Ţica); project administration, O.Ţ. (Ovidiu Ţica); funding acquisition, M.I.Ş. O.Ţ. (Otilia Ţica), O.Ţ. (Ovidiu Ţica), and M.I.Ş. have equally contributed to the manuscript. All authors have read and agreed to the published version of the manuscript.

**Funding:** This research received no external funding. The APC for this article was funded by the University of Oradea.

**Institutional Review Board Statement:** The study was conducted in accordance with the Declaration of Helsinki and approved by the Institutional Review Board and Ethics Committee of SCJU Bh (number 22223/24 June 2025).

**Informed Consent Statement:** Patient consent was waived due to the retrospective nature of the study and the use of anonymized data, in accordance with the approval granted by the Ethics Committee of the Emergency County Clinical Hospital of Bihor, Oradea, Romania.

**Data Availability Statement:** The data supporting the findings of this study are available from the corresponding authors upon reasonable request. Due to ethical and privacy considerations, the data are not publicly available.

**Acknowledgments:** The authors would like to thank the Emergency County Clinical Hospital of Bihor and the University of Oradea for their institutional support in conducting this study. We also acknowledge the assistance provided by the Faculty of Medicine and Pharmacy, University of Oradea, covering the Article Processing Charges (APC) for this publication. The authors have reviewed and edited the output and take full responsibility for the content of this publication.

**Conflicts of Interest:** The authors declare no conflicts of interest. The Article Processing Charges (APCs) were funded by the University of Oradea. The funder had no role in the design of the study; in the collection, analysis, or interpretation of data; in the writing of the manuscript; or in the decision to publish the results.

## Abbreviations

The following abbreviations are used in this manuscript:

| | |
|---|---|
| AUC | Area Under the Curve |
| BMI | Body Mass Index |
| CI | Confidence Interval |
| CVD | Cardiovascular Disease |
| DCM | Dilated Cardiomyopathy |
| DM | Diabetes Mellitus |
| EHR | Electronic Health Record |
| GCP | Good Clinical Practice |
| HTN | Hypertension |
| IRB | Institutional Review Board |

| | |
|---|---|
| NYHA | New York Heart Association |
| OR | Odds Ratio |
| SGLT2 | Sodium–Glucose Cotransporter-2 |
| T2DM | Type 2 Diabetes Mellitus |
| TSH | Thyroid-Stimulating Hormone |
| VIF | Variance Inflation Factor |

## References

1. Kahaly, G.J.; Liu, Y.; Persani, L. Hypothyroidism: Playing the cardiometabolic risk concerto. *Thyroid Res.* **2025**, *18*, 20. [CrossRef]
2. Ahmad, A.; Lim, L.L.; Morieri, M.L.; Tam, C.H.; Cheng, F.; Chikowore, T.; Dudenhöffer-Pfeifer, M.; Fitipaldi, H.; Huang, C.; Kanbour, S.; et al. Precision prognostics for cardiovascular disease in Type 2 diabetes: A systematic review and meta-analysis. *Commun. Med.* **2024**, *4*, 11. [CrossRef]
3. Țica, O.; Țica, O. Molecular Diagnostics in Heart Failure: From Biomarkers to Personalized Medicine. *Diagnostics* **2025**, *15*, 1807. [CrossRef]
4. Marfella, R.; Sardu, C.; Mansueto, G.; Napoli, C.; Paolisso, G. Evidence for human diabetic cardiomyopathy. *Acta Diabetol.* **2021**, *58*, 983–988. [CrossRef]
5. Tan, Y.; Zhang, Z.; Zheng, C.; Wintergerst, K.A.; Keller, B.B.; Cai, L. Mechanisms of diabetic cardiomyopathy and potential therapeutic strategies: Preclinical and clinical evidence. *Nat. Rev. Cardiol.* **2020**, *17*, 585–607. [CrossRef]
6. Bernardo, B.C.; Faulkner, A.N.; Wang, T.; Huynh, K.; Meikle, P.J.; Weeks, K.L.; Tham, Y.K. Lipidomic Profiling of a Preclinical Model of Streptozotocin-Induced Diabetic Cardiomyopathy Reveals Potential Plasma Biomarkers. *Heart Lung Circ.* **2025**, *34*, 739–742. [CrossRef]
7. Fan, G.; Lin, L.; Xu, C. Causal association between body mass index and dilated cardiomyopathy: A Mendelian randomization study. *Arch. Med. Sci.* **2024**, *20*, 2040–2042. [CrossRef]
8. Michaelidou, M.; Pappachan, J.M.; Jeeyavudeen, M.S. Management of diabesity: Current concepts. *World J. Diabetes* **2023**, *14*, 396–411. [CrossRef]
9. Țica, O.; Țica, O.; Bunting, K.V.; deBono, J.; Gkoutos, G.V.; Popescu, M.I.; Kotecha, D. Post-mortem examination of high mortality in patients with heart failure and atrial fibrillation. *BMC Med.* **2022**, *20*, 331. [CrossRef] [PubMed]
10. Kittleson, M.M.; Benjamin, E.J.; Blumer, V.; Harrington, J.; Januzzi, J.L.; McMurray, J.J.V.; Vest, A.R. 2025 ACC Scientific Statement on the Management of Obesity in Adults With Heart Failure: A Report of the American College of Cardiology. *J. Am. Coll. Cardiol.* **2025**. [CrossRef]
11. Xu, Y.; Derakhshan, A.; Hysaj, O.; Wildisen, L.; Ittermann, T.; Pingitore, A.; Abolhassani, N.; Medici, M.; Kiemeney, L.; Riksen, N.P.; et al. The optimal healthy ranges of thyroid function defined by the risk of cardiovascular disease and mortality: Systematic review and individual participant data meta-analysis. *Lancet Diabetes Endocrinol.* **2023**, *11*, 743–754. [CrossRef]
12. Țica, O.; Țica, O. Anemia in Heart Failure: Diagnostic Insights and Management Patterns Across Ejection Fraction Phenotypes. *Diagnostics* **2025**, *15*, 2079. [CrossRef]
13. Țica, O.; Sabău, M.; Venter, A.; Beiușanu, C.; Berechet, M.; Huniadi, A.; Șandor, M.I.; Țica, O. Stroke in Dilated Cardiomyopathy: An Autopsy-Based Study of Mechanisms, Topography, and Clinical Implications. *Diagnostics* **2025**, *15*, 2287. [CrossRef]
14. Petersen, J.K.; Selmer, C.; Kristensen, S.L.; Rørth, R.; Yafasova, A.; Vejlstrup, L.G.B.; Ali, S.A.; Schou, M.; Køber, L.; Fosbøl, E.L.; et al. Prognostic Impact of Thyroid Function in Patients With Incident Heart Failure: A Nationwide Study. *J. Card. Fail.* **2025**, *31*, 1118–1126. [CrossRef]
15. Hershberger, R.E.; Zareba, K.M. Dilated Cardiomyopathy: New Distinct Phenotypes or Temporal Phases of Disease? *J. Am. Coll. Cardiol.* **2022**, *79*, 2233–2235. [CrossRef]
16. Zhao, X.; Liu, S.; Wang, X.; Chen, Y.; Pang, P.; Yang, Q.; Lin, J.; Deng, S.; Wu, S.; Fan, G.; et al. Diabetic cardiomyopathy: Clinical phenotype and practice. *Front. Endocrinol.* **2022**, *13*, 1032268. [CrossRef]
17. Abdullah, A.R.; Seliem, M.A.; Khidr, E.G.; Sobhy, A.M.; El-Shiekh, R.A.; Hafeez, M.; El-Husseiny, A.A. A comprehensive review on diabetic cardiomyopathy (DCM): Histological spectrum, diagnosis, pathogenesis, and management with conventional treatments and natural compounds. *Naunyn Schmiedebergs Arch. Pharmacol.* **2025**, *398*, 9929–9969. [CrossRef]
18. Huo, J.L.; Feng, Q.; Pan, S.; Fu, W.J.; Liu, Z.; Liu, Z. Diabetic cardiomyopathy: Early diagnostic biomarkers, pathogenetic mechanisms, and therapeutic interventions. *Cell Death Discov.* **2023**, *9*, 256. [CrossRef]
19. Li, J.; Liu, J.; Wang, Y.; Yan, H.; Li, Q.; Wen, W. Global research trends in the relationship between diabetic cardiomyopathy and mitochondria: A bibliometric analysis. *Hereditas* **2025**, *162*, 114. [CrossRef]
20. Yalçın, N.; Aktaş, S.; Uyar, S.; Koca, N. Impact of SGLT2 Inhibitors on Cardiovascular Risk Scores, Metabolic Parameters, and Laboratory Profiles in Type 2 Diabetes. *Life* **2025**, *15*, 722. [CrossRef]

21. Stanton, A.M.; Vaduganathan, M.; Chang, L.S.; Turchin, A.; Januzzi, J.L., Jr.; Aroda, V.R. Asymptomatic Diabetic Cardiomyopathy: An Underrecognized Entity in Type 2 Diabetes. *Curr. Diab. Rep.* **2021**, *21*, 41. [CrossRef]
22. Jelavic, M.M.; Țica, O.; Pintaric, H.; Țica, O. Circulating Neuropeptide Y May Be a Biomarker for Diagnosing Atrial Fibrillation. *Cardiology* **2023**, *148*, 593–595. [CrossRef]
23. Ruan, W.; Zhou, X.; Li, J.; Wang, T.; Liu, H.; Zhang, G.; Lin, K. Thyroid function effect on cardiac structure, cardiac function, and disease risk: Evidence of causal associations in European ancestry. *Heart Rhythm.* **2024**, *21*, 2272–2281. [CrossRef]
24. Shah, R.; Orlov, S.; Paré, G.; Gerstein, H.C. Thyroid Dysfunction and Cardiovascular Events in Patients With Dysglycemia. *Can. J. Diabetes* **2023**, *47*, 257–262. [CrossRef]
25. Yang, G.; Wang, Y.; Ma, A.; Wang, T. Subclinical thyroid dysfunction is associated with adverse prognosis in heart failure patients with reduced ejection fraction. *BMC Cardiovasc. Disord.* **2019**, *19*, 83. [CrossRef]
26. Țica, O.; Khamboo, W.; Kotecha, D. Breaking the Cycle of Heart Failure With Preserved Ejection Fraction and Atrial Fibrillation. *Card. Fail. Rev.* **2022**, *8*, e32. [CrossRef]
27. Țica, O.; Teodorovich, N.; Champsi, A.; Swissa, M. Are the Four Pillars the Ideal Treatment for the Elderly? *Cardiology* **2023**, *148*, 296–299. [CrossRef]
28. Gopal, K.; Chahade, J.J.; Kim, R.; Ussher, J.R. The Impact of Antidiabetic Therapies on Diastolic Dysfunction and Diabetic Cardiomyopathy. *Front. Physiol.* **2020**, *11*, 603247. [CrossRef]
29. Zhang, N.; Liu, X.; Wang, L.; Zhang, Y.; Xiang, Y.; Cai, J.; Xu, H.; Xiao, X.; Zhao, X. Lifestyle factors and their relative contributions to longitudinal progression of cardio-renal-metabolic multimorbidity: A prospective cohort study. *Cardiovasc. Diabetol.* **2024**, *23*, 265. [CrossRef]
30. Bellemare, M.; Bourcier, L.; Iglesies-Grau, J.; Boulet, J.; O'Meara, E.; Bouabdallaoui, N. Mechanisms of diabetic cardiomyopathy: Focus on inflammation. *Diabetes Obes. Metab.* **2025**, *27*, 2326–2338. [CrossRef]
31. Seferović, P.M.; Paulus, W.J.; Rosano, G.; Polovina, M.; Petrie, M.C.; Jhund, P.S.; Tschöpe, C.; Sattar, N.; Piepoli, M.; Papp, Z.; et al. Diabetic myocardial disorder. A clinical consensus statement of the Heart Failure Association of the ESC and the ESC Working Group on Myocardial & Pericardial Diseases. *Eur. J. Heart Fail.* **2024**, *26*, 1893–1903. [CrossRef] [PubMed]
32. Robertson, J.; Schaufelberger, M.; Lindgren, M.; Adiels, M.; Schiöler, L.; Torén, K.; McMurray, J.; Sattar, N.; Åberg, M.; Rosengren, A. Higher Body Mass Index in Adolescence Predicts Cardiomyopathy Risk in Midlife. *Circulation* **2019**, *140*, 117–125. [CrossRef]
33. Weerts, J.; Țica, O.; Aranyo, J.; Basile, C.; Borizanova-Petkova, A.; Borovac, J.A.; Camilli, M.; Eichenlaub, M.; Fiori, E.; Van Loon, T.; et al. Atrial cardiomyopathy: From healthy atria to atrial failure. A clinical consensus statement of the Heart Failure Association of the ESC. *Eur. J. Heart Fail.* 2025, *Online ahead of print*. [CrossRef]
34. Kaimori, R.; Nishida, H.; Tamura, M.; Kuroki, K.; Murata, K.; Kawamura, K.; Mori, S.; Daa, T. Obesity cardiomyopathy could contribute to sudden cardiac death: A Japanese epidemiological morphological study. *Cardiovasc. Diabetol.* **2024**, *23*, 378. [CrossRef]
35. Lisco, G.; Giagulli, V.A.; Iovino, M.; Zupo, R.; Guastamacchia, E.; De Pergola, G.; Iacoviello, M.; Triggiani, V. Endocrine system dysfunction and chronic heart failure: A clinical perspective. *Endocrine* **2022**, *75*, 360–376. [CrossRef]
36. Soetedjo, N.N.M.; Agustini, D.; Permana, H. The impact of thyroid disorder on cardiovascular disease: Unraveling the connection and implications for patient care. *Int. J. Cardiol. Heart Vasc.* **2024**, *55*, 101536. [CrossRef]
37. Akbari, T.; Hammersley, D.J.; May, C.Y.; Halliday, B.P.; Prasad, S.K. The Impact of Cardio-Renal-Metabolic Profile in Dilated Cardiomyopathy. *Curr. Cardiol. Rep.* **2025**, *27*, 89. [CrossRef]
38. Powell-Wiley, T.M.; Poirier, P.; Burke, L.E.; Després, J.P.; Gordon-Larsen, P.; Lavie, C.J.; Lear, S.A.; Ndumele, C.E.; Neeland, I.J.; Sanders, P.; et al. Obesity and Cardiovascular Disease: A Scientific Statement From the American Heart Association. *Circulation* **2021**, *143*, e984–e1010. [CrossRef]
39. Prasad, B.; Bjourson, A.J.; Shukla, P. Data-driven patient stratification of UK Biobank cohort suggests five endotypes of multimorbidity. *Brief Bioinform.* **2022**, *23*, bbac410. [CrossRef]
40. GBD 2021 Adult BMI Collaborators. Global, regional, and national prevalence of adult overweight and obesity, 1990–2021, with forecasts to 2050: A forecasting study for the Global Burden of Disease Study 2021. *Lancet* **2025**, *405*, 813–838. [CrossRef] [PubMed]

biomedicines

MDPI

*Review*

# A Clinical Review of the Connections Between Diabetes Mellitus, Periodontal Disease, and Cardiovascular Pathologies

**Otilia Țica [1,*,†], Ioana Romanul [2,*,†], Gabriela Ciavoi [2], Vlad Alin Pantea [2], Ioana Scrobota [2], Lucian Șipoș [2], Cristian Marius Daina [3] and Ovidiu Țica [4,5,†]**

1 Cardiology Clinic, Emergency County Clinical Hospital of Bihor, 410165 Oradea, Romania
2 Department of Dental Medicine, Faculty of Medicine and Pharmacy, University of Oradea, 410068 Oradea, Romania; gciavoi@uoradea.ro (G.C.); vlad_pantea@uoradea.ro (V.A.P.); ioana_scrobota@uoradea.ro (I.S.); lsipos@uoradea.ro (L.Ș.)
3 Department of Surgical Disciplines, Faculty of Medicine and Pharmacy, University of Oradea, 410073 Oradea, Romania; cdaina@uoradea.ro
4 Department of Morphological Disciplines, Faculty of Medicine and Pharmacy, University of Oradea, 410073 Oradea, Romania; ovidiu.tica@gmail.com
5 Department of Pathology, Emergency County Clinical Hospital of Bihor, 410165 Oradea, Romania
* Correspondence: otilia_cristea01@yahoo.com (O.Ț.); ioana_romanul@uoradea.ro (I.R.)
† These authors contributed equally to this work.

## Abstract

**Background**: Diabetes mellitus (DM), periodontal disease (PD), and cardiovascular disease (CVD) are highly prevalent global health conditions with overlapping pathophysiological mechanisms. Emerging evidence suggests a bidirectional and synergistic relationship among them, driven by chronic inflammation, immune dysregulation, oxidative stress, and microbial dysbiosis. **Objective**: This review synthesizes current literature on the interconnectedness of DM, PD, and CVD, emphasizing shared molecular pathways, clinical implications, and opportunities for integrated management. **Methods**: A systematic review and narrative synthesis of recent clinical trials, observational studies, and multi-omics investigations was conducted to explore the mechanisms linking these three conditions. A structured literature search was performed across PubMed, Scopus, and Web of Science from database inception until 30 June 2025. Key findings were contextualized within systems biology, precision medicine, and real-world clinical strategies. **Results**: DM exacerbates periodontal inflammation and accelerates tissue destruction via hyperglycemia-induced inflammatory mediators, while periodontitis worsens glycemic control and insulin resistance. Both conditions independently elevate cardiovascular risk, and their co-occurrence significantly amplifies the incidence of adverse cardiovascular events. Shared biomarkers such as Interleukin (IL)-6, Tumor Necrosis Factor (TNF)-α, and CRP, as well as overlapping genetic and epigenetic signatures, underscore a common inflammatory axis. Periodontal therapy has demonstrated modest but meaningful benefits on glycemic control and endothelial function, while cardiometabolic therapies (e.g., statins, Glucagon-Like Peptide (GLP-1) receptor agonists, SGLT2 inhibitors) show potential to improve periodontal outcomes. Probiotics, microbiome-targeted therapies, and AI-based risk models are emerging as future tools. **Conclusions**: DM, PD, and CVD form a mutually reinforcing triad mediated by systemic inflammation and metabolic dysregulation. Integrated, multidisciplinary care models and precision health strategies are essential to address this inflammatory burden and improve long-term outcomes. Further large-scale interventional trials and mechanistic human studies are needed to establish causal links and optimize combined therapeutic approaches.

 https://doi.org/10.3390/biomedicines13092309

**Keywords:** diabetes mellitus; cardiovascular disease; endothelial dysfunction; insulin resistance; atherosclerosis; periodontal disease

## 1. Introduction

Diabetes mellitus (DM), periodontal disease (PD), and cardiovascular disease (CVD) represent three major global health concerns that not only share common risk factors but also appear to be pathophysiologically interconnected. DM currently affects over 500 million individuals globally and is expected to rise steadily due to aging populations and lifestyle changes [1]. Periodontal disease, a chronic inflammatory condition of the gums and supporting tissues, affects approximately 50% of adults worldwide, with severe forms impacting up to 10% [2]. CVD remains the leading cause of morbidity and mortality worldwide, responsible for an estimated 18 million deaths annually [3,4].

Emerging evidence underscores a bidirectional relationship between DM and PD, wherein each condition exacerbates the other. Simultaneously, both DM and PD have been independently associated with increased cardiovascular risk, hinting at shared systemic inflammatory pathways. Chronic inflammation, immune dysregulation, oxidative stress, and endothelial dysfunction are recurrent themes across these conditions [5]. In this comprehensive systematic review, we aim to synthesize current clinical and mechanistic evidence linking diabetes mellitus, periodontal disease, and cardiovascular disease, with particular emphasis on their overlapping mechanisms of inflammation, metabolic dysregulation, and microbial dysbiosis. The main objectives are: (i) to synthesize current clinical and mechanistic evidence connecting these three conditions, (ii) to highlight the shared inflammatory, metabolic, and microbial pathways that underpin their interactions, and (iii) to explore translational, therapeutic, and multidisciplinary strategies for integrated management. Unlike previous reviews that have examined these associations in isolation, our work integrates data across the three axes to highlight their triangular interplay. By systematically presenting clinical, mechanistic, and therapeutic perspectives, this review contributes to the field by framing periodontal disease as a cardiometabolic risk enhancer and by outlining future translational opportunities, including multi-omics, AI-based tools, and integrated care models.

## 2. Materials and Methods

We conducted this review in accordance with the Preferred Reporting Items for Systematic Reviews and Meta-Analyses (PRISMA) 2020 statement [6]. The protocol was not registered in PROSPERO because this was designed as a narrative clinical review; however, the systematic framework was applied to ensure transparency and reproducibility.

### 2.1. Information Sources and Search Strategy

The literature search was conducted with the explicit aim of identifying studies that addressed at least one of the following: mechanistic links between DM, PD, and CVD; shared biomarkers or inflammatory pathways; and clinical or therapeutic strategies relevant to their integrated management."

A comprehensive literature search was carried out across PubMed/MEDLINE, Scopus, and Web of Science databases, covering the period from database inception until 30 June 2025. The search combined the following keywords and Medical Subject Headings (MeSH): "periodontal disease", "diabetes mellitus", "cardiovascular disease", "atherosclerosis", "inflammation", "oral-systemic connection", and "endothelial dysfunction." Boolean operators (AND/OR) were used to refine the strategy.

In addition, the reference lists of relevant articles and recent reviews were screened manually to identify further eligible studies.

### 2.2. Eligibility Criteria

Studies were included if they met the following criteria: (i) peer-reviewed articles published in English; (ii) human clinical studies (observational studies, randomized controlled trials, or meta-analyses); and (iii) mechanistic or translational studies directly linking DM, PD, and/or CVD through inflammatory, microbial, or metabolic pathways.

We excluded (i) non-peer-reviewed publications, case reports, letters, or conference abstracts without full data; (ii) purely animal or in vitro studies unless they offered direct mechanistic insight relevant to humans; and (iii) articles without accessible full texts.

### 2.3. Study Selection and Data Extraction

After duplicate removal, two reviewers (O.T. and I.R.) independently screened titles and abstracts, followed by full-text evaluation for potentially eligible records. Data were extracted using a predefined form, collecting information on study design, sample characteristics, outcomes, biomarkers, and therapeutic implications. Any disagreements were resolved through discussion with a third author (O.A.T).

The database searches identified a total of 2145 records. After removal of 345 duplicates, 1800 records were screened by title and abstract. Of these, 1420 were excluded as not relevant to the scope of this review. The remaining 380 full-text articles were retrieved and assessed for eligibility. Following full-text review, 270 studies were excluded for the following reasons: insufficient mechanistic or clinical relevance (n = 145), case reports or letters only (n = 68), non-English language (n = 32), or inaccessible full text (n = 25). Ultimately, 110 studies met all inclusion criteria and were incorporated into the qualitative synthesis.

### 2.4. Methodological Note

Given the narrative clinical review design, no new clinical or laboratory data were collected. All searches, screening, and data extraction were performed manually by two independent reviewers (O.T. and I.R.), with disagreements resolved by a third author (O.A.T.). No automated text-mining, keyword scraping, or AI-based tools were used in this review. The synthesis represents a qualitative integration of evidence. No statistical pooling (meta-analysis) was conducted; findings were integrated through narrative synthesis.

## 3. Diabetes Mellitus and Periodontal Disease: Bidirectional Interactions

This review synthesized evidence from 110 eligible studies that investigated the relationships between diabetes mellitus (DM), periodontal disease (PD), and cardiovascular disease (CVD). The findings are presented in four domains corresponding to the three established axes and their triangular interaction.

The relationship between diabetes mellitus and periodontal disease has been extensively studied and is widely regarded as bidirectional. Patients with poorly controlled diabetes exhibit a higher prevalence and severity of periodontal disease, while those with periodontitis often present with poorer glycaemic control [7].

This review synthesized evidence from 110 eligible studies identified through our structured literature search across PubMed, Scopus, and Web of Science. The included works comprised observational studies, randomized controlled trials, mechanistic analyses, and meta-analyses, all of which addressed at least one of the predefined objectives of this review. For clarity, the findings are presented systematically according to the three established disease axes and their triangular interaction: (i) the relationship between diabetes mellitus and periodontal disease (DM–PD), (ii) the association between periodontal disease and cardiovascular disease (PD–CVD), (iii) the well-established axis of diabetes mellitus and cardiovascular disease (DM–CVD), and (iv) the triangular relationship that integrates all three conditions.

*3.1. Pathways and Clinical Outcomes of the Diabetes–Periodontitis Interaction*

The literature consistently supports a bidirectional association between DM and PD, wherein hyperglycemia-driven inflammation worsens periodontal outcomes, while periodontal disease contributes to poor glycemic control.

3.1.1. Inflammatory Pathways

One of the primary mechanisms underpinning this association involves systemic inflammation. Hyperglycaemia leads to increased production of advanced glycation end-products (AGEs), which bind to RAGE (receptors for AGEs) on immune cells, enhancing the secretion of pro-inflammatory cytokines such as TNF-α, IL-1β, and IL-6 [8]. These cytokines exacerbate periodontal tissue destruction and impair wound healing in gingival tissues.

These inflammation-related mechanisms are supported by clinical studies showing significant reductions in HbA1c and CRP following periodontal therapy in T2DM patients [9], and in cross-sectional comparisons demonstrating higher HbA1c in diabetic individuals with periodontitis compared to those without.

The interplay between metabolic dysregulation and microbial inflammation is depicted in Figure 1, which outlines how hyperglycaemia-induced AGEs, bacterial LPS (notably from *P. gingivalis*), and TLR activation synergistically contribute to insulin resistance, systemic cytokine release, and ultimately, the progression of vascular inflammation and atherogenesis.

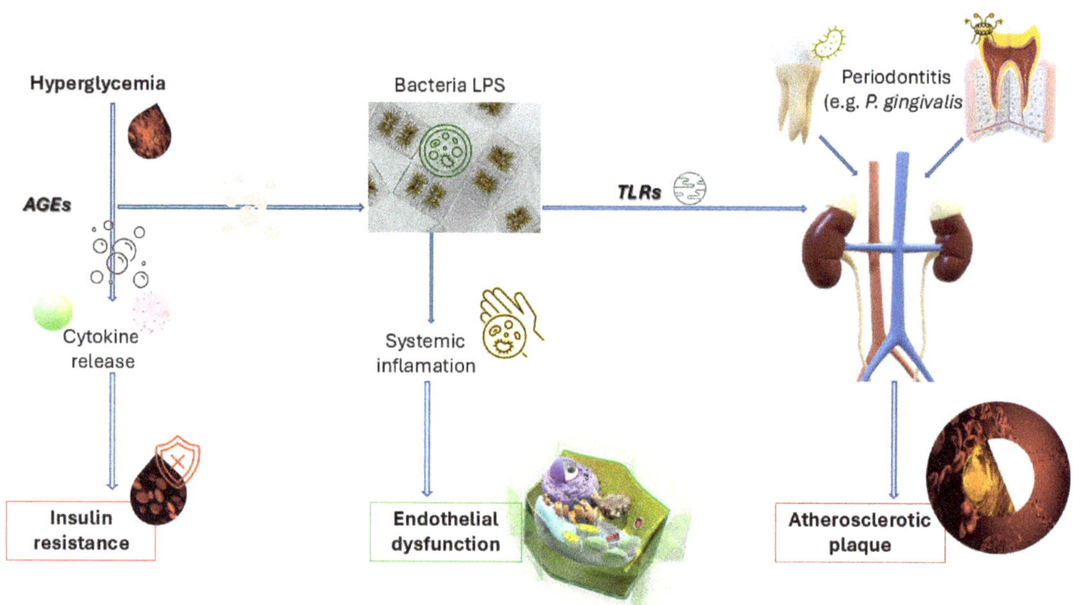


**Figure 1.** Biological cascade from hyperglycemia to atherosclerosis via periodontitis. Biological cascade from hyperglycemia to atherosclerosis via periodontitis. The flowchart illustrates the connections between hyperglycemia, bacterial lipopolysaccharide (LPS) from *Porphyromonas gingivalis*, advanced glycation end-products (AGEs), and toll-like receptor (TLR) activation, and their downstream effects on insulin resistance, systemic inflammation, and atherosclerotic plaque development. Evidence supporting these pathways has been reported in mechanistic and clinical studies linking hyperglycaemia, periodontal inflammation, and vascular end-products (AGEs), and toll-like receptor

(TLR) activation, and their downstream effects on insulin resistance, systemic inflammation, and atherosclerotic plaque development. Evidence supporting these pathways has been reported in mechanistic and clinical studies linking hyperglycaemia, periodontal inflammation, and vascular dysfunction [4,5,7]. Abbreviation: AGEs—Advanced Glycation End-products: harmful compounds formed from sugar–protein reactions; LPS—Lipopolysaccharide: a bacterial endotoxin that triggers immune responses; *P. gingivalis*—*Porphyromonas gingivalis*: a key periodontal pathogen; TLRs—Toll-like Receptors: receptors involved in pathogen recognition and immune activation.

The figures included in this review are evidence-based conceptual models designed to synthesize published findings into accessible visual frameworks. Each pathway or mediator depicted is directly supported by peer-reviewed studies, systematic reviews, or consensus reports, as detailed in the revised figure legends. Thus, the figures are not speculative but rather serve as integrative summaries of current mechanistic and clinical evidence.

In periodontal disease, inflammation is primarily localized to the gingival and periodontal tissues, leading to connective tissue breakdown and alveolar bone loss. In diabetes mellitus, chronic low-grade inflammation affects pancreatic islets, liver, and adipose tissue, contributing to impaired insulin secretion and systemic insulin resistance. Cardiovascular disease is characterized by inflammation within the vascular endothelium and atherosclerotic plaques, where activated immune cells and cytokines promote plaque instability. Adipose tissue is increasingly recognized as a key metabolic and inflammatory organ in DM. Recent studies [10,11] show that adipose tissue in T2D exhibits inflammatory cell infiltration, altered adipokine profiles, and contributes to systemic inflammation, which links PD and CVD.

Periodontal pathogens like *Porphyromonas gingivalis* also contribute to systemic inflammation by stimulating toll-like receptors (TLRs) and promoting the release of systemic inflammatory mediators, which worsen insulin resistance [12]. This vicious cycle amplifies both periodontal and diabetic pathologies. These inflammatory processes provide a mechanistic rationale for the clinical outcomes described in Sections 3.1.2 and 3.1.3, where improved glycemic control (reflected by HbA1c reduction) following periodontal therapy can be partly explained by reductions in systemic cytokines such as IL-6 and TNF-α, and by improvements in endothelial function.

#### 3.1.2. Clinical Evidence and Outcomes

Meta-analyses have shown that individuals with diabetes are 2 to 3 times more likely to develop periodontitis than those without diabetes [13]. Conversely, interventional trials demonstrate that nonsurgical periodontal therapy (NSPT) improves glycemic control, with reported reductions in HbA1c ranging from 0.3% to 0.6% over a 3- to 6-month period [4].

A recent randomized controlled trial involving over 1000 patients with type 2 diabetes reported that periodontal therapy was associated with significant improvements in HbA1c, independent of medication changes, highlighting the systemic benefits of oral healthcare [14].

For instance, Hayashi et al. [15]. observed that non-surgical periodontal treatment not only improved periodontal health but also reduced HbA1c levels and systemic markers of inflammation including TNF-α, linking improved glycemic control to reduced inflammatory burden.

#### 3.1.3. Glycemic Control as a Modifying Factor

Glycemic control directly influences periodontal outcomes. Patients with poorly controlled diabetes (HbA1c > 8%) exhibit more rapid periodontal destruction, including alveolar bone loss, compared to those with better metabolic control [16]. This suggests that strict glycemic regulation should be a cornerstone in periodontal management for diabetic individuals.

### 3.2. *Periodontal Disease and Cardiovascular Disease: Inflammation as a Common Denominator*

Multiple studies demonstrate that periodontal disease is linked with increased cardiovascular risk via bacteremia, systemic inflammation, and endothelial dysfunction.

The connection between periodontal disease (PD) and cardiovascular disease (CVD) has garnered significant attention in recent years, with increasing evidence supporting a causal or at least contributory relationship. The hypothesis centers on shared mechanisms of chronic inflammation, bacteremia, and immune dysregulation, which link oral infections to systemic vascular pathologies [17].

#### 3.2.1. Epidemiological Associations

Multiple observational studies and meta-analyses have demonstrated a consistent association between periodontitis and increased cardiovascular risk, including coronary artery disease (CAD), myocardial infarction (MI), and stroke [18,19]. A large-scale cohort study found that individuals with severe periodontitis had a 1.6-fold higher risk of major adverse cardiovascular events (MACE), even after adjusting for traditional risk factors such as smoking, diabetes, and hypertension [20].

Furthermore, a meta-analysis concluded that periodontitis increases the risk of future cardiovascular events by 20% to 25%, particularly in individuals with comorbid conditions [21].

#### 3.2.2. Mechanistic Insights: From Mouth to Heart

The underlying pathophysiological mechanisms that connect PD and CVD involve both direct and indirect pathways:

##### Bacteremia and Endothelial Dysfunction

Periodontal pathogens such as *Porphyromonas gingivalis*, *Aggregatibacter actinomycetemcomitans*, and *Tannerella forsythia* can enter the systemic circulation during daily oral activities or dental procedures. These microbes have been detected in atherosclerotic pla ques, suggesting a direct role in vascular pathology [22]. While acute bacteremia may rarely precipitate sepsis, the more clinically relevant mechanism in the DM–PD–CVD axis is chronic low-grade bacteremia, which sustains systemic inflammation and contributes to progressive atherosclerosis and plaque instability [9,15,23,24]. Observations of periodontal pathogens in atheromatous plaques and observations of monocyte adhesion, endothelial disruption, and immune activation [24] further support the hypothesis that chronic low-grade bacteremia contributes to atherosclerotic progression rather than merely acute sepsis. A recent study confirmed the presence of DNA from *P. gingivalis* and its lipopolysaccharide (LPS) in coronary artery samples obtained from patients undergoing coronary artery bypass grafting [25].

This translocation triggers an inflammatory cascade, leading to endothelial activation, increased expression of adhesion molecules (e.g., Vascular Cell Adhesion Molecule (VCAM)-1, Intercellular Adhesion Molecule (ICAM)-1), and subsequent leukocyte infiltration—hallmarks of early atherogenesis.

As illustrated in Figure 2, the translocation of *P. gingivalis* into systemic circulation leads to endothelial activation and the upregulation of adhesion molecules such as VCAM-1 and ICAM-1. This cascade fosters leukocyte recruitment, chronic inflammation, and increased vulnerability of atherosclerotic plaques—offering a visual summary of the oral-systemic interface in vascular pathology.

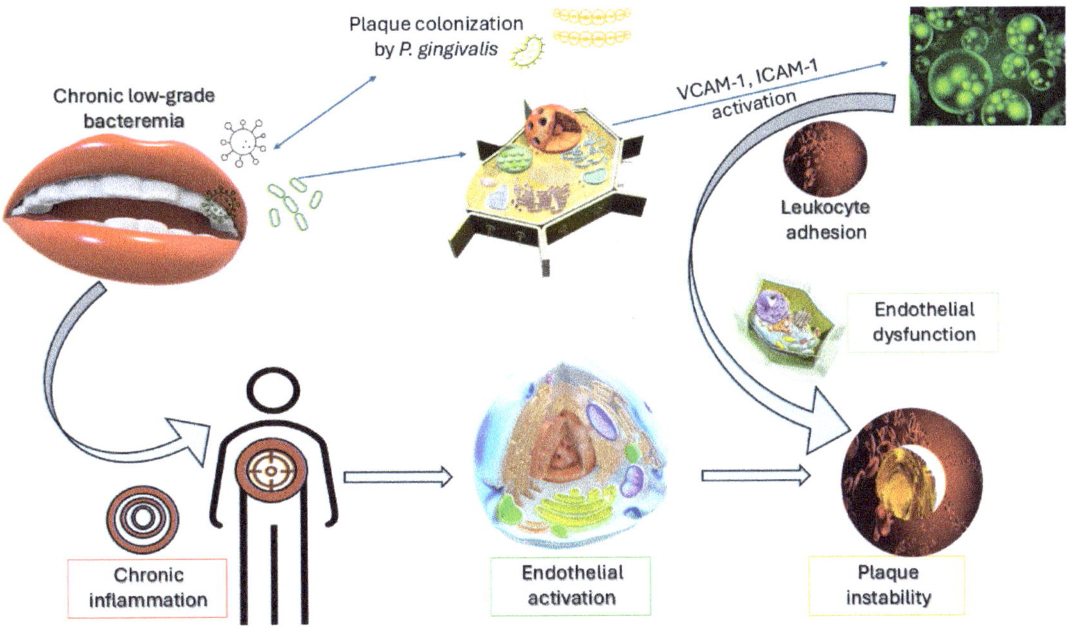


**Figure 2.** Pathway linking bacteremia and atherosclerotic plaque instability. The diagram illustrates how periodontal pathogens such as *Porphyromonas gingivalis* enter the circulation, activate endothelial adhesion molecules (VCAM-1, ICAM-1), and promote leukocyte recruitment, systemic inflammation, and plaque vulnerability. While acute bacteremia can rarely result in sepsis, the more clinically relevant pathway is chronic low-grade bacteremia, which sustains vascular inflammation and contributes to atherosclerosis. This figure is based on evidence showing the detection of periodontal bacteria in coronary plaques and their role in endothelial dysfunction and plaque vulnerability [12,26]. Abbreviation: ICAM-1—Intercellular Adhesion Molecule-1: another key adhesion molecule in vascular inflammation; *P. gingivalis—Porphyromonas gingivalis*: periodontal bacterium implicated in systemic inflammation; VCAM-1—Vascular Cell Adhesion Molecule-1: facilitates leukocyte adhesion to endothelial cells.

Systemic Inflammation and Cytokine Release

Chronic periodontal infection results in the sustained release of inflammatory mediators such as interleukin-6 (IL-6), tumor necrosis factor-alpha (TNF-$\alpha$), and C-reactive protein (CRP), which circulate systemically and contribute to vascular inflammation and plaque instability [27]. Elevated CRP levels have been independently associated with both periodontitis and cardiovascular events, reinforcing the inflammation-driven model of CVD [28].

In a recent prospective cohort study of patients with type 2 diabetes, individuals with high periodontal inflammatory burden demonstrated significantly elevated serum IL-6 and CRP levels, which correlated with increased coronary artery calcification scores [25]. This primary patient-based study provides clinical evidence linking systemic inflammatory mediators from PD with subclinical atherosclerosis [29].

#### 3.2.3. Clinical Impact of Periodontal Treatment on Cardiovascular Outcomes

Several interventional studies have examined whether treating periodontitis may reduce systemic inflammation and improve cardiovascular health. In a randomized controlled trial including patients with moderate-to-severe periodontitis and established cardiovascular disease, intensive periodontal therapy significantly improved brachial artery flow-mediated dilation (FMD) at four weeks and was associated with reductions in circulating CRP and IL-6 levels [17,30,31]. These biomarker changes indicate a short-term anti-inflammatory effect, but it is important to clarify that lowering inflammatory markers does not necessarily equate to a complete resolution of systemic inflammation.

Although long-term cardiovascular endpoints remain to be conclusively proven, these findings suggest that periodontal treatment may serve as a complementary strategy for managing systemic inflammation in high-risk cardiovascular patients.

#### 3.2.4. The Oral-Systemic Inflammatory Axis

Emerging research underscores the concept of an "oral-systemic inflammatory axis," where chronic oral inflammation acts as a reservoir for systemic immune activation. A consensus report published in the *European Heart Journal* emphasized the importance of the oral–systemic inflammatory axis in a systems biology framework, proposing that PD may serve as both a marker and a mediator of cardiovascular risk, particularly in individuals with preexisting metabolic disturbances [26].

Additionally, oral dysbiosis—disruption of the normal oral microbiota—has been implicated in endothelial dysfunction and systemic inflammation. Studies using shotgun metagenomic sequencing have found increased abundance of pro-atherogenic bacteria in the oral cavities of patients with coronary artery disease [32].

### 3.3. *Diabetes and Cardiovascular Disease: A Well-Established Axis*

The relationship between DM and CVD is robust and extensively documented, with hyperglycemia, insulin resistance, endothelial dysfunction, and dyslipidemia as central mediators. Cardiovascular complications are the leading cause of morbidity and mortality in individuals with diabetes, accounting for more than 65% of deaths in this population [33]. The connection is multifactorial, involving hyperglycemia, insulin resistance, endothelial dysfunction, dyslipidemia, and systemic inflammation.

#### 3.3.1. Hyperglycemia and Endothelial Dysfunction

Chronic hyperglycemia plays a pivotal role in the development of cardiovascular pathology by impairing endothelial function. Elevated glucose levels promote the generation of reactive oxygen species (ROS), reduce nitric oxide bioavailability, and increase the formation of advanced glycation end-products (AGEs) [34]. These biochemical changes lead to increased vascular permeability, leukocyte adhesion, and smooth muscle proliferation—hallmarks of atherosclerosis.

In a recent study using transcriptomic profiling of human coronary artery endothelial cells (HCAECs) obtained from patients with poorly controlled diabetes, investigators reported significant upregulation of both pro-inflammatory and pro-thrombotic gene networks [35]. Specifically, increased expression of IL-6, TNF-α, and NF-κB–related signaling pathways was observed, alongside enhanced transcription of pro-thrombotic mediators such as tissue factor (F3), plasminogen activator inhibitor-1 (PAI-1/SERPINE1), and von Willebrand factor (vWF). These findings provide direct molecular evidence that hyperglycemia induces endothelial dysfunction by simultaneously driving inflammation and thrombosis in the vascular wall.

#### 3.3.2. Insulin Resistance, Dyslipidemia, and Atherosclerosis

Insulin resistance, central to the development of type 2 diabetes, also contributes to atherogenesis through multiple mechanisms. It leads to increased free fatty acid release, hepatic overproduction of very-low-density lipoprotein (VLDL), and reduced high-density lipoprotein (HDL), promoting an atherogenic lipid profile [36]. At the molecular level, these changes are associated with activation of NF-κB signaling and upregulation of adhesion molecules such as ICAM-1 and VCAM-1, which facilitate leukocyte recruitment to the endothelium. Pro-inflammatory cytokines including IL-6 and TNF-α further amplify vascular inflammation, while impaired endothelial nitric oxide synthase (eNOS) activity reduces nitric oxide (NO) bioavailability. In parallel, pro-thrombotic mediators such as plasminogen activator inhibitor-1 (PAI-1) and tissue factor are upregulated, enhancing thrombosis risk. Together, these molecular alterations explain how insulin resistance promotes endothelial dysfunction, inflammation, and thrombotic complications in patients with diabetes and cardiovascular disease. A review noted that insulin resistance is associated with increased carotid intima-media thickness and coronary artery calcification, both strong predictors of future cardiovascular events [37].

#### 3.3.3. Role of HbA1c and Glycemic Variability

Hemoglobin A1c (HbA1c) is a well-established marker for long-term glycemic control and cardiovascular risk. Elevated HbA1c is directly correlated with an increased incidence of myocardial infarction, stroke, and heart failure. A meta-analysis found that each 1% increase in HbA1c was associated with an 18% increase in cardiovascular events [38].

However, emerging data suggest that glycemic variability—fluctuations in blood glucose levels—is an independent risk factor for cardiovascular outcomes, possibly due to oxidative stress and vascular inflammation. This was demonstrated in a review examining continuous glucose monitoring data in high-risk diabetic populations [39].

#### 3.3.4. Evidence from Cardiovascular Outcome Trials (CVOTs)

Several large-scale cardiovascular outcome trials (CVOTs) have transformed our understanding of the interplay between diabetes treatment and cardiovascular risk:

SGLT2 inhibitors: Trials such as EMPA-REG OUTCOME (empagliflozin) [40] and DAPA-HF (dapagliflozin) [41] have shown significant reductions in hospitalization for heart failure and cardiovascular mortality [42,43].

GLP-1 receptor agonists: Liraglutide (LEADER trial) [44] and semaglutide (SUSTAIN-6) [45] reduced the incidence of major adverse cardiovascular events (MACE) in type 2 diabetic patients [46].

A comprehensive meta-analysis covering 14 CVOTs concluded that both SGLT2 inhibitors and GLP-1 receptor agonists offer consistent cardiovascular protection in diabetic populations, particularly against heart failure [47] and atherosclerotic events [48,49].

#### 3.3.5. Genetic and Epigenetic Factors

Recent studies have uncovered shared genetic and epigenetic pathways between DM and CVD. A multi-omics integrative analysis identified common polymorphisms in inflammatory genes such as *IL6*, *TNF*, and *NFKB1*, which modulate both glycemic control and cardiovascular risk [50].

Moreover, epigenetic modifications like DNA methylation and histone acetylation have been implicated in the regulation of pro-inflammatory and metabolic pathways. These findings open new avenues for precision medicine targeting the shared biology of diabetes and cardiovascular disease.

### 3.4. *Triangular Relationship: DM, PD, and CVD—A Shared Inflammatory Burden*

Emerging evidence indicates that DM, PD, and CVD form a mutually reinforcing triad, underpinned by systemic inflammation, oxidative stress, microbial dysbiosis, and shared genetic susceptibilities.

While the individual links between diabetes mellitus (DM), periodontal disease (PD), and cardiovascular disease (CVD) are well documented, growing evidence suggests these three conditions form a mutually reinforcing triad mediated by chronic low-grade systemic inflammation, oxidative stress, immune dysregulation, and microbial dysbiosis. Understanding this triangular interplay is essential for advancing both pathophysiological insights and integrated clinical management strategies.

#### 3.4.1. Systems Biology Perspective

Systems biology approaches have been instrumental in delineating the interconnectedness of DM, PD, and CVD. A 2022 integrative omics study analyzed blood and tissue transcriptomes from patients with co-existing diabetes and periodontitis and found overlapping gene expression signatures related to inflammation, innate immunity, and endothelial activation [51]. Pathways involving nuclear factor (NF)-κB signaling, Toll-like receptor pathways, and oxidative phosphorylation were especially prominent in patients who later developed cardiovascular events.

Similarly, multi-omics network modeling revealed common nodes—such as IL-6, TNF-α, and ICAM-1—across the three disease states, underscoring the pivotal role of inflammation in driving systemic disease progression [52].

#### 3.4.2. Shared Biomarkers: Inflammatory and Immune Mediators

The inflammatory burden in DM, PD, and CVD arises from distinct yet overlapping tissue sites. In periodontal disease, gingival and periodontal connective tissues are the primary sites of chronic inflammation. In diabetes mellitus, inflammation occurs not only in pancreatic islets but also in adipose tissue and the liver, where it drives systemic insulin resistance. In cardiovascular disease, the vascular endothelium and atherosclerotic plaques serve as key inflammatory niches, where leukocyte infiltration and cytokine release accelerate plaque instability.

Several biomarkers consistently emerge across studies investigating DM, PD, and CVD [53]. Notably:

- C-reactive protein (CRP): Elevated in all three conditions; serves as a marker for systemic inflammation and cardiovascular risk [54].
- Interleukin-6 (IL-6): Linked to insulin resistance, periodontal inflammation, and endothelial dysfunction [55].
- Tumor necrosis factor-alpha (TNF-α): Involved in pancreatic β-cell dysfunction, periodontal tissue degradation, and atherosclerosis [56].
- Fibrinogen: A coagulation factor elevated in chronic inflammation and predictive of cardiovascular events [21].

Recent studies [57,58] confirm that in type 2 diabetes pancreatic islets undergo low-grade but persistent inflammation, with macrophage infiltration and activation of NF-κB and IL-1β signaling contributing to β-cell dysfunction. The single-cell transcriptomics work [10] further clarifies that pancreatic duct/duct-acinar cells express pro-inflammatory chemokines (e.g., CXCL9, CXCL10, etc.), suggesting that non-β endocrine cells also contribute to islet inflammation.

L-6 and TNF-α act as upstream drivers of insulin resistance and endothelial activation, amplifying vascular inflammation. NF-κB serves as a central transcriptional hub integrating microbial and metabolic signals [59]. Within the DM–PD–CVD triad, NF-κB activation drives pro-inflammatory cytokine release (IL-6, TNF-α), endothelial adhesion molecule expression (ICAM-1, VCAM-1), and amplifies systemic low-grade inflammation. While NF-κB is indeed implicated in numerous other chronic diseases, in this context it functions as a pivotal convergence point linking periodontal dysbiosis, hyperglycemia, and vascular dysfunction. CRP, produced in the liver in response to IL-6, is not only a marker but also an active participant in endothelial dysfunction, promoting monocyte adhesion and plaque progression. Together, these pathways highlight endothelial dysfunction and systemic cytokine signaling as pivotal mechanisms that unify local periodontal inflammation with metabolic and cardiovascular pathology.

Key molecular pathways operate across all three disease states. IL-6 and TNF-α promote insulin resistance in adipose tissue, impair periodontal tissue repair, and drive vascular inflammation. NF-κB functions as a transcriptional regulator that integrates microbial LPS from periodontal pathogens with metabolic stress signals, leading to sustained cytokine release in gingival, pancreatic, and vascular cells [60]. CRP, produced in the liver under IL-6 stimulation, actively contributes to endothelial dysfunction by enhancing leukocyte adhesion. NLRP3 inflammasome activation [61] and IL-1β release [62] amplify tissue destruction in periodontal pockets, β-cell apoptosis in pancreatic islets, and plaque instability in atherosclerosis [63]. Collectively, these circuits converge on endothelial dysfunction, which emerges as the key link between local periodontal inflammation, systemic metabolic dysregulation, and cardiovascular complications.

These mechanistic interactions have clear clinical implications. A prospective cohort study demonstrated that individuals with high levels of both IL-6 and CRP, along with moderate-to-severe periodontitis and type 2 diabetes, had a 2.3-fold increased risk of major adverse cardiovascular events compared to individuals with none of these risk factors [54].

These overlapping pathophysiological mechanisms can be visually represented through a systems-level view of shared inflammatory mediators among diabetes mellitus, periodontal disease, and cardiovascular disease. Figure 3 illustrates the intersection of these three conditions, highlighting key molecules such as IL-6, TNF-α, CRP, and the roles of oxidative stress and endothelial dysfunction as central drivers of disease synergy.

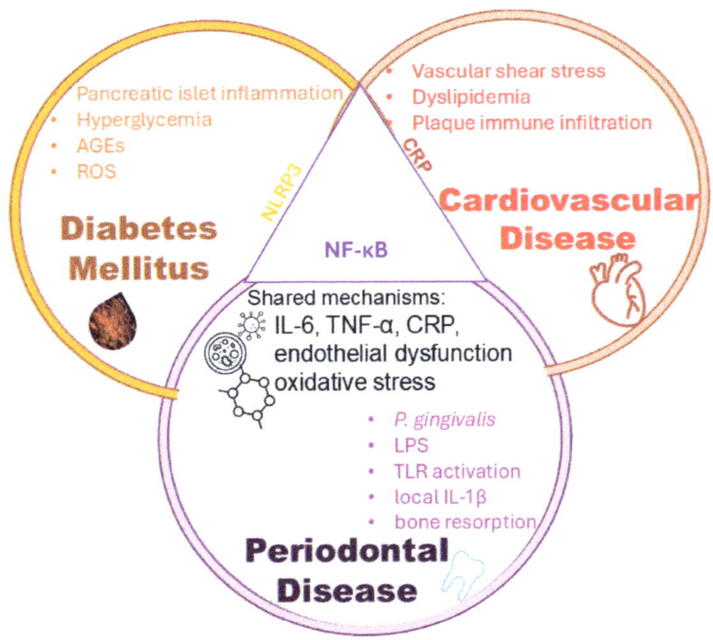


**Figure 3.** Shared inflammatory mechanisms linking diabetes, cardiovascular disease, and periodontal disease. A Venn diagram illustrates overlapping molecular pathways among DM, PD, and CVD. The shared circuits include IL-6, TNF-α, CRP, NF-κB, NLRP3 inflammasome activation, IL-1β signaling, oxidative stress, and endothelial dysfunction, which together drive systemic inflammation and cardiovascular risk. Evidence supporting these pathways is drawn from recent mechanistic and clinical studies. These mediators have been consistently reported as common drivers of systemic inflammation

across the three conditions [54,55,64]. Abbreviations: CRP—C-reactive Protein; IL—Interleukin; NF-κB—Nuclear Factor kappa-light-chain-enhancer of activated B cells; NLRP3—NOD-like Receptor Family Pyrin domain-containing 3 (inflammasome); ROS—Reactive Oxygen Species; TNF-α—Tumor Necrosis Factor-alpha TLR—Toll-Like Receptor.

Although inflammatory mediators such as IL-6, TNF-α, and CRP are elevated across all three conditions, endothelial dysfunction and systemic cytokine signaling represent central converging mechanisms, linking local periodontal inflammation and metabolic dysregulation to cardiovascular complications. The vascular endothelium itself is a primary site of inflammation in both CVD and the DM-PD axis. A recent study [65] reviewed how periodontal disease correlates with endothelial dysfunction (measured by markers such as flow-mediated dilation, VCAM/ICAM), and how periodontal therapy reduces endothelial inflammatory biomarkers. Such evidence strengthens the argument that endothelial tissue is a common denominator across all three diseases.

Key molecular pathways such as NF-κB, NLRP3 inflammasome activation, and IL-1β signaling are consistently implicated in inflammation in pancreatic islets [58], in adipose tissue, and in endothelial cells [65]. These pathways appear to operate both locally (e.g., periodontal tissue) and systemically, mediating crosstalk among tissues.

#### 3.4.3. The Oral–Gut–Systemic Axis

Beyond the local effects of oral inflammation, emerging data suggest a more complex microbiome-mediated axis that links the oral cavity, gut, and systemic circulation. Dysbiosis of the oral microbiota—characterized by overgrowth of pathogens such as *P. gingivalis* and *F. nucleatum*—has been shown to influence gut microbial composition and promote intestinal permeability ("leaky gut"), which facilitates systemic endotoxemia [66].

A recent study found that periodontitis altered gut microbial diversity and abundance in patients with diabetes, particularly increasing the abundance of pro-inflammatory Enterobacteriaceae species [42]. These alterations can amplify insulin resistance and systemic inflammation, contributing to the pathogenesis of both CVD and metabolic disease [67,68].

#### 3.4.4. Genetic and Epigenetic Links

Recent advances in genomics and epigenomics have uncovered shared genetic predispositions. For example, polymorphisms in *CDKN2A/B*, Peroxisome Proliferator-Activated Receptor Gamma (*PPARG*), and *IL6* genes have been implicated in both DM and CVD risk and have also shown associations with severe periodontitis [51].

In a Mendelian randomization study, genetically determined periodontitis was causally associated with increased risk of ischemic heart disease in patients with diabetes, suggesting a potential genetic mediation in this triad [69].

Furthermore, shared epigenetic mechanisms, including methylation changes in inflammatory gene promoters (e.g., IL6, TNF, and ICAM1), histone acetylation/deacetylation, and altered microRNA expression, have been reported in patients with PD, DM, and CVD. For example, Kang et al. [69]. demonstrated distinct DNA methylation and histone modification profiles in periodontitis patients with type 2 diabetes, while Patrick et al. [50]. identified shared genetic and epigenetic risk factors between chronic inflammatory diseases and coronary artery diseases. These findings provide evidence that epigenetic regulation contributes to the persistence and co-progression of these conditions despite conventional therapy.

#### 3.4.5. Clinical Clustering and Risk Synergy

Patients with comorbid diabetes and periodontitis are significantly more likely to develop cardiovascular complications than those with only one condition. In a longitudinal study, the presence of both diabetes and periodontitis was associated with a 3.1-fold higher risk of myocardial infarction and a 2.7-fold higher risk of stroke compared to those without either condition [12]. Importantly, these findings highlight that the elevated cardiovascular risk is not attributable to diabetes or periodontitis alone, but to their synergistic interaction, whereby chronic hyperglycemia and periodontal inflammation jointly amplify systemic cytokine release, endothelial dysfunction, and atherosclerotic progression [7,12].

This synergistic effect highlights the importance of early identification and integrated management of at-risk individuals. It also raises the potential for using PD as an early indicator or risk enhancer in cardiovascular risk stratification models for diabetic patients.

Taken together, these results demonstrate that diabetes mellitus, periodontal disease, and cardiovascular disease are linked through overlapping inflammatory, metabolic, and microbial pathways, with each axis reinforcing the others. While the evidence strongly supports these associations, their clinical and therapeutic implications require further integration and critical evaluation. These aspects are addressed in detail in the following Discussion section.

## 4. Discussion

This review highlights three well-established axes—diabetes mellitus and periodontal disease (DM–PD), periodontal disease and cardiovascular disease (PD–CVD), and diabetes mellitus and cardiovascular disease (DM–CVD)—that converge into a triangular model of shared systemic inflammation, immune dysregulation, and microbial dysbiosis. Together, these overlapping mechanisms explain why individuals with comorbid DM and PD are at substantially higher risk of cardiovascular complications compared to those with a single condition. The evidence presented in the Results demonstrates that hyperglycemia-driven inflammation, periodontal pathogen–induced bacteremia, and systemic endothelial dysfunction represent the key pathways uniting these disorders.

### *4.1. Therapeutic Implications and Multidisciplinary Management*

Because gingival tissues, pancreatic islets, adipose depots, and the vascular endothelium all represent active inflammatory sites in DM, PD, and CVD, integrated management must address both local and systemic inflammation. The convergence of DM, PD, and CVD is mediated not simply by systemic inflammation but by overlapping molecular circuits, particularly NF-κB, NLRP3 inflammasome activation, and IL-1β signaling. These pathways explain how periodontal dysbiosis, hyperglycemia, and vascular stress converge on a common axis of endothelial dysfunction. Therapeutically, interventions that target these circuits—such as IL-1β inhibitors, statins with NF-κB–modulating activity, or experimental inflammasome modulators—represent promising avenues for cross-disease management. Endothelial dysfunction, in particular, emerges as a unifying pathway, explaining how localized oral infection can amplify cardiovascular risk in susceptible diabetic patients.

Given that endothelial tissue emerges as the most critical site of convergence, therapies that reduce systemic inflammation (e.g., statins, SGLT2 inhibitors, GLP-1 receptor agonists) or directly target endothelial activation have particular translational relevance.

The interwoven pathophysiology of diabetes mellitus (DM), periodontal disease (PD), and cardiovascular disease (CVD) demands a paradigm shift in how these conditions are managed. Rather than treating each in isolation, an integrated, multidisciplinary model is emerging—one that aligns oral health with systemic disease prevention and management. This approach holds promise for improved patient outcomes and reduced healthcare burdens.

#### 4.1.1. Periodontal Therapy: Beyond Oral Health

Periodontal therapy, particularly nonsurgical interventions such as scaling and root planing (SRP), has demonstrated systemic benefits. Multiple meta-analyses have confirmed that effective periodontal treatment leads to modest but statistically significant reductions in HbA1c (−0.3% to −0.6%) in patients with type 2 diabetes [70].

Moreover, improvements in endothelial function have been documented after intensive periodontal therapy. A randomized trial showed that endothelial function, as measured by brachial artery flow-mediated dilation (FMD), improved significantly four weeks after treatment in patients with moderate-to-severe periodontitis [17].

These findings support the idea that periodontal treatment is not merely a local intervention but a systemic anti-inflammatory strategy, especially for individuals with cardiometabolic comorbidities.

#### 4.1.2. Anti-Inflammatory and Cardiometabolic Medications

Sodium–glucose cotransporter-2 (SGLT2) inhibitors are a class of oral antidiabetic agents that lower blood glucose by promoting urinary glucose excretion and have demonstrated robust benefits in reducing heart failure hospitalization and cardiovascular mortality in type 2 diabetes [71–73]. Glucagon-like peptide-1 (GLP-1) receptor agonists are injectable incretin-based therapies that enhance glucose-dependent insulin secretion, delay gastric emptying, and promote weight reduction; beyond glycemic control, large cardiovascular outcome trials such as LEADER and SUSTAIN-6 have shown significant reductions in major adverse cardiovascular events [46,74–76]. These pharmacological backgrounds provide important context for understanding why SGLT2 inhibitors and GLP-1 receptor agonists may also exert beneficial effects on periodontal inflammation and systemic inflammatory pathways.

Drugs that reduce systemic inflammation and improve metabolic profiles may also benefit periodontal health:

- SGLT2 inhibitors (e.g., empagliflozin, dapagliflozin) and GLP-1 receptor agonists (e.g., liraglutide, semaglutide) have been shown to reduce systemic inflammation and oxidative stress, which could theoretically benefit periodontal tissues [17].
- Statins, primarily used for lipid-lowering and cardiovascular protection, possess anti-inflammatory and antibacterial properties. It was reported that statin use was associated with reduced periodontitis progression, particularly in patients with elevated hs-CRP [17,77].
- Low-dose doxycycline, a Matrix Metalloproteinase (MMP)-inhibitor, has been U.S. Food and Drug Administration (FDA)-approved as an adjunctive periodontal therapy and shows promise in reducing systemic inflammatory burden, including CRP and IL-6 levels [78].

Recent evidence indicates that certain pharmacological therapies may exert synergistic effects when combined with periodontal treatment. For example, statin therapy, in addition to lowering lipids and reducing cardiovascular risk, has been shown to enhance the outcomes of nonsurgical periodontal therapy by attenuating periodontal inflammation and reducing probing depth progression [33,77]. Similarly, GLP-1 receptor agonists and SGLT2 inhibitors have been validated in cardiovascular outcome trials [40,45,48], also reduce systemic inflammation and oxidative stress, which may indirectly benefit periodontal health. These data suggest that combining periodontal therapy with cardiometabolic drugs could provide additive anti-inflammatory benefits.

These pharmacologic strategies highlight the potential for cross-benefits when targeting systemic inflammation.

#### 4.1.3. Probiotics and Microbiome Modulation

Given the emerging role of oral and gut microbiota in the pathogenesis of DM, PD, and CVD, microbiome-based therapies are under investigation. Probiotic lozenges containing *Lactobacillus reuteri* have demonstrated some efficacy in reducing periodontal inflammation [79].

Pilot studies are also examining whether modulating the gut microbiome in diabetic patients with prebiotic fibers or probiotic supplements could reduce systemic endotoxemia and improve oral and cardiovascular health. While results are preliminary, this area holds potential for integrative and preventive medicine [80].

#### 4.1.4. Lifestyle Interventions: Diet, Exercise, and Smoking Cessation

Lifestyle modifications remain foundational. Nutritional approaches like the Mediterranean diet and DASH diet have been shown to improve glycemic control, reduce cardiovascular risk, and lower inflammatory biomarkers, while also improving periodontal parameters [81,82].

Beyond pharmacological approaches, nutritional interventions may work synergistically with periodontal therapy and systemic treatments. The Mediterranean and DASH diets, rich in fruits, vegetables, whole grains, polyphenols, and omega-3 fatty acids, have been shown to improve glycemic control, reduce systemic inflammation, and lower cardiovascular risk, while also improving periodontal parameters [36]. Polyphenol supplementation has demonstrated modulatory effects on periodontal bacteria and matrix metalloproteinase activity, providing an additional anti-inflammatory pathway that may complement both systemic drugs and local periodontal therapies [70]. These findings highlight the potential of coordinated drug–diet–oral health strategies to achieve more robust clinical benefits. Table 1 summarizes pharmacological and dietary strategies that not only provide primary benefits in DM or CVD management but also exert synergistic effects on periodontal health through shared anti-inflammatory and metabolic pathways.

**Table 1.** Examples of pharmacological and nutritional strategies with synergistic effects across DM, PD, and CVD.

| Intervention | Primary Clinical Effect | Additional/Synergistic Effect on DM–PD–CVD Axis | References |
|---|---|---|---|
| Statins | Lipid-lowering, cardiovascular protection | Reduce periodontal inflammation and disease progression; enhance outcomes of nonsurgical periodontal therapy | [33,77] |
| SGLT2 inhibitors (e.g., empagliflozin, dapagliflozin) | Improve glycemic control, reduce heart failure risk and CV mortality | Anti-inflammatory and antioxidative effects may benefit periodontal health | [40,41,48] |
| GLP-1 receptor agonists (e.g., liraglutide, semaglutide) | Improve glycemic control, reduce MACE in T2DM | Reduce systemic inflammation and improve endothelial function; potential positive effects on PD outcomes | [44–46] |
| Polyphenols (e.g., resveratrol, flavonoids) | Antioxidant and anti-inflammatory dietary compounds | Modulate periodontal bacteria and matrix metalloproteinase (MMP) activity; support metabolic and vascular health | [70] |
| Omega-3 fatty acids (e.g., EPA, DHA) | Improve lipid profile, reduce CV risk | Enhance resolution of periodontal inflammation; improve insulin sensitivity | [36] |
| Mediterranean/ DASH diet | Improve glycemic control, lower blood pressure, reduce CV events | Reduce periodontal inflammation, improve periodontal indices, and lower systemic inflammatory markers | [36] |

Pharmacological and nutritional interventions with synergistic effects across diabetes mellitus (DM), periodontal disease (PD), and cardiovascular disease (CVD). The table highlights how drugs and dietary strategies provide their primary benefits while simultaneously exerting secondary effects that reduce systemic inflammation, improve periodontal outcomes, and mitigate cardiovascular risk. Abbreviations: CVD—cardiovascular disease; CV—Cardiovascular; DASH—Dietary Approaches to Stop Hypertension; DM—Diabetes Mellitus; EPA—Eicosapentaenoic Acid; DHA—Docosahexaenoic Acid; GLP-1—Glucagon-Like Peptide-1; MACE—Major Adverse Cardiovascular Events; MMP—Matrix Metalloproteinase; NSPT—Nonsurgical Periodontal Therapy; PD—Periodontal Disease; SGLT2—Sodium-Glucose Cotransporter-2; T2DM—Type 2 Diabetes Mellitus.

Smoking cessation is particularly critical, as smoking amplifies all three conditions. Smokers with diabetes and periodontitis are at exponentially higher risk for cardiovascular events, making cessation programs essential in any integrated care strategy [77].

#### 4.1.5. Integrated Care Models

Several models of integrated care have been proposed and tested:

The Centre for Oral-Systemic Health (United States of America) and National Health Service NHS-integrated diabetes-periodontal clinics (UK) represent institutional examples of interdisciplinary care teams including dentists, endocrinologists, cardiologists, and nutritionists [83].

A pilot program demonstrated that embedding dental professionals within diabetes management clinics led to improved oral health literacy, higher periodontal treatment uptake, and better glycemic control at 6 months [77].

These programs not only improve clinical outcomes but may also reduce long-term healthcare costs through prevention and early intervention.

#### 4.1.6. Digital Health and Artificial Intelligence (AI)

AI-powered risk stratification tools and telemedicine platforms are beginning to play a role in identifying patients at the intersection of DM, PD, and CVD. Machine learning models trained on electronic health record (EHR) data have successfully predicted cardiovascular events in diabetic patients using periodontal status as a variable, suggesting that oral health indicators can enhance risk models [84].

Apps that integrate oral hygiene tracking with glucose monitoring and heart rate variability analysis are also under development, supporting patient self-management across domains.

### 4.2. Limitations in Current Research and Gaps in Knowledge

Despite compelling evidence supporting the associations between diabetes mellitus (DM), periodontal disease (PD), and cardiovascular disease (CVD), several limitations persist in the current literature that constrain our ability to establish causality and implement unified care strategies.

#### 4.2.1. Heterogeneity in Study Designs and Populations

Many existing studies vary in methodology, including the diagnostic criteria for PD, definitions of glycemic control, and endpoints for cardiovascular outcomes. This heterogeneity reduces the comparability of results and weakens the ability to generate robust meta-analytic conclusions.

For example, a review noted that more than 40% of trials assessing PD and CVD failed to standardize periodontal disease classification, leading to inconsistencies in findings [85].

Additionally, many cohort studies are conducted in high-income countries, limiting generalizability to low- and middle-income populations where the burden of all three diseases is rising [86].

#### 4.2.2. Confounding and Reverse Causality

Many observational studies are prone to confounding factors such as smoking, obesity, diet, and socioeconomic status, all of which influence DM, PD, and CVD. While multivariate adjustment helps, residual confounding remains a challenge. Moreover, the bidirectional nature of the relationships—especially between DM and PD—complicates interpretation of causality.

Mendelian randomization (MR) studies are emerging as tools to overcome confounding, but even MR analyses are limited by the availability and specificity of genetic instruments [87].

#### 4.2.3. Underpowered or Short-Term Interventional Studies

Though some interventional trials suggest periodontal therapy may improve glycemic control or endothelial function, most are short (e.g., $\leq$6 months) and lack hard cardiovascular outcomes like myocardial infarction or stroke. A meta-analysis emphasized the need for long-term randomized controlled trials (RCTs) with clinical endpoints [30].

The absence of large-scale, longitudinal interventional studies hinders the development of evidence-based integrated treatment guidelines.

#### 4.2.4. Lack of Mechanistic Human Studies

Most mechanistic insights into inflammation, dysbiosis, or immune pathways are derived from in vitro or animal studies. While useful, these models do not fully replicate the human biological complexity. There is a need for well-designed human studies that utilize tissue biopsies, single-cell sequencing, and proteomic profiling to map real-time interactions between oral and systemic inflammation [88,89].

Addressing these limitations will be crucial to translating current evidence into actionable clinical practice. Standardization of definitions, long-term interventional data, and inclusion of diverse populations must be prioritized. Additionally, mechanistic human studies will strengthen causal inference and facilitate biomarker development. Overcoming these research gaps is essential for moving from correlation to causation, and from fragmented care to unified disease prevention.

It should also be noted that comparable inflammatory and microbial pathways are implicated in other diabetes-related complications. For example, diabetic nephropathy is characterized by endothelial dysfunction and NF-κB–driven cytokine release within renal tissues, while diabetic retinopathy involves chronic low-grade inflammation, oxidative stress, and vascular leakage in the retina [7,68,90,91]. Beyond diabetes, systemic inflammatory diseases such as rheumatoid arthritis share overlapping cytokine profiles (e.g., elevated IL-6 and TNF-α) and are associated with increased periodontal and cardiovascular risk. Recognizing these parallels situates the DM–PD–CVD triad within a broader network of chronic inflammatory conditions, highlighting the systemic relevance of these mechanisms.

### 4.3. *Future Directions*

Addressing current gaps will require a shift toward more integrative, precise, and equitable research frameworks. Several promising directions are emerging. Although not applied in the present review, AI-driven natural language processing and automated literature-mining platforms are emerging as valuable tools to support systematic evidence synthesis and risk stratification, and they may play an important role in future integrative analyses of DM, PD, and CVD [84,92].

#### 4.3.1. Precision Medicine and Risk Stratification

As multi-omics and machine learning tools mature, individualized risk prediction models incorporating genetic, inflammatory, microbiome, and clinical data will become feasible. Precision strategies [64] could stratify patients based on inflammatory profiles, enabling personalized periodontal and metabolic therapies.

A study described an AI-integrated model combining oral microbiome signatures with HbA1c and hs-CRP levels to predict future cardiovascular events with 85% accuracy in diabetic patients—a proof-of-concept for precision prevention [93].

#### 4.3.2. Integrated Clinical Trials

Large-scale, multicenter RCTs that evaluate the impact of combined dental and medical interventions on systemic outcomes are urgently needed. Ongoing trials and studies in Europe are evaluating whether routine periodontal therapy in diabetic patients reduces long-term CVD risk [94–96].

Such trials should use standardized periodontal assessment tools and include long-term follow-up for hard outcomes, enabling definitive conclusions.

#### 4.3.3. Microbiome and Metabolome Research

Future research should focus on oral-gut-systemic microbial interactions using metagenomics, metabolomics, and lipidomics. Understanding how microbial metabolites like trimethylamine-N-oxide (TMAO) or short-chain fatty acids modulate systemic inflammation could open new therapeutic avenues, including dietary or probiotic interventions [67].

#### 4.3.4. Therapeutic Targets and Emerging Mechanistic Strategies

This subsection highlights therapeutic targets that act on common molecular circuits linking DM, PD, and CVD, including pharmacological agents, microbiome-modulating therapies, nutritional bioactives, and AI-driven strategies.

New pharmacological agents such as sodium-glucose co-transporter-2 (SGLT2) inhibitors and glucagon-like peptide-1 (GLP-1) receptor agonists, already validated in large cardiovascular outcome trials, have demonstrated systemic anti-inflammatory effects and improved endothelial function. Preliminary evidence suggests that these agents may also reduce periodontal inflammation, although larger dedicated trials are needed. Similarly, statins, traditionally used for lipid lowering, have shown promise in slowing periodontal disease progression due to their anti-inflammatory and antibacterial properties [97,98].

Microbiome-modulating therapies are another emerging field. Probiotics (e.g., Lactobacillus reuteri lozenges) and prebiotics have been studied as adjuncts to periodontal therapy, with early results suggesting reductions in gingival inflammation and systemic inflammatory markers. Recent patents have also described novel anti-inflammatory and microbiome-targeted formulations, including polyphenol-based compounds and engineered bacterial strains, designed to simultaneously modulate oral and gut dysbiosis with potential benefit across DM, PD, and CVD [10,99].

Novel mechanistic insights have been provided by multi-omics analyses that integrate genomics, transcriptomics, proteomics, and metabolomics. These studies have identified shared signatures of inflammation, including IL-6, TNF-$\alpha$, and NF-$\kappa$B signaling, as well as alterations in lipid and amino acid metabolism, which may represent therapeutic targets. The concept of an "oral–gut–systemic axis" has gained attention, suggesting that periodontal dysbiosis can influence gut microbiota composition and systemic immune activation, thereby linking PD to both metabolic and cardiovascular dysfunction [25].

AI-driven approaches represent an additional frontier. Machine-learning models that integrate periodontal status, glycemic indices, and systemic biomarkers have shown promise in predicting cardiovascular events in diabetic populations. These tools may support personalized risk stratification and highlight oral health as a variable of clinical importance in cardiometabolic care [100–102].

Together, these therapeutic targets demonstrate how interventions directed at shared molecular pathways (NF-$\kappa$B, NLRP3 inflammasome, IL-1$\beta$, IL-6/CRP axis) may yield synergistic benefits across DM, PD, and CVD, and should be prioritized in future translational trials.

#### 4.3.5. Policy and Practice Integration

Healthcare systems must evolve to incorporate dental care as part of chronic disease management. Policies that incentivize interdisciplinary collaboration, integrated electronic health records, and co-location of dental and medical services could significantly improve outcomes.

Moreover, educating clinicians—especially in primary care—about the oral-systemic disease connection should be prioritized in medical and dental curricula [103].

The convergence of digital health, systems biology, and interdisciplinary care is paving the way for a new era in chronic disease management. The future lies in data-driven, integrated models that detect disease earlier, intervene more precisely, and monitor more holistically. Success in this endeavor will depend not only on scientific innovation but also on policy reform, clinician education, and patient empowerment.

Taken together, these findings underscore the need to shift from a siloed approach to a truly integrated model of care. Periodontal health should be recognized as a cardiometabolic risk factor, and its management incorporated into diabetes and cardiovascular care pathways. At the same time, the convergence of novel therapies, microbiome-targeted interventions, and AI-driven precision tools offers new opportunities to reduce the inflammatory burden that unites these conditions. By embedding oral-systemic health within multidisciplinary care and policy frameworks, clinicians and researchers can help reshape prevention and management strategies for chronic diseases worldwide.

## 5. Conclusions

The evidence connecting diabetes mellitus, periodontal disease, and cardiovascular pathologies is compelling and continues to grow. These conditions, long viewed as distinct entities, are now recognized as components of a complex, interconnected disease network driven by chronic inflammation, immune dysregulation, microbial dysbiosis, and metabolic dysfunction.

Bidirectional relationships between diabetes and periodontitis, and between periodontitis and cardiovascular disease, underscore the need for a systemic view of oral health. Furthermore, diabetes itself significantly increases cardiovascular risk through a host of overlapping mechanisms, including endothelial dysfunction, oxidative stress, and pro-inflammatory states—all of which are further exacerbated by coexisting periodontitis.

Emerging research, including omics-based systems biology and advanced machine learning, is beginning to unravel the shared molecular pathways that drive this triad. Clinically, the integration of periodontal care into cardiometabolic risk management has already shown promise, with evidence supporting modest improvements in glycemic control, endothelial function, and inflammatory biomarkers.

However, significant gaps remain—particularly the lack of large, long-term interventional trials with definitive cardiovascular outcomes. Moving forward, precision medicine, integrated care models, and microbiome-focused interventions may reshape how these diseases are diagnosed, treated, and prevented.

Despite promising evidence for the systemic benefits of periodontal therapy and cardiometabolic drugs, significant gaps remain—particularly the lack of large, long-term interventional trials with definitive cardiovascular outcomes. Moving forward, multidisciplinary care models that bridge dentistry, endocrinology, and cardiology, combined with precision medicine and microbiome-targeted therapies, will be essential to reduce the global burden of these interconnected conditions and improve long-term outcomes.

**Author Contributions:** Conceptualization: O.Ţ. (Otilia Ţica), I.R. and O.Ţ. (Ovidiu Ţica); methodology: O.Ţ. (Otilia Ţica), G.C. and V.A.P.; Investigation: I.R., G.C., V.A.P. and I.S.; resources: C.M.D., L.Ş. and I.S.; writing—Original Draft Preparation: O.Ţ. (Otilia Ţica) and I.R.; writing—Review and Editing: O.Ţ. (Ovidiu Ţica), C.M.D. and I.S.; visualization: G.C. and V.A.P.; supervision: O.Ţ. (Ovidiu Ţica) and C.M.D.; project administration: O.Ţ. (Otilia Ţica) and I.R. All authors have read and agreed to the published version of the manuscript.

**Funding:** The APC for this article was funded by the University of Oradea.

**Institutional Review Board Statement:** Not applicable. This study is a narrative clinical review and did not involve the collection of new data from human or animal subjects.

**Informed Consent Statement:** Not applicable. No individual data or patient information was collected or reported in this review.

**Data Availability Statement:** No new data were created or analyzed in this study. Data sharing does not apply to this article.

**Acknowledgments:** The authors would like to thank the University of Oradea and the Emergency County Clinical Hospital of Bihor for academic support and institutional access to medical literature. We also acknowledge the collaborative insights provided by colleagues in the Departments of Cardiology, Dental Medicine, and Pathology. The authors have reviewed and edited the output and take full responsibility for the content of this publication.

**Conflicts of Interest:** The authors declare no conflicts of interest. The funders had no role in the design of the study; in the collection, analyses, or interpretation of data; in the writing of the manuscript; or in the decision to publish the results.

## Abbreviations

The following abbreviations are used in this manuscript:

| | |
|---|---|
| AI | Artificial Intelligence |

| | |
|---|---|
| CAD | Coronary Artery Disease |
| CRP | C-Reactive Protein |
| CVD | Cardiovascular Disease |
| DAPA-HF | Dapagliflozin and Prevention of Adverse Outcomes in Heart Failure |
| DASH | Dietary Approaches to Stop Hypertension |
| DM | Diabetes Mellitus |
| DNA | Deoxyribonucleic Acid |
| EHR | Electronic Health Record |
| EMPA-REG | Empagliflozin Cardiovascular Outcome Event Trial in Type 2 Diabetes Mellitus Patients—Removing Excess Glucose |
| FDA | U.S. Food and Drug Administration |
| FMD | Flow-Mediated Dilation |
| GLP | Glucagon-Like Peptide |
| HDL | High-Density Lipoprotein |
| ICAM | Intercellular Adhesion Molecule |
| IL | Interleukin |
| LEADER | Liraglutide Effect and Action in Diabetes: Evaluation of Cardiovascular Outcome Results |
| LPS | Lipopolysaccharide |
| MACE | Major Adverse Cardiovascular Events |
| MI | Myocardial Infarction |
| MMP | Matrix Metalloproteinase |
| MR | Mendelian Randomization |
| NF | Nuclear Factor |
| NHS | National Health Service |
| NSPT | Nonsurgical Periodontal Therapy |
| OUTCOME | Outcome Trial (e.g., EMPA-REG OUTCOME) |
| PD | Periodontal Disease |
| PPARG | Peroxisome Proliferator-Activated Receptor Gamma |
| RAGE | Receptor for Advanced Glycation End-products |
| ROS | Reactive Oxygen Species |
| SRP | Scaling and Root Planing |
| SUSTAIN | Semaglutide Unabated Sustainability in Treatment of Type 2 Diabetes |
| TMAO | Trimethylamine N-oxide |
| TNF | Tumor Necrosis Factor |
| UK | United Kingdom |
| U.S. | United States of America |
| VCAM | Vascular Cell Adhesion Molecule |
| VLDL | Very-Low-Density Lipoprotein |

## References


1. Saeedi, P.; Petersohn, I.; Salpea, P.; Malanda, B.; Karuranga, S.; Unwin, N.; Colagiuri, S.; Guariguata, L.; Motala, A.A.; Ogurtsova, K.; et al. Global and regional diabetes prevalence estimates for 2019 and projections for 2030 and 2045: Results from the International Diabetes Federation Diabetes Atlas, 9(th) edition. *Diabetes Res. Clin. Pract.* **2019**, *157*, 107843. [CrossRef]
2. Tonetti, M.S.; Jepsen, S.; Jin, L.; Otomo-Corgel, J. Impact of the global burden of periodontal diseases on health, nutrition and wellbeing of mankind: A call for global action. *J. Clin. Periodontol.* **2017**, *44*, 456–462. [CrossRef]
3. Roth, G.A.; Mensah, G.A.; Johnson, C.O.; Addolorato, G.; Ammirati, E.; Baddour, L.M.; Barengo, N.C.; Beaton, A.Z.; Benjamin, E.J.; Benziger, C.P.; et al. Global Burden of Cardiovascular Diseases and Risk Factors, 1990-2019: Update From the GBD 2019 Study. *J. Am. Coll. Cardiol.* **2020**, *76*, 2982–3021. [CrossRef]
4. Simpson, T.C.; Clarkson, J.E.; Worthington, H.V.; MacDonald, L.; Weldon, J.C.; Needleman, I.; Iheozor-Ejiofor, Z.; Wild, S.H.; Qureshi, A.; Walker, A.; et al. Treatment of periodontitis for glycaemic control in people with diabetes mellitus. *Cochrane Database Syst. Rev.* **2022**, *4*, CD004714. [CrossRef]
5. Baeza, M.; Morales, A.; Cisterna, C.; Cavalla, F.; Jara, G.; Isamitt, Y.; Pino, P.; Gamonal, J. Effect of periodontal treatment in patients with periodontitis and diabetes: Systematic review and meta-analysis. *J. Appl. Oral Sci.* **2020**, *28*, e20190248. [CrossRef]

6. Page, M.J.; McKenzie, J.E.; Bossuyt, P.M.; Boutron, I.; Hoffmann, T.C.; Mulrow, C.D.; Shamseer, L.; Tetzlaff, J.M.; Akl, E.A.; Brennan, S.E.; et al. The PRISMA 2020 statement: An updated guideline for reporting systematic reviews. *BMJ* **2021**, *372*, n71. [CrossRef]
7. Schulze-Späte, U.; Wurschi, L.; van der Vorst, E.P.C.; Hölzle, F.; Craveiro, R.B.; Wolf, M.; Noels, H. Crosstalk between periodontitis and cardiovascular risk. *Front. Immunol.* **2024**, *15*, 1469077. [CrossRef] [PubMed]
8. Sanghvi, M.M.; Ramírez, J.; Chadalavada, S.; Aung, N.; Munroe, P.B.; Donos, N.; Petersen, S.E. The Association Between Periodontal Disease and Cardiovascular Disease: Insights From Imaging, Observational, and Genetic Data. *JACC Adv.* **2024**, *3*, 101241. [CrossRef] [PubMed]
9. Umezaki, Y.; Yamashita, A.; Nishimura, F.; Naito, T. The role of periodontal treatment on the reduction of hemoglobinA1c, comparing with existing medication therapy: A systematic review and meta-analysis. *Front. Clin. Diabetes Healthc.* **2025**, *6*, 1541145. [CrossRef]
10. Wei, X.; Qian, S.; Yang, Y.; Mo, J. Microbiome-based therapies for periodontitis and peri-implantitis. *Oral Dis.* **2024**, *30*, 2838–2857. [CrossRef] [PubMed]
11. Muñoz-Carrillo, J.L.; Palomeque-Molina, P.I.; Villacis-Valencia, M.S.; Gutiérrez-Coronado, O.; Chávez-Ruvalcaba, F.; Vázquez-Alcaraz, S.J.; Villalobos-Gutiérrez, P.T.; Palomeque-Molina, J. Relationship between periodontitis, type 2 diabetes mellitus and COVID-19 disease: A narrative review. *Front. Cell Infect. Microbiol.* **2025**, *15*, 1527217. [CrossRef]
12. Norhammar, A.; Näsman, P.; Buhlin, K.; de Faire, U.; Ferrannini, G.; Gustafsson, A.; Kjellström, B.; Kvist, T.; Jäghagen, E.L.; Lindahl, B.; et al. Does Periodontitis Increase the Risk for Future Cardiovascular Events? Long-Term Follow-Up of the PAROKRANK Study. *J. Clin. Periodontol.* **2025**, *52*, 16–23. [CrossRef] [PubMed]
13. Xie, H.; Qin, Z.; Ling, Z.; Ge, X.; Zhang, H.; Guo, S.; Liu, L.; Zheng, K.; Jiang, H.; Xu, R. Oral pathogen aggravates atherosclerosis by inducing smooth muscle cell apoptosis and repressing macrophage efferocytosis. *Int. J. Oral Sci.* **2023**, *15*, 26. [CrossRef]
14. Wojtkowska, A.; Zapolski, T.; Wysokińska-Miszczuk, J.; Wysokiński, A.P. The inflammation link between periodontal disease and coronary atherosclerosis in patients with acute coronary syndromes: Case-control study. *BMC Oral Health* **2021**, *21*, 5. [CrossRef]
15. Hayashi, J.; Hasegawa, A.; Hayashi, K.; Suzuki, T.; Ishii, M.; Otsuka, H.; Yatabe, K.; Goto, S.; Tatsumi, J.; Shin, K. Effects of periodontal treatment on the medical status of patients with type 2 diabetes mellitus: A pilot study. *BMC Oral Health* **2017**, *17*, 77. [CrossRef] [PubMed]
16. Oe, Y.; Nomoto, H.; Cho, K.Y.; Yokozeki, K.; Ono, T.; Miya, A.; Kameda, H.; Nakamura, A.; Arimura, Y.; Atsumi, T. Efficacy and safety of oral semaglutide in older patients with type 2 diabetes: A retrospective observational study (the OTARU-SEMA study). *BMC Endocr. Disord.* **2024**, *24*, 124. [CrossRef]
17. Molina, A.; Ambrosio, N.; Molina, M.; Montero, E.; Virto, L.; Herrera, D.; Figuero, E.; Sanz, M. Effect of periodontal therapy on endothelial function and serum biomarkers in patients with periodontitis and established cardiovascular disease: A pilot study. *Front. Oral Health* **2025**, *6*, 1488941. [CrossRef]
18. Czerniuk, M.R.; Surma, S.; Romańczyk, M.; Nowak, J.M.; Wojtowicz, A.; Filipiak, K.J. Unexpected Relationships: Periodontal Diseases: Atherosclerosis-Plaque Destabilization? From the Teeth to a Coronary Event. *Biology* **2022**, *11*, 272. [CrossRef] [PubMed]
19. Țica, O.; Sabău, M.; Venter, A.; Beiușanu, C.; Berechet, M.; Huniadi, A.; Șandor, M.I.; Țica, O. Stroke in Dilated Cardiomyopathy: An Autopsy-Based Study of Mechanisms, Topography, and Clinical Implications. *Diagnostics* **2025**, *15*, 2287. [CrossRef]
20. Sayols-Baixeras, S.; Dekkers, K.F.; Baldanzi, G.; Jönsson, D.; Hammar, U.; Lin, Y.T.; Ahmad, S.; Nguyen, D.; Varotsis, G.; Pita, S.; et al. Streptococcus Species Abundance in the Gut Is Linked to Subclinical Coronary Atherosclerosis in 8973 Participants From the SCAPIS Cohort. *Circulation* **2023**, *148*, 459–472. [CrossRef]
21. Sinning, C.; Makarova, N.; Völzke, H.; Schnabel, R.B.; Ojeda, F.; Dörr, M.; Felix, S.B.; Koenig, W.; Peters, A.; Rathmann, W.; et al. Association of glycated hemoglobin A(1c) levels with cardiovascular outcomes in the general population: Results from the BiomarCaRE (Biomarker for Cardiovascular Risk Assessment in Europe) consortium. *Cardiovasc. Diabetol.* **2021**, *20*, 223. [CrossRef] [PubMed]
22. Park, M.J.; Choi, K.M. Association between Variability of Metabolic Risk Factors and Cardiometabolic Outcomes. *Diabetes Metab. J.* **2022**, *46*, 49–62. [CrossRef]
23. Kara, G.; Cifcibasi, E.; Karsidag, K.; Cintan, S. Short term effects of periodontal therapy on inflammatory markers in patients with type-2 diabetes. *Saudi Med. J.* **2015**, *36*, 469–476. [CrossRef]
24. Huang, X.; Xie, M.; Lu, X.; Mei, F.; Song, W.; Liu, Y.; Chen, L. The Roles of Periodontal Bacteria in Atherosclerosis. *Int. J. Mol. Sci.* **2023**, *24*, 12861. [CrossRef]
25. Wang, Z.; Kaplan, R.C.; Burk, R.D.; Qi, Q. The Oral Microbiota, Microbial Metabolites, and Immuno-Inflammatory Mechanisms in Cardiovascular Disease. *Int. J. Mol. Sci.* **2024**, *25*, 12337. [CrossRef]
26. Sanz, M.; Marco Del Castillo, A.; Jepsen, S.; Gonzalez-Juanatey, J.R.; D'Aiuto, F.; Bouchard, P.; Chapple, I.; Dietrich, T.; Gotsman, I.; Graziani, F.; et al. Periodontitis and cardiovascular diseases: Consensus report. *J. Clin. Periodontol.* **2020**, *47*, 268–288. [CrossRef]
27. Gandhi, M.; Haider, S.; Chang, H.Z.Y.; Kazlauskas, A. The Role of Endothelial Senescence in the Pathogenesis of Diabetic Retinopathy. *Int. J. Mol. Sci.* **2025**, *26*, 5211. [CrossRef]

28. Lamri, A.; De Paoli, M.; De Souza, R.; Werstuck, G.; Anand, S.; Pigeyre, M. Insight into genetic, biological, and environmental determinants of sexual-dimorphism in type 2 diabetes and glucose-related traits. *Front. Cardiovasc. Med.* **2022**, *9*, 964743. [CrossRef]
29. Chen, Y.F.; Zhan, Q.; Wu, C.Z.; Yuan, Y.H.; Chen, W.; Yu, F.Y.; Li, Y.; Li, L.J. Baseline HbA1c Level Influences the Effect of Periodontal Therapy on Glycemic Control in People with Type 2 Diabetes and Periodontitis: A Systematic Review on Randomized Controlled Trails. *Diabetes Ther.* **2021**, *12*, 1249–1278. [CrossRef] [PubMed]
30. Meng, R.; Xu, J.; Fan, C.; Liao, H.; Wu, Z.; Zeng, Q. Effect of non-surgical periodontal therapy on risk markers of cardiovascular disease: A systematic review and meta-analysis. *BMC Oral Health* **2024**, *24*, 692, Correction in *BMC Oral Health* **2024**, *24*, 819. [CrossRef] [PubMed]
31. Truong, H.P.; Tran, H.M.; Huynh, T.; Nguyen, D.N.Q.; Ho, D.T.; Tran, C.C.; Nguyen, S.V.; Vo, T.M. Correlation Between SCORE2-Diabetes and Coronary Artery Calcium Score in Patients with Type 2 Diabetes Mellitus: A Cross-Sectional Study in Vietnam. *J. Imaging* **2025**, *11*, 130. [CrossRef]
32. Ramos, T.C.S.; Boas, M.L.V.; Nunes, C.M.M.; Ferreira, C.L.; Pannuti, C.M.; Santamaria, M.P.; Jardini, M.A.N. Effect of systemic antibiotic and probiotic therapies as adjuvant treatments of subgingival instrumentation for periodontitis: A randomized controlled clinical study. *J. Appl. Oral Sci.* **2022**, *30*, e20210583. [CrossRef]
33. de Carvalho, R.D.P.; Casarin, R.C.V.; de Lima, P.O.; Cogo-Müller, K. Statins with potential to control periodontitis: From biological mechanisms to clinical studies. *J. Oral Biosci.* **2021**, *63*, 232–244. [CrossRef]
34. Gatto, A.; Liu, K.; Milan, N.; Wong, S. The Effects of GLP-1 Agonists on Musculoskeletal Health and Orthopedic Care. *Curr. Rev. Musculoskelet. Med.* **2025**, *18*, 469–480. [CrossRef] [PubMed]
35. Yang, D.R.; Wang, M.Y.; Zhang, C.L.; Wang, Y. Endothelial dysfunction in vascular complications of diabetes: A comprehensive review of mechanisms and implications. *Front. Endocrinol.* **2024**, *15*, 1359255. [CrossRef] [PubMed]
36. Aalizadeh, Y.; Khamisi, N.; Asghari, P.; Safari, A.; Mottaghi, M.; Taherkhani, M.H.; Alemi, A.; Ghaderi, M.; Rahmanian, M. The Mediterranean diet and periodontitis: A systematic review and meta-analysis. *Heliyon* **2024**, *10*, e35633. [CrossRef] [PubMed]
37. Fuster, V.; García-Álvarez, A.; Devesa, A.; Mass, V.; Owen, R.; Quesada, A.; Fuster, J.J.; García-Lunar, I.; Pocock, S.; Sánchez-González, J.; et al. Influence of Subclinical Atherosclerosis Burden and Progression on Mortality. *J. Am. Coll. Cardiol.* **2024**, *84*, 1391–1403. [CrossRef]
38. Han, F.; Shi, X.L.; Pan, J.J.; Wu, K.Y.; Zhu, Q.T.; Yuan, C.C.; Xiao, W.M.; Ding, Y.B.; Yu, X.P.; Jiao, X.P.; et al. Elevated serum HbA1c level, rather than previous history of diabetes, predicts the disease severity and clinical outcomes of acute pancreatitis. *BMJ Open Diabetes Res. Care* **2023**, *11*, e003070. [CrossRef]
39. Lee, S.; Liu, T.; Chung, C.T.; Reinhold, J.; Vassiliou, V.S.; Tse, G. PowerAI-Diabetes: Review of glycemic and lipid variability to predict cardiovascular events in Chinese diabetic population. *npj Metab. Health Dis.* **2024**, *2*, 14. [CrossRef]
40. Fitchett, D.; Inzucchi, S.E.; Wanner, C.; Mattheus, M.; George, J.T.; Vedin, O.; Zinman, B.; Johansen, O.E. Relationship between hypoglycaemia, cardiovascular outcomes, and empagliflozin treatment in the EMPA-REG OUTCOME® trial. *Eur. Heart J.* **2020**, *41*, 209–217. [CrossRef]
41. Zannad, F.; Ferreira, J.P.; Pocock, S.J.; Anker, S.D.; Butler, J.; Filippatos, G.; Brueckmann, M.; Ofstad, A.P.; Pfarr, E.; Jamal, W.; et al. SGLT2 inhibitors in patients with heart failure with reduced ejection fraction: A meta-analysis of the EMPEROR-Reduced and DAPA-HF trials. *Lancet* **2020**, *396*, 819–829. [CrossRef]
42. Ţica, O.; Ţica, O.; Bunting, K.V.; deBono, J.; Gkoutos, G.V.; Popescu, M.I.; Kotecha, D. Post-mortem examination of high mortality in patients with heart failure and atrial fibrillation. *BMC Med.* **2022**, *20*, 331. [CrossRef] [PubMed]
43. Ţica, O.; Khamboo, W.; Kotecha, D. Breaking the Cycle of Heart Failure with Preserved Ejection Fraction and Atrial Fibrillation. *Card. Fail. Rev.* **2022**, *8*, e32. [CrossRef]
44. Buse, J.B.; Bain, S.C.; Mann, J.F.E.; Nauck, M.A.; Nissen, S.E.; Pocock, S.; Poulter, N.R.; Pratley, R.E.; Linder, M.; Monk Fries, T.; et al. Cardiovascular Risk Reduction with Liraglutide: An Exploratory Mediation Analysis of the LEADER Trial. *Diabetes Care* **2020**, *43*, 1546–1552. [CrossRef]
45. Nauck, M.A.; Quast, D.R. Cardiovascular Safety and Benefits of Semaglutide in Patients with Type 2 Diabetes: Findings From SUSTAIN 6 and PIONEER 6. *Front. Endocrinol.* **2021**, *12*, 645566. [CrossRef]
46. Marx, N.; Husain, M.; Lehrke, M.; Verma, S.; Sattar, N. GLP-1 Receptor Agonists for the Reduction of Atherosclerotic Cardiovascular Risk in Patients with Type 2 Diabetes. *Circulation* **2022**, *146*, 1882–1894. [CrossRef] [PubMed]
47. Ţica, O.; Teodorovich, N.; Champsi, A.; Swissa, M. Are the Four Pillars the Ideal Treatment for the Elderly? *Cardiology* **2023**, *148*, 296–299. [CrossRef]
48. Neuen, B.L.; Fletcher, R.A.; Heath, L.; Perkovic, A.; Vaduganathan, M.; Badve, S.V.; Tuttle, K.R.; Pratley, R.; Gerstein, H.C.; Perkovic, V.; et al. Cardiovascular, Kidney, and Safety Outcomes with GLP-1 Receptor Agonists Alone and in Combination with SGLT2 Inhibitors in Type 2 Diabetes: A Systematic Review and Meta-Analysis. *Circulation* **2024**, *150*, 1781–1790. [CrossRef] [PubMed]

49. Norgard, N.B.; Lopez-Candales, A. The rise of SGLT2 inhibitors: The time is now for cardiovascular specialists to lead the charge. *Postgrad. Med.* **2022**, *134*, 11–13. [CrossRef]
50. Patrick, M.T.; Li, Q.; Wasikowski, R.; Mehta, N.; Gudjonsson, J.E.; Elder, J.T.; Zhou, X.; Tsoi, L.C. Shared genetic risk factors and causal association between psoriasis and coronary artery disease. *Nat. Commun.* **2022**, *13*, 6565. [CrossRef]
51. Wei, X.; Zhang, X.; Chen, R.; Li, Y.; Yang, Y.; Deng, K.; Cai, Z.; Lai, H.; Shi, J. Impact of periodontitis on type 2 diabetes: A bioinformatic analysis. *BMC Oral Health* **2024**, *24*, 635. [CrossRef]
52. Wang, X.; Chen, L.; Wei, J.; Zheng, H.; Zhou, N.; Xu, X.; Deng, X.; Liu, T.; Zou, Y. The immune system in cardiovascular diseases: From basic mechanisms to therapeutic implications. *Signal Transduct. Target. Ther.* **2025**, *10*, 166. [CrossRef] [PubMed]
53. Jelavic, M.M.; Ţica, O.; Pintaric, H.; Ţica, O. Circulating Neuropeptide Y May Be a Biomarker for Diagnosing Atrial Fibrillation. *Cardiology* **2023**, *148*, 593–595. [CrossRef]
54. Pink, C.; Holtfreter, B.; Völzke, H.; Nauck, M.; Dörr, M.; Kocher, T. Periodontitis and systemic inflammation as independent and interacting risk factors for mortality: Evidence from a prospective cohort study. *BMC Med.* **2023**, *21*, 430, Correction in *BMC Med.* **2025**, *23*, 16. [CrossRef]
55. Herrera, D.; Sanz, M.; Shapira, L.; Brotons, C.; Chapple, I.; Frese, T.; Graziani, F.; Hobbs, F.D.R.; Huck, O.; Hummers, E.; et al. Periodontal diseases and cardiovascular diseases, diabetes, and respiratory diseases: Summary of the consensus report by the European Federation of Periodontology and WONCA Europe. *Eur. J. Gen. Pract.* **2024**, *30*, 2320120. [CrossRef]
56. Martínez-García, M.; Hernández-Lemus, E. Periodontal Inflammation and Systemic Diseases: An Overview. *Front. Physiol.* **2021**, *12*, 709438. [CrossRef] [PubMed]
57. Cuenco, J.; Dalmas, E. Islet Inflammation and β Cell Dysfunction in Type 2 Diabetes. *Handb. Exp. Pharmacol.* **2022**, *274*, 227–251. [CrossRef] [PubMed]
58. Chen, J.; Fei, S.; Chan, L.W.C.; Gan, X.; Shao, B.; Jiang, H.; Li, S.; Kuang, P.; Liu, X.; Yang, S. Inflammatory signaling pathways in pancreatic β-cell: New insights into type 2 diabetes pathogenesis. *Pharmacol. Res.* **2025**, *216*, 107776. [CrossRef] [PubMed]
59. Easter, Q.T.; Fernandes Matuck, B.; Beldorati Stark, G.; Worth, C.L.; Predeus, A.V.; Fremin, B.; Huynh, K.; Ranganathan, V.; Ren, Z.; Pereira, D.; et al. Single-cell and spatially resolved interactomics of tooth-associated keratinocytes in periodontitis. *Nat. Commun.* **2024**, *15*, 5016. [CrossRef]
60. Li, H.; Sun, L.; Wang, Y. Inhibition of LPS-induced NLRP3 inflammasome activation by stem cell-conditioned culture media in human gingival epithelial cells. *Mol. Med. Rep.* **2023**, *27*, 106. [CrossRef]
61. Didilescu, A.C.; Chinthamani, S.; Scannapieco, F.A.; Sharma, A. NLRP3 inflammasome activity and periodontal disease pathogenesis-A bidirectional relationship. *Oral Dis.* **2024**, *30*, 4069–4077. [CrossRef]
62. Dhungana, G.; Srisai, D.; Sampath, C.; Soliman, J.; Kelly, R.M.; Saleh, H.Y.; Sedik, A.; Raynes, E.; Ferguson, A.; Alluri, L.S.C.; et al. Unveiling the Molecular Crosstalk Between Periodontal and Cardiovascular Diseases: A Systematic Review. *Dent. J.* **2025**, *13*, 98. [CrossRef]
63. Wang, Z.; Bao, J.; Yang, Y.; Wei, Y.; Li, C.; Pan, Y.; Chen, L. The Impact of Periodontitis on Cardiovascular Disease: Mechanisms, Evidence, and Therapeutic Implications. *Int. J. Dent.* **2025**, *2025*, 3694736. [CrossRef] [PubMed]
64. Perez, M.V.; Mahaffey, K.W.; Hedlin, H.; Rumsfeld, J.S.; Garcia, A.; Ferris, T.; Balasubramanian, V.; Russo, A.M.; Rajmane, A.; Cheung, L.; et al. Large-Scale Assessment of a Smartwatch to Identify Atrial Fibrillation. *N. Engl. J. Med.* **2019**, *381*, 1909–1917. [CrossRef] [PubMed]
65. Angjelova, A.; Jovanova, E.; Polizzi, A.; Laganà, L.; Santonocito, S.; Ragusa, R.; Isola, G. Impact of Periodontitis on Endothelial Risk Dysfunction and Oxidative Stress Improvement in Patients with Cardiovascular Disease. *J. Clin. Med.* **2024**, *13*, 3781. [CrossRef] [PubMed]
66. Corbi, S.C.T.; de Vasconcellos, J.F.; Bastos, A.S.; Bussaneli, D.G.; da Silva, B.R.; Santos, R.A.; Takahashi, C.S.; Rocha, C.d.S.; Carvalho, B.d.S.; Maurer-Morelli, C.V.; et al. Circulating lymphocytes and monocytes transcriptomic analysis of patients with type 2 diabetes mellitus, dyslipidemia and periodontitis. *Sci. Rep.* **2020**, *10*, 8145. [CrossRef]
67. Peña-Durán, E.; García-Galindo, J.J.; López-Murillo, L.D.; Huerta-Huerta, A.; Balleza-Alejandri, L.R.; Beltrán-Ramírez, A.; Anaya-Ambriz, E.J.; Suárez-Rico, D.O. Microbiota and Inflammatory Markers: A Review of Their Interplay, Clinical Implications, and Metabolic Disorders. *Int. J. Mol. Sci.* **2025**, *26*, 1773. [CrossRef]
68. Zhao, Y.; Shen, Q.Q. Link between periodontitis and diabetic retinopathy: Inflammatory pathways and clinical implications. *World J. Diabetes* **2024**, *15*, 1842–1846. [CrossRef]
69. Kang, J.; Lee, H.; Joo, J.Y.; Song, J.M.; Kim, H.J.; Kim, Y.H.; Park, H.R. Comparison of genetic and epigenetic profiles of periodontitis according to the presence of type 2 diabetes. *MedComm* **2024**, *5*, e620. [CrossRef]
70. Thouvenot, K.; Turpin, T.; Taïlé, J.; Clément, K.; Meilhac, O.; Gonthier, M.P. Links between Insulin Resistance and Periodontal Bacteria: Insights on Molecular Players and Therapeutic Potential of Polyphenols. *Biomolecules* **2022**, *12*, 378. [CrossRef]
71. Patel, S.M.; Kang, Y.M.; Im, K.; Neuen, B.L.; Anker, S.D.; Bhatt, D.L.; Butler, J.; Cherney, D.Z.I.; Claggett, B.L.; Fletcher, R.A.; et al. Sodium-Glucose Cotransporter-2 Inhibitors and Major Adverse Cardiovascular Outcomes: A SMART-C Collaborative Meta-Analysis. *Circulation* **2024**, *149*, 1789–1801. [CrossRef] [PubMed]

72. Usman, M.S.; Siddiqi, T.J.; Anker, S.D.; Bakris, G.L.; Bhatt, D.L.; Filippatos, G.; Fonarow, G.C.; Greene, S.J.; Januzzi, J.L., Jr.; Khan, M.S.; et al. Effect of SGLT2 Inhibitors on Cardiovascular Outcomes Across Various Patient Populations. *J. Am. Coll. Cardiol.* **2023**, *81*, 2377–2387. [CrossRef]
73. Țica, O.; Țica, O. Anemia in Heart Failure: Diagnostic Insights and Management Patterns Across Ejection Fraction Phenotypes. *Diagnostics* **2025**, *15*, 2079. [CrossRef]
74. Apperloo, E.M.; Neuen, B.L.; Fletcher, R.A.; Jongs, N.; Anker, S.D.; Bhatt, D.L.; Butler, J.; Cherney, D.Z.I.; Herrington, W.G.; Inzucchi, S.E.; et al. Efficacy and safety of SGLT2 inhibitors with and without glucagon-like peptide 1 receptor agonists: A SMART-C collaborative meta-analysis of randomised controlled trials. *Lancet Diabetes Endocrinol.* **2024**, *12*, 545–557. [CrossRef]
75. Rivera, F.B.; Cruz, L.L.A.; Magalong, J.V.; Ruyeras, J.; Aparece, J.P.; Bantayan, N.R.B.; Lara-Breitinger, K.; Gulati, M. Cardiovascular and renal outcomes of glucagon-like peptide 1 receptor agonists among patients with and without type 2 diabetes mellitus: A meta-analysis of randomized placebo-controlled trials. *Am. J. Prev. Cardiol.* **2024**, *18*, 100679. [CrossRef]
76. Giugliano, D.; Longo, M.; Scappaticcio, L.; Bellastella, G.; Maiorino, M.I.; Esposito, K. SGLT-2 inhibitors and cardiorenal outcomes in patients with or without type 2 diabetes: A meta-analysis of 11 CVOTs. *Cardiovasc. Diabetol.* **2021**, *20*, 236. [CrossRef]
77. Zemedikun, D.T.; Chandan, J.S.; Raindi, D.; Rajgor, A.D.; Gokhale, K.M.; Thomas, T.; Falahee, M.; De Pablo, P.; Lord, J.M.; Raza, K.; et al. Burden of chronic diseases associated with periodontal diseases: A retrospective cohort study using UK primary care data. *BMJ Open* **2021**, *11*, e048296. [CrossRef] [PubMed]
78. Amersfoort, J.; Eelen, G.; Carmeliet, P. Immunomodulation by endothelial cells-partnering up with the immune system? *Nat. Rev. Immunol.* **2022**, *22*, 576–588. [CrossRef]
79. Luppi, S.; Aldegheri, L.; Azzalini, E.; Pacetti, E.; Barucca Sebastiani, G.; Fabiani, C.; Robino, A.; Comar, M. Unravelling the Role of Gut and Oral Microbiota in the Pediatric Population with Type 1 Diabetes Mellitus. *Int. J. Mol. Sci.* **2024**, *25*, 10611. [CrossRef] [PubMed]
80. Lafleur, S.; Bodein, A.; Mbuya Malaïka Mutombo, J.; Mathieu, A.; Joly Beauparlant, C.; Minne, X.; Chandad, F.; Droit, A.; Houde, V.P. Multi-Omics Data Integration Reveals Key Variables Contributing to Subgingival Microbiome Dysbiosis-Induced Inflammatory Response in a Hyperglycemic Microenvironment. *Int. J. Mol. Sci.* **2023**, *24*, 8832. [CrossRef]
81. Henein, M.Y.; Vancheri, S.; Longo, G.; Vancheri, F. The Role of Inflammation in Cardiovascular Disease. *Int. J. Mol. Sci.* **2022**, *23*, 12906. [CrossRef]
82. Țica, O.; Țica, O. Molecular Diagnostics in Heart Failure: From Biomarkers to Personalized Medicine. *Diagnostics* **2025**, *15*, 1807. [CrossRef]
83. Borgnakke, W.S. Current scientific evidence for why periodontitis should be included in diabetes management. *Front. Clin. Diabetes Healthc.* **2023**, *4*, 1257087. [CrossRef]
84. Wang, X.; Mobley, A.R.; Tica, O.; Okoth, K.; Ghosh, R.E.; Myles, P.; Williams, T.; Haynes, S.; Nirantharakumar, K.; Shukla, D.; et al. Systematic approach to outcome assessment from coded electronic healthcare records in the DaRe2THINK NHS-embedded randomized trial. *Eur. Heart J. Digit. Health* **2022**, *3*, 426–436. [CrossRef]
85. West, N.X.; Gormley, A.; Pollard, A.J.; Izzetti, R.; Marruganti, C.; Graziani, F. Evaluating the Performance and Implementation of the 2018 Classification of Periodontal Diseases: A Systematic Review and Survey. *J. Clin. Periodontol.* **2025**, *52*, 34–57. [CrossRef]
86. Wu, L.; Huang, C.M.; Wang, Q.; Wei, J.; Xie, L.; Hu, C.Y. Burden of severe periodontitis: New insights based on a systematic analysis from the Global Burden of Disease Study 2021. *BMC Oral Health* **2025**, *25*, 861. [CrossRef] [PubMed]
87. Chaudhry, M.; Kumar, M.; Singhal, V.; Srinivasan, B. Metabolic health tracking using Ultrahuman M1 continuous glucose monitoring platform in non- and pre-diabetic Indians: A multi-armed observational study. *Sci. Rep.* **2024**, *14*, 6490. [CrossRef] [PubMed]
88. Fang, J.; Wu, Y.; Wang, H.; Zhang, J.; You, L. Oral Health and Diabetic Cardiomyopathy: Mechanisms, Biomarkers, and Early Screening Approaches. *J. Inflamm. Res.* **2025**, *18*, 8689–8704. [CrossRef] [PubMed]
89. Weerts, J.; Țica, O.; Aranyo, J.; Basile, C.; Borizanova-Petkova, A.; Borovac, J.A.; Camilli, M.; Eichenlaub, M.; Fiori, E.; Van Loon, T.; et al. Atrial cardiomyopathy: From healthy atria to atrial failure. A clinical consensus statement of the Heart Failure Association of the ESC. *Eur. J. Heart Fail.* **2025**. [CrossRef]
90. Tattar, R.; da Costa, B.D.C.; Neves, V.C.M. The interrelationship between periodontal disease and systemic health. *Br. Dent. J.* **2025**, *239*, 103–108. [CrossRef]
91. Martínez Nieto, M.; De Leon Rodríguez, M.L.; Anaya Macias, R.D.C.; Lomelí Martínez, S.M. Periodontitis and chronic kidney disease: A bidirectional relationship based on inflammation and oxidative stress. *World J. Clin. Cases* **2024**, *12*, 6775–6781. [CrossRef] [PubMed]
92. Karwath, A.; Bunting, K.V.; Gill, S.K.; Tica, O.; Pendleton, S.; Aziz, F.; Barsky, A.D.; Chernbumroong, S.; Duan, J.; Mobley, A.R.; et al. Redefining β-blocker response in heart failure patients with sinus rhythm and atrial fibrillation: A machine learning cluster analysis. *Lancet* **2021**, *398*, 1427–1435. [CrossRef]
93. Li, Y.; Liu, Y.; Liu, S.; Gao, M.; Wang, W.; Chen, K.; Huang, L.; Liu, Y. Diabetic vascular diseases: Molecular mechanisms and therapeutic strategies. *Signal Transduct. Target. Ther.* **2023**, *8*, 152. [CrossRef]

94. Sanz, M.; Del Castillo, A.M.; Jepsen, S.; Gonzalez-Juanatey, J.R.; D'Aiuto, F.; Bouchard, P.; Chapple, I.; Dietrich, T.; Gotsman, I.; Graziani, F.; et al. Periodontitis and Cardiovascular Diseases. Consensus Report. *Glob. Heart* **2020**, *15*, 1. [CrossRef]
95. Bunting, K.V.; Stanbury, M.; Tica, O.; Kotecha, D. Transforming clinical research by involving and empowering patients- the RATE-AF randomized trial. *Eur. Heart J.* **2021**, *42*, 2411–2414. [CrossRef] [PubMed]
96. Patel, M.R.; Armstrong, P.W.; Bhatt, D.L.; Braunwald, E.; Camm, A.J.; Fox, K.A.; Harrington, R.A.; Hiatt, W.R.; James, S.K.; Kirtane, A.J.; et al. Sharing Data from Cardiovascular Clinical Trials—A Proposal. *N. Engl. J. Med.* **2016**, *375*, 407–409. [CrossRef] [PubMed]
97. Inoue, M.; Sakanaka, A.; Katakami, N.; Furuno, M.; Nishizawa, H.; Omori, K.; Taya, N.; Ishikawa, A.; Mayumi, S.; Tanaka Isomura, E.; et al. Periodontal tissue susceptibility to glycaemic control in type 2 diabetes. *Diabetes Obes. Metab.* **2024**, *26*, 4684–4693. [CrossRef]
98. Langeveld, M.; Balfoort, B.M.; Wortmann, S.B. SGLT2 inhibitors for periodontitis in glycogen storage disease. *Mol. Genet. Metab.* **2025**, *144*, 109057. [CrossRef] [PubMed]
99. Pacheco-Yanes, J.; Reynolds, E.; Li, J.; Mariño, E. Microbiome-targeted interventions for the control of oral-gut dysbiosis and chronic systemic inflammation. *Trends Mol. Med.* **2023**, *29*, 912–925. [CrossRef]
100. Walter, E.; Brock, T.; Lahoud, P.; Werner, N.; Czaja, F.; Tichy, A.; Bumm, C.; Bender, A.; Castro, A.; Teughels, W.; et al. Predictive modeling for step II therapy response in periodontitis-model development and validation. *npj Digit. Med.* **2025**, *8*, 445. [CrossRef]
101. Rebeiz, T.; Lawand, G.; Martin, W.; Gonzaga, L.; Revilla-León, M.; Khalaf, S.; Megarbané, J.M. Development of an artificial intelligence model for assisting periodontal therapy decision-making: A retrospective longitudinal cohort study. *J. Dent.* **2025**, *159*, 105780. [CrossRef] [PubMed]
102. Sarakbi, R.M.; Varma, S.R.; Muthiah Annamma, L.; Sivaswamy, V. Implications of artificial intelligence in periodontal treatment maintenance: A scoping review. *Front. Oral Health* **2025**, *6*, 1561128. [CrossRef] [PubMed]
103. Nymberg, P.; Milos-Nymberg, V.; Grundberg, A.; Oscarson, N.; Stenman, E.; Sundquist, K. Exploring the link between self-rated poor oral health and cardiovascular risk: A cross-sectional study using SCORE2. *BMC Oral Health* **2025**, *25*, 298. [CrossRef] [PubMed]

*biomedicines*

MDPI

*Review*

# Tertiary Hyperparathyroidism in Diabetic Nephropathy: An Underrecognized Complication—A Narrative Review

**Mirona Costea [1,2], Dana-Mihaela Tilici [1,2,*], Diana Loreta Paun [2,3], Vanda Roxana Nimigean [3,4], Sorin Constantin Paun [3,5] and Rucsandra Elena Danciulescu-Miulescu [3,6]**

1 Doctoral School, University of Medicine and Pharmacy "Carol Davila", 020021 Bucharest, Romania; mirona.costea@drd.umfcd.ro
2 Endocrinology Department, Bucharest University Emergency Hospital, 050098 Bucharest, Romania; diana.paun@umfcd.ro
3 Faculty of Medicine, University of Medicine and Pharmacy "Carol Davila", 020021 Bucharest, Romania; vanda.nimigian@umfcd.ro (V.R.N.); sorin.paun@umfcd.ro (S.C.P.); rucsandra.danciulescu@umfcd.ro (R.E.D.-M.)
4 Department of Oral Rehabilitation, Faculty of Dentistry, University of Medicine and Pharmacy "Carol Davila", 020021 Bucharest, Romania
5 Bucharest Emergency Clinical Hospital, Calea Floreasca 8, 014461 Bucharest, Romania
6 Department of Diabetes, Nutrition and Metabolic Diseases, "Carol Davila" University of Medicine and Pharmacy, 020021 Bucharest, Romania
* Correspondence: dana-mihaela.tilici@drd.umfcd.ro

**Abstract**

Tertiary hyperparathyroidism (THPT) arises in patients with chronic kidney disease (CKD) as a consequence of prolonged secondary hyperparathyroidism and is marked by autonomous parathyroid hormone (PTH) secretion. In some cases, parathyroid hyperplasia persists even after successful renal transplantation, resulting in sustained PTH elevation and hypercalcaemia. These alterations contribute to bone loss, vascular calcification, and increased cardiovascular risk. Management includes medical therapy with calcimimetics or vitamin D analogues and surgical intervention via parathyroidectomy. However, optimal timing and treatment strategy remain uncertain. This review examines the pathophysiology, clinical manifestations, and current management paradigms of THPT, with an emphasis on areas that require further research and consensus.

**Keywords:** parathyroid hormone (PTH); tertiary hyperparathyroidism (THPT); chronic kidney disease (CKD); diabetic nephropathy

## 1. Introduction

Tertiary hyperparathyroidism (THPT) is a condition mainly seen in patients with end-stage renal disease (ESRD), especially after prolonged secondary hyperparathyroidism (SHPT). It involves autonomous parathormone (PTH) secretion and ongoing hypercalcemia even after the initial cause is addressed—usually following successful kidney transplantation. THPT signifies a failure of the parathyroid glands to return to normal function, often due to parathyroid hyperplasia or nodular changes caused by chronic stimulation during SHPT.

Diabetic nephropathy, a microvascular complication of diabetes mellitus, has become the leading cause of ESRD globally. The increasing prevalence of diabetes and its renal complications has significantly shifted the epidemiological landscape of chronic kidney disease (CKD) and related mineral and bone disorders. Patients with diabetic nephropathy


https://doi.org/10.3390/biomedicines13112604

experience a combination of metabolic derangements that affect calcium, phosphate, vitamin D, and PTH homeostasis. These patients often develop SHPT earlier and may have a more rapid progression toward tertiary hyperparathyroidism due to more severe and persistent alterations in mineral metabolism.

Despite the well-recognized association between CKD and mineral bone disease, the specific incidence and clinical trajectory of THPT in individuals with diabetic nephropathy remain under-investigated. It is unclear whether diabetic nephropathy confers an independent risk for the development of THPT beyond its role as a cause of CKD. Some studies suggest that diabetic patients may have a blunted PTH response in early CKD stages, while others report a higher prevalence of parathyroid hyperplasia in diabetic ESRD populations. Furthermore, chronic inflammation, oxidative stress, and insulin resistance—hallmarks of diabetes—may further exacerbate parathyroid gland dysfunction.

The management of THPT becomes particularly challenging in diabetic individuals, who commonly present with comorbidities, vascular calcifications, and elevated cardiovascular risk, necessitating a personalized approach to medical versus surgical treatment. Parathyroidectomy, while often curative, may be associated with increased perioperative risk in patients with diabetes and cardiovascular disease. On the other hand, medical therapies such as calcimimetics may be less effective in cases of nodular or autonomous hyperplasia.

This narrative review aims to explore the incidence and clinical implications of THPT in patients with diabetic nephropathy, highlighting the unique pathophysiological mechanisms, risk factors, diagnostic challenges, and treatment considerations specific to this population. By synthesizing available evidence from clinical studies, guidelines, and expert opinions, we aim to delineate the scope of the problem and identify potential directions for future research and clinical practice improvement. A deeper understanding of how diabetic nephropathy influences the development and course of THPT may ultimately lead to better stratification of at-risk patients and optimization of therapeutic interventions.

## 2. Materials and Methods

Study design and search strategy: For this narrative review, a comprehensive literature search was conducted. We have used major medical databases, including PubMed and Google Scholar. The search primarily focused on articles published between 2019 and 2025. Keywords used in the search included "tertiary hyperparathyroidism", "diabetic nephropathy", and "chronic kidney disease". Manual screening of the bibliographies of included studies and clinical guidelines was performed to uncover further relevant literature.

Eligibility criteria and study selection: We included clinical studies, guidelines, and systematic reviews. Priority was given to publications that addressed the pathophysiology, clinical presentation, diagnosis, and management. Exclusion criteria: animal/in vitro studies, non-relevant reports, pediatric population. Single case reports were also excluded.

Data extraction and synthesis: Relevant data, including study design, diagnostic definitions, clinical findings, and treatment modalities, were systematically extracted. The data were qualitatively synthesized and interpreted in the context of existing guideline recommendations. No meta-analysis was performed.

## 3. Diabetic Nephropathy and THPT–Pathophysiological Mechanisms

Diabetes mellitus, a chronic metabolic disorder marked by hyperglycemia, is associated with a wide range of systemic complications, among which renal involvement represents a major cause of morbidity and mortality. Current projections estimate that over

12% of the global population will be affected by diabetes by 2045, with type 2 diabetes accounting for the vast majority of the cases [1–3].

The natural evolution of diabetes mellitus is characterized by numerous microvascular complications, including diabetic nephropathy. This is the most common complication for type 2 diabetes mellitus and has become the main cause of ESRD globally [4]. Over 25–40% of patients with diabetes mellitus can experience renal damage after 20 years of disease [3].

Diabetic nephropathy is characterized by persistent albuminuria, measured at least twice within 3–6 months, and a decreased glomerular filtration rate (GFR). The ultimate result of nephropathy is ESRD [3,5]. Usually, the decreased GFR occurs in association with hypertension, accelerating the progression of renal deterioration [4].

It is important for clinicians to detect as early as possible the presence of diabetic nephropathy. Various biomarkers have been investigated, with albuminuria generally chosen as the primary tool for assessing renal function. The detection of albumin in the urine reflects the increased glomerular permeability to macromolecules. However, this is not a perfect way of screening because albuminuria has limitations—it is not sensitive to diabetic nephropathy, may not be present in the early stages, and it has larger variability [4]. The gold standard method is a kidney biopsy, an invasive procedure only available in specialized centers. The future screening method is expected to be more specific, like urinary microRNA markers [6].

The pathogenetic mechanisms of diabetic nephropathy are extremely complex and need further investigation. The well-described mechanisms of developing diabetic nephropathy are the accumulation of extracellular matrix, the vascular thickening and hyalinization, and the perturbations of intraglomerular hemodynamics. The intrinsic mechanisms responsible for these changes include advanced glycation products, activation of protein kinase C, and reactive oxygen species. High-glucose flux leads to advanced glycation end products (AGEs). Their presence plays a major role in the activation of the renin-angiotensin system (RAS). Increased renin secretion due to alterations in the juxtaglomerular apparatus ultimately leads to the production of angiotensin II. This molecule is responsible for adrenal aldosterone secretion, promoting sodium retention. As a result, arterial hypertension develops, exacerbating renal damage [5].

The switch between microalbuminuria, defined by 30–300 mg/day, and macroalbuminuria, defined by more than 300 mg/day, represents a critical stage for the evolution of diabetic nephropathy. This is a marker of severe renal dysfunction and represents a critical stage of the disease. It is accompanied by glomerular filtration rate deterioration and represents a sign of ESRD becoming imminent [7].

The long evolution of kidney disease frequently triggers additional metabolic disturbances, among which phosphate retention is particularly significant. Fibroblast Growth Factor 23 (FGF23) and its cofactor Klotho are important key regulators in phosphate metabolism. While the renal dysfunction continues, the level of FGF23 rises. FGF23 is the product of bone cells—osteocytes and osteoblasts and is responsible for decreasing plasma phosphate. This process is the result of reducing tubular phosphate reabsorption in a similar way to PTH. Also, FGF23 secretion decreased the renal synthesis of calcitriol, playing a crucial role in developing hypocalcemia. The underlying mechanism for FGF23 to activate its receptors is the presence of Klotho, a cofactor of the process. FGF23 alone cannot express its actions, as the activation of its receptors needs Klotho. Despite advances in understanding, the precise mechanisms governing the FGF23-Klotho axis remain incompletely elucidated and warrant further investigation [8,9].

In CKD patients, hyperphosphatemia associated with calcitriol deficiency leads to hypocalcemia, which in turn stimulates PTH secretion, ultimately resulting in secondary hyperparathyroidism. This form of hyperparathyroidism is characterized by adequate

PTH secretation due to hypocalcemia. In the parathyroid glands, oxidative stress has been shown to disrupt calcium-sensing receptor signaling and the parathyroid cell proliferation control mechanism. Inflammatory cytokines can also modulate PTH gene expression and secretion, potentially accelerating the onset of secondary hyperparathyroidism. This highlights a more direct pathophysiological link between diabetes-related inflammation and parathyroid dysfunction, beyond the traditional CKD-mediated pathway [10].

Secondary hyperparathyroidism in the context of CKD often represents a transitional state. With disease progression, PTH secretion becomes persistent and dysregulated. This advanced stage is defined as THPT, characterized by autonomous parathyroid gland function, hypercalcemia, and progressive enlargement of all four parathyroid glands—Figure 1 [10].

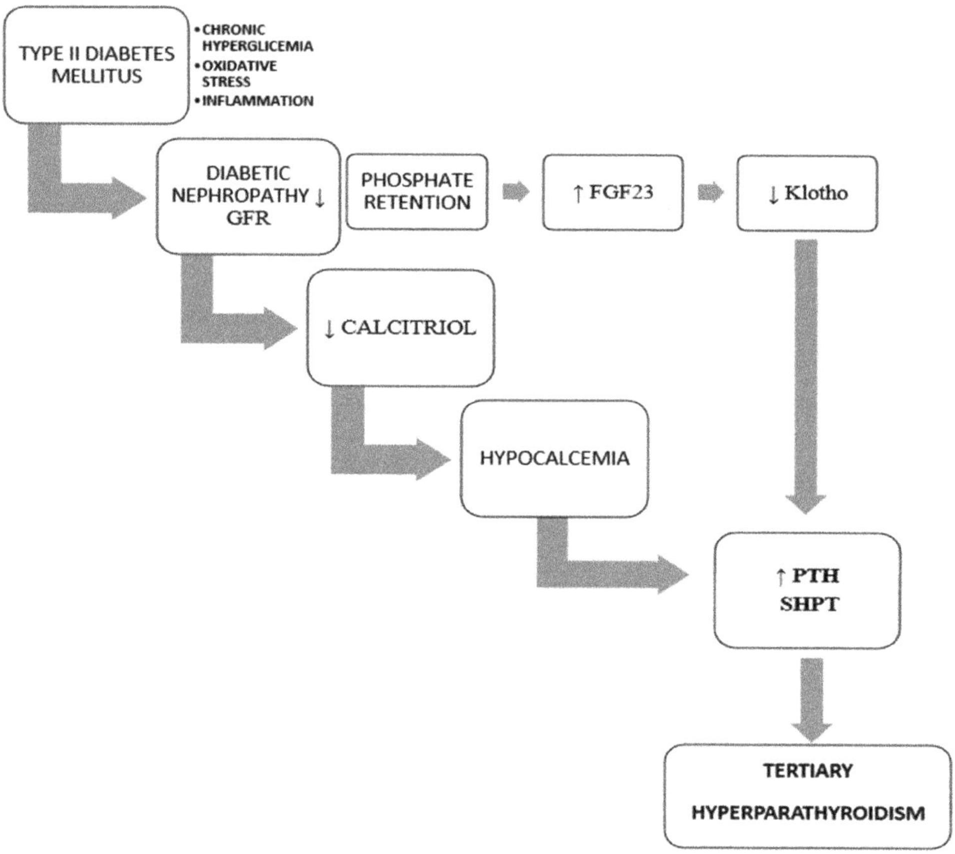


**Figure 1.** Pathophysiological Mechanism Leading to THP.

The development of TPTH is multifactorial and primarily associated with CKD progression. The risk is considered to be maximal once the GFR drops below 45 mL/min/1.73 m$^2$, reflecting significant loss of renal function and disrupted calcium-phosphate homeostasis. Several clinical and biochemical risk factors have been associated with the progression of TPHT. These are summarized in Table 1 below [11].

**Table 1.** Summary of risk factors associated with TPTH.

| Risk Factor | Mechanism |
| --- | --- |
| eGFR < 45 mL/min/ 1.73 m$^2$ | Reduced phosphate excretion, vitamin D deficiency |
| Young age | Higher bone turnover and sensitivity to mineral imbalance |
| Male sex | Possibly linked to differences in skeletal and hormonal profiles |
| Hypoalbuminemia | Affects calcium binding and PTH regulation |
| Use of diuretics | May increase calcium excretion and contribute to chronic PTH stimulation |
| Long dialysis duration | Prolonged SPTH |
| Poor phosphate control | Hyperphosphatemia promotes PTH secretion |
| Vitamin D deficiency | Leads to hypocalcemia and compensatory PTH increase |

eGFR = estimated glomerular filtration rate; PTH = parathormone; SPTH = secondary hyperparatyroidism.

Given the overlapping clinical and biochemical features, distinguishing between secondary and tertiary hyperparathyroidism can be challenging. To aid clinicians in this differentiation, the following table, Table 2 summarizes the key distinguishing characteristics of these two entities [8].

**Table 2.** Distinguishing SHPT and THPT.

| | SHPT | THPT |
| --- | --- | --- |
| Biochemical Assessment | ↑ PTH<br>N/↓ Calcium | ↑ PTH<br>↑ Calcium |
| Clinical Context | Low CKD stage | Renal transplantation history, chronic dialysis |
| Evaluate response to calcimimetics and vitamin D analogs | Responsive | Poor response |
| Imaging | No visible parathyroid adenoma | Parathyroid adenoma |
| Management | Monitor biochemical markers<br>Optimize medical therapy | Optimize medical therapy<br>Refer to surgery |

N = Normal Level; SHPT = Secondary hyperparathyroidism; THPT = Tertiary hyperparathyroidism; PTH = parathormone.

PTH actions are modulated by the Calcium Sensing Receptor (CaSR), localized at parathyroid and kidney levels. It is capable of detecting slight variations in ionized calcium levels and stimulating PTH secretation when the ionized calcium level is low. PTH is a hypercalcemic hormone that acts on various organs—bone, kidney, and gastrointestinal tract. Via PTH Receptor 1, PTH increases tubular absorption of calcium, decreases phosphate concentration, and stimulates the activation of vitamin D. PTH also applies its actions on the bone and intestines, leading to calcium absorption [12].

THPT is a condition that affects the patient's quality of life, in addition to the renal disease [13]. It is vital to have a multidisciplinary approach in managing THPT [7].

## 4. THPT and Kidney Transplant

The role of kidney transplantation in managing end-stage renal disease (ESRD) should be emphasized. The majority of patients with THPT are kidney transplant recipients who, despite restoration of renal function and interruption of the initial stimulus for PTH secretion, continue to exhibit elevated PTH concentrations. This persistence indicates autonomous parathyroid secretion, thereby characterizing TPTH [14].

After a successful kidney transplant, PTH and FGF23 levels should drop off in the first 3 months [15,16]. High levels of PTH after this period are responsible for hypercalcemia and hypophosphatemia and define THPT. Well-known risk factors for developing THPT after transplant are high PTH values before transplant, long-time dialyzed patients, and nodular parathyroid hyperplasia. Vitamin D in its active form also plays a role in the calcium and PTH axes. It is crucial to measure the levels of 25-OH-Vitamin D, as the deficiency of this metabolite may contribute to high PTH levels [16].

In a cohort study where 849 patients were followed 1 year after a kidney transplant, 21.5% of them developed THPT, defined by elevated PTH levels (>70 pg/mL) and hypercalcemia (>10 mg/dL). Pre-transplant PTH levels greater than 300 pg/mL were established to be a risk factor for developing TPHT. The study shows that CKD is a complex disease, and its effects can persist long after a renal transplant. Clinicians must be aware of this fact and ensure continuous patient monitoring [17].

Other data suggest even a greater percentage of patients suffering from THPT after renal transplant, up to 50% [18].

It is important to screen, recognize, and treat THPT, as it is a condition associated with renal allograft dysfunction and chronic allograft nephropathy [19]. THPT is also a risk factor for graft failure and ultimately death of the patience [20].

There is an urgent need to standardize the diagnostic approach. We must take into account biochemic findings—including high PTH, bone-specific alkaline phosphatase, vitamin D, and calcium levels– the preferred formula involves using the corrected calcium-to-albumin ratio [21].

To distinguish residual secondary HPTH from autonomous TPTH post-transplant, clinicians should consider the following practical criteria:

- Time course: persistent PTH elevation beyond 6–12 months post-transplant may suggest tertiary hyperparathyroidism
- Calcium levels: hypercalcemia supports the diagnosis of TPTH; in contrast, residual SPTH is typically associated with normocalcemia or even hypocalcemia
- Phosphate levels: hypophosphatemia is more common in TPTH due to excessive PTH secretion
- PTH response to treatment: lack of improvement with correction of vitamin D deficiency and withdrawal of calcimimetics or phosphate binders may indicate autonomous secretion

Applying their criteria, clinicians can appropriately guide further diagnostic steps [11,14,16].

## 5. Clinical Manifestations

Chronic hypercalcemia presents with a wide spectrum of clinical manifestations. It can vary from no symptoms to severe bone affection. Non-specific symptoms include abdominal pain, headache, fatigue, irritability, and weakness. Pruritus may also be present in some cases. Bone disease may take the form of severe osteoarthritis, known as renal osteopathy [22]. Skeletal anomalies are found in the majority of stage 5 CKD patients [15].

The gold standard for diagnosis and assessment of renal osteodystrophy (ROD) remains bone biopsy, which allows histomorphometric analysis. The TMV classification (bone turnover, mineralization, and volume) categorizes ROD into osteitis fibrosa, adynamic bone disease, osteomalacia, and mixed uremic osteodystrophy [15]. Despite being the gold standard, bone biopsy is infrequently utilized in clinical practice due to its invasive nature and substantial cost. As a result, non-invasive imaging modalities such as 18 F-sodium fluoride positron emission tomography (18F-NaFPET) have been investigated and demonstrated strong correlations with histomorphometric indices. This investigation offers high diagnostic accuracy in the classification of ROD subtypes [23]. Unfortunately, this method is currently not available in Romania, limiting its clinical application in our country.

Osteitis fibrosa is the result of prolonged hyperparathyroidism [15,24]. PTH increases bone resorption and leads to negative bone balance. The osteoid formed under the elevated PTH status is not structurally normal, and it is called woven bone. The bones lose their physiological laminar disposition. Also, PTH is responsible for the appearance of peritrabecular fibrosis. In the end, cortical bone is thinned, and the laminar osteoid is replaced by fibrous tissue, accumulating cysts [24].

Adynamic bone disease is defined by low bone turnover and normal mineralization. It is the main bone affection among dialysis patients [15]. The activity of both osteoblasts and osteoclasts is severely reduced [25]. Adynamic bone disease manifests with bone pain and fracture. There are biochemical findings that are suggestive of adynamic bone disease: low levels of PTH, low bone-specific alkaline phosphatase, hypercalcemia, and vascular calcification, with the golden standard for diagnosis remaining bone biopsy. Bone turnover markers, such as bone-specific alkaline phosphatase, can aid in the assessment of bone metabolic activity [26].

Osteomalacia is often characterized by severe diffuse pain, in contrast to osteoporosis [25]. Osteomalacia is characterized by low bone turnover and abnormal mineralization. Its prevalence is now decreasing, as it was historically caused by the use of aluminum-based phosphate binders [15].

The metabolism of the skeletal system is different in CKD patients compared to the general population. Factors that contribute to the bone alteration in CKD patients are hemodialysis, peritoneal dialysis, or a history of kidney transplantation [24]. When discussing bone diseases, osteoporosis cannot be excluded. Osteoporosis is a condition characterized by low mineral density of the bones and is defined as a T-score below −2.5 SD. It is important to use DXA measurement for CKD patients, but we must consider the fact that it is not a helpful tool when wanting to capture the quality of the bone [24,27]. The bone mineral density is expected to be lower than in the general population, especially when measuring the density of the femoral neck and total hip. Factors involved in low mineral density are age, low body mass index, use of tobacco, and high PTH levels [24]. Although it has certain limitations, DXA remains a useful non-invasive tool for monitoring bone health in CKD patients. Bone quality can be measured in specific centers by using high-resolution quantitative peripheral computed tomography (HR-pQCT) [27].

It is well-known that fracture risk is significantly higher in patients with CKD than in the general population. Various studies have shown that the risk of fragility fractures is 5 times higher in patients with eGFR below 15 mL/min/1.73 $m^2$ compared to individuals with eGFR higher than 60 mL/min/1.73 $m^2$ [24]. Besides the fracture risk, bone disease is a well-known negative factor for developing cardiovascular comorbidities and mortality [15].

Chronic kidney-disease-related bone and mineral disease (CKD-MBD) is a complex clinical syndrome defined by a systemic disorder of mineral and bone metabolism due to CKD, which is manifested by abnormalities in bone and mineral metabolism and/or extraskeletal calcification. CKD-MBD arises as a result of the interaction among three distinct pathophysiological mechanisms: high PTH levels, abnormalities in bone metabolism, and diffuse soft tissue calcification. This may represent a potential background for patients to develop osteoporosis. Patients who develop this clinical entity are at high risk for fragility fractures, especially hip fractures [28]. The clinical findings include, besides bone affection, vascular calcification, neurological occurrences—dementia, cognitive decline, small vessel abnormalities, gastrointestinal events—constipation, liver inflammation, microbiome affection, infections, and malnutrition [29]. Several studies suggest that the development of CKD-MBD may be driven by a complex interaction between the immune system and the microbiome of the patients. This interplay serves as a foundation for ongoing research and the development of new therapeutic strategies [30].

While bone pain may be present in both diabetic neuropathy and TPTH, the latter is often associated with radiographic evidence of bone resorption, which is not characteristic of diabetic complications. Hypercalcemia in TPTH is typically persistent and biochemically distinct, whereas hypercalcemia is not a usual finding in diabetic neuropathy or nephropathy. Elevated PTH levels, in conjunction with hypercalcemia and imaging findings, support the diagnosis of TPTH [15].

Another important clinical aspect is related to the cardiovascular pathology of CKD patients. There are several risk factors for developing cardiovascular disease. They can be related to diabetes mellitus, such as insulin resistance, dyslipidemia, arterial stiffness, and arterial hypertension. Also, there are CKD-related mechanisms that can lead to cardiovascular injury due to the accumulation of uremic toxins, sodium retention, hyperphosphatemia, vascular calcifications, and elevated levels of renin and aldosterone [31]. Atherosclerosis is another important feature of this group of patients, and this can lead to severe events—stroke, coronary artery disease, or transient ischemic attack [21].

Soft tissue complications can also occur, with calciphylaxis (calcific uremic arteriolopathy) being the most severe manifestation. This affects small vessels, and it may trigger thrombosis, non-healing wounds, secondary infections, and skin necrosis, a potentially lethal condition [21].

## 6. Management Strategies

### 6.1. Medical Therapies

The optimal therapeutic approach varies according to the stage of the disease. If the patient is being carefully watched, the phase of hypocalcemia and hyperphosphatemia that will later trigger the secretion of PTH needs immediate and specific intervention [21]. Phosphorus levels typically increase when GFR goes under 30 mL/min/1.73 $m^2$ [32].

Therapeutic options for CKD-MBD include vitamin D sterols, active vitamin D analogs, calcimimetics, and phosphate binders. Phosphate binders act by binding dietary phosphate in the gastrointestinal tract to prevent absorption and are classified as calcium-based or calcium-free. Calcium-based binders (calcium carbonate, calcium acetate) effectively lower serum phosphate but increase the risk of hypercalcemia, positive calcium balance, and vascular calcifications. Calcium-free binders (sevelamer hydrochloride/carbonate, lanthanum carbonate) are equally or slightly less effective but reduce the risk of hypercalcemia and vascular calcifications. However, they present specific adverse effects: sevelamer is associated with gastrointestinal intolerance, while lanthanum carbonate, though generally well tolerated, may lead to vomiting, tissue accumulation in bone and liver, and rare gastric mucosal deposition, with its long-term hepatic safety yet to be fully established [33].

The therapeutic response to agents such as calcimimetics, vitamin D analogues, and phosphate binders may offer important diagnostic clues in differentiating secondary from tertiary hyperparathyroidism. In secondary hyperparathyroidism, PTH levels usually respond to medical management—especially with calcimimetics, resulting in reduced PTH secretion. In contrast, THPT is characterized by autonomous PTH secretetation, and patients often exhibit a minimal response to this therapy. Persistently elevated PTH and hypercalcemia despite optimal medical management may suggest the transition to THPT, thereby guiding consideration for surgical intervention. Recognizing these response patterns can aid clinicians in timely diagnosis and management [8,21].

The last line of therapy is dialysis, an efficient way to reduce phosphate, but difficult for the patient [21].

Sevelamer carbonate is a calcium-free agent designed to lower the levels of phosphorus. Besides this, it is also capable of binding bile acids and resulting in reduced LDL-cholesterol and preventing bone loss [32,34]. The target group for the use of this drug is patients who require dialysis or patients with phosphate levels higher than 1.78 mmol/L who do not require dialysis. The guidelines suggest a starting dose of 0.8–1.6 g/day, administered three times a day with meals. The typical adverse effects were related to the gastrointestinal tract—vomiting, nausea, dyspepsia, diarrhea, or constipation [34,35].

Although there are many therapies on the market, the perfect chelator has not yet been found—it should bind excess phosphorus in the gastrointestinal tract but without adverse effects and be cost-effective [32].

The involvement of cardiology is a critical component in the management of patients with CKD. Optimal pharmacological control of hypertension remains essential, with recommended target values below 120/80 mmHg. Anemia requires correction through iron supplementation or the administration of erythropoiesis-stimulating agents, aiming for hemoglobin concentrations of 10–12 g/dL in patients undergoing dialysis. Dyslipidemia also necessitates intervention, with therapeutic strategies including statins and sodium–glucose cotransporter-2 (SGLT2) inhibitors. Current guidelines additionally recommend the use of novel agents, such as monoclonal antibodies directed against dyslipidemia (e.g., alirocumab, evolocumab) and glucagon-like peptide-1 (GLP-1) receptor agonists [31]. Besides these facts, we must keep in mind that the risk of atrial fibrillation is higher in CKD patients and prevent embolic events by using anticoagulants [36].

### 6.2. Vitamin D Analogues

It was shown that 56% of the patients with stage 5 CKD lack vitamin D, a secosteroid hormone [37]. Vitamin D analogs target their receptors and contribute to the suppression of PTH synthesis [38]. Active forms of vitamin D include calcitriol, paricalcitrol, 22-oxacalcitriol, and others [37]. Synthetic calcitriol analogs include its prodrug, named Alfacalcidol, the most powerful and fastest analog for treating vitamin D deficiency at the moment. There are more analogs currently being the subject of various studies, like Falecalcitriol, which is potentially more efficient in PTH suppression [39].

These medications have the role of decreasing PTH secretion, but it was more recently shown that they also reduce the activation of the renin-angiotensin axis and reduce inflammation [37,40]. There is a proven reduction in PTH levels as a result of vitamin D analogs, but unfortunately, there are no well-defined target levels of PTH [41].

Several studies have failed to demonstrate a clear reduction in systemic calcifications and, in some cases, have even reported a potential increase in vascular calcification risk [38,42]. On the other hand, certain findings suggest that these interventions may exert favorable effects on vascular function; however, the evidence remains insufficient [40]. The role of vitamin D analogs in vascular calcification remains controversial. These conflicting data likely stem from differences in study design, dosing regimens, and patient characteristics such as the presence of diabetes and pre-existing calcifications. Moreover, the lack of standardized endpoints complicates comparison across studies. Further research is required to establish consistent, reliable biochemical targets and to clarify whether improvements in vascular function can occur independently of, or despite, progression in vascular calcification. Current guidelines offer limited and sometimes conflicting recommendations [38,40,42].

### 6.3. Calcimimetics

Calcimimetics represent a key category of drugs that revolutionized treating HPTH and CKD.

Calcimimetics function as allosteric modulators of the CaSR, a G protein–coupled receptor that governs PTH synthesis and secretion in accordance with fluctuations in extracellular calcium within the parathyroid glands. Practically, they imitate serum calcium's effect on CaSR, inducing a decline in PTH secretation. They not only contribute to lowering PTH concentrations but also promote a reduction in parathyroid gland volume [38]. Regardless of whether the patient is receiving concomitant vitamin D therapy or not, calcimimetics managed to reduce calcium and phosphate levels of patients with HPTH undergoing dialysis [43].

Currently, two orally administered calcimimetics—cinacalcet, the first agent of this class to receive approval, and evocalcet—are available, alongside an intravenously administered formulation, etelcalcetide [38,44]. Each of these three agents has its own disadvantages, and the choice of therapy should be individualized based on clinical and biochemical characteristics of the patient [38].

Cinacalcet was shown to have an anabolic effect on the bone, promoting bone formation and serving as a key ally in combating adynamic bone disease [45]. Cinacalcet is the most effective PTH-suppressing agent in the prevention of vascular events and bone fractures [36]. Its side effects are usually minor, including nausea and vomiting, but hypocalcemia may also occur. This remains the most prescribed calcimetic agent [44].

Evocalcet, the more recent oral calcimimetic therapy, has promising results and may become the first-line oral therapy. It has been shown to have better gastrointestinal tolerability compared to cinacalcet, but the incidence of hypocalcemia was similar [46].

Etelcalcetide, a novel second-generation intravenous calcimimetic, allows for thrice-weekly dosing during hemodialysis and was developed to improve treatment efficacy, patient adherence, and minimize gastrointestinal side effects compared with cinacalcet [47].

Etelcalcetide is more effective than cicalcet in lowering PTH levels. This fact tends to induce a higher rate of hypocalcemia (calcium <7.5 mg/dL). Beneficial effects on bone structure and a potential reduction in fracture incidence were described when using etelcalcetide [48]. It is crucial to mention that the progression of left ventricular hypertrophy was inhibited by the use of etelcalcetide, demonstrating cardiovascular benefits; however, further studies are needed in this area [49].

Upacicalcet is one of the new injectable calcimimetic agents, currently being used only in Japan. It is used for patients undergoing dialysis, being injected three times a week into the venous side of the dialysis machine. However, it is important to note that most of the evidence supporting the use of Upicalcet comes from early-phase clinical trials and observational data, predominantly conducted in Japan. Furthermore, long-term safety and efficacy data are still limited, especially in diverse patient populations. Thus, while the results available at the moment appear promising in terms of safety and tolerability, particularly the lower incidence of hypocalcemia, the overall quality of evidence remains moderate, and further validation is required [50]. It is considered to be effective and safe, with a low rate of induced hypocalcemia, in only 2% of patients from a 2023 study [51].

### 6.4. Surgical Interventions

Despite the availability of numerous pharmacological therapies, some patients persist with significantly high levels of calcium and PTH. Furthermore, even one year following renal transplantation, some patients do not attain adequate biochemical control, highlighting the potential necessity for radical surgical intervention [52].

At present, no standardized criteria have been established to define the optimal timing of surgical intervention. In clinical practice, PTH levels ranging from 2- to 9-fold above the upper reference limit are generally regarded as indicative of surgical necessity [10,16].

Although no universally standardized criteria exist, several clinical and biochemical thresholds are widely accepted in guiding surgical referral. Current guidelines, such as KDIGO and the European Society of Endocrinology, suggest considering surgery in the following scenarios:

- PTH levels persistently elevated at 2–9 times the upper normal limit despite optimal medical therapy
- Chronic hypercalcemia, particularly if >11 mg/dL

- Persistent hypercalcemia for 3 to 12 months post-renal transplant
- Hyperphosphatemia refractory to phosphate binders
- Severe bone disease
- Uncontrolled symptoms, such as pruritus, fatigue, nephrolithiasis, or nephrocalcinosis

The main goal of surgery is to reestablish normal calcium concentration while reducing parathyroid gland mass [16,53,54].

According to the Miami criteria, a successful surgical intervention is indicated by an intraoperative PTH decrease of over 50% measured at least 10 min post-resection [16,55]. Other authors consider that a reduction in PTH levels >7-% 20 min after surgery is a good indicator for a successful surgical act [56]. It is important to mention the difference in clearance of intact PTH in CKD patients, as it is much slower. In renal failure, the iPTH decay is around 6.6 min. If the levels of PTH do not drop as expected, the surgeon must take into consideration the existence of ectopic parathyroid glands [55].

It is recommended to use imagistic methods in order to localize the parathyroid glands. The routine use of parathyroid imaging remains a topic of debate. We suggest that imaging should not be routinely performed in all cases but rather reserved for those being considered for surgery or in whom the biochemical diagnosis is uncertain. The most accessible approach is by using ultrasonography—an evaluation best used in cases of over 1 g adenomas. Besides that, we can detect the presence of all four hyperplastic parathyroid glands by using Tc−99 m sestamibi SPECT-CT. Clinicians must guide the use of imagistic methods and adjust them to every patient's needs [52]. On the other hand, some authors suggest that it is not mandatory to localize the parathyroid glands before surgery, and the Miami criteria are the most efficient indicator, as an experienced surgeon can find ectopic glands without preoperative localization [57]. Preoperative localization remains a useful tool, especially in cases of recurrent HPTH [56].

Globally, several operative approaches are implemented in the treatment of THPT. These include: subtotal parathyroidectomy with bilateral cervical thymectomy, entailing removal of three glands and partial excision of the fourth, with a residual parathyroid remnant of approximately 40–80 mg; total parathyroidectomy, with or without autologous parathyroid tissue transplantation, usually in conjunction with bilateral cervical thymectomy; and total parathyroidectomy without autotransplantation or associated thymectomy [16]. Thymectomy may be used, as the majority of ectopic parathyroid glands are situated there [56]. Potential sites for reimplantation include the sternocleidomastoid muscle, the brachioradial muscle, or, more appropriately, the subcutaneous adipose tissue of the forearm [16,56].

Total parathyroidectomy is generally not considered a first-line approach due to the high likelihood of chronic hypoparathyroidism, although it offers the benefit of a low recurrence rate. In contrast, subtotal parathyroidectomy is associated with a higher risk of disease recurrence, but postoperative hypocalcemia occurs less frequently. Total parathyroidectomy combined with autotransplantation has been shown to more effectively improve patients' quality of life, although there is the potential for regrowth of the transplanted parathyroid tissue. According to some authors, this procedure remains the treatment of choice [56].

Given the complexity of the cervical region, various intraoperative assistive methods were developed to facilitate the identification of parathyroid glands. We mention noncarbon suspension-assistance with negative parathyroid imaging, near-infrared fluorescence, and the use of indocyanine green [56]. The technique of injecting indocyanine green dye followed by detection using fluorescence has been used for years in several centers. This proved to be safe, effective, and had a key role in the preservation of the parathyroid glands [58].

As surgery practice moves toward minimal methods, we have to mention ablation. This is a new way to approach tumors, and whether it is using microwave or radiofrequency, some studies show benefits when used for HPTH. While ablation techniques (including microwave and radiofrequency ablation) offer a minimally invasive alternative for patients with contraindications to surgery, most available data are derived from retrospective studies or small prospective cohorts [38,56]. This technique offers a valid alternative for patients with contraindications for anesthesia. Although a meta-analysis found no significant differences between parathyroidectomy and ablation in terms of PTH, calcium, or phosphate levels. Ablation has shown a lower rate of hypocalcemia after the procedure, a shorter hospitalization time, but an increased risk of recurrence. The heterogeneity of patient selection, procedural protocols, and outcome measures limits the generalizability of the findings [38]. In addition, the risk of recurrence and the lack of long-term follow-up in many studies suggest that ablation should currently be considered as an alternative option in selected cases, rather than a standard first-line therapy. The overall quality of evidence is moderate, emphasizing the need for randomized, controlled, and adequately powered studies [58].

The most clinically relevant complications of parathyroidectomy are hypocalcemia and hungry bone syndrome, a condition that can become lethal. Among patients undergoing dialysis, hungry bone syndrome has a high prevalence, ranging from 26% to 95%. This condition usually presents with a rapid onset of hypocalcemia and often lasts more than four days after surgery. The standard treatment includes continuous calcium infusion [59]. Elevated levels of PTH pre-operatively, with a cutoff of over 166 pmol/L, are a powerful predictor of post-operative hypocalcemia and require higher doses of replacement [60].

Besides this, hypoparathyroidism may appear as transitory or permanent, and it is accompanied by hypocalcemia [52]. Hypocalcemia may be asymptomatic or present as tingling and numbness of the extremities. Clinical indicators of hypocalcemia include tetany, a positive Trousseau or Chvostek sign, and a prolonged QT interval. In cases of permanent hypoparathyroidism, long-term complications may arise, such as cataracts, cardiovascular involvement, and basal ganglia calcification. The treatment involves chronic calcium substitution, although ongoing research is exploring more effective therapeutic alternatives [61].

## 7. Conclusion and Future Directions

THPT complicating diabetic nephropathy is marked by a pronounced clinical heterogeneity and poses a multifaceted therapeutic challenge. The absence of standardized diagnostic criteria delays identification and consistent treatment selection. Management must therefore integrate targeted metabolic and cardiovascular measures—rigorous blood-pressure control, correction of anemia, treatment of dyslipidemia—alongside therapies directed at mineral metabolism, including vitamin D analogues, calcimimetics (cinacalcet, etelcalcetide, upacicalcet) and phosphate-lowering strategies; these interventions can favorably modify PTH levels, parathyroid gland volume and bone structure, but require careful monitoring for hypocalcemia and vascular calcification.

Patients who remain refractory to medical therapy may benefit from surgical approaches (subtotal or total parathyroidectomy with or without autotransplantation) or minimally invasive ablation. However, these options also carry risks—such as hungry bone syndrome or persistent hypocalcemia, demanding meticulous perioperative planning.

Future research should prioritize the identification and validation of specific biomarkers that can better stratify disease severity, predict therapeutic response, and monitor disease progression in SPTH and TPTH. At the moment, there is a lack of validated, non-invasive biomarkers beyond serum PTH and calcium-phosphate levels, which show high variability and limited predictive value.

Moreover, consensus is still lacking regarding optimal PTH targets in dialysis patients, with current guidelines offering wide reference ranges without solid evidence for improved outcomes. There is a critical need for large-scale, longitudinal studies to determine the PTH thresholds that correlate with reduced cardiovascular calcification and improve bone turnover, particularly in high-risk subgroups such as diabetic patients.

In addition, further investigation into the long-term cardiovascular and skeletal outcomes of emerging pharmacologic agents is needed, along with personalized approaches that account for comorbidities such as diabetes.

Ultimately, developing individualized treatment algorithms that integrate pharmacologic, interventional, and surgical modalities—tailored to patient phenotype and comorbidity profile- will be essential to improving quality of life and long-term survival in this complex patient population.

**Author Contributions:** Conceptualization, M.C. and D.-M.T.; methodology, D.L.P. and V.R.N.; software, M.C.; validation, M.C., D.L.P., and D.-M.T.; formal analysis, M.C.; investigation, M.C.; resources, M.C. and S.C.P.; data curation, R.E.D.-M.; writing—original draft preparation, M.C.; writing—review and editing, M.C.; visualization, R.E.D.-M.; supervision, R.E.D.-M.; project administration, M.C. and S.C.P. All authors have read and agreed to the published version of the manuscript.

**Funding:** This research received no external funding.

**Institutional Review Board Statement:** Not applicable.

**Data Availability Statement:** No new data were created.

**Acknowledgments:** Publication of this paper was supported by the University of Medicine and Pharmacy Carol Davila, through the institutional program Publish not Perish.

**Conflicts of Interest:** The authors declare no conflicts of interest.

## Abbreviations

The following abbreviations are used in this manuscript:

| | |
|---|---|
| THPT | Tertiary Hyperparathyroidism |
| CKD | Chronic Kidney Disease |
| PTH | Parathyroid Hormone |
| ESRD | End-stage Renal Disease |
| SHPT | Secondary Hyperparathyroidism |
| GFR | Glomerular Filtration Rate |
| AGEs | Advanced Glycation End products |
| RAS | Renin-angiotensin System |
| FGF23 | Fibroblast Growth Factor 23 |
| CaSR | Calcium Sensing Receptor |
| ROD | Renal Osteodystrophy |
| DXA | Dual-energy X-ray Absorptiometry |
| CKD-MBD | Chronic kidney-disease-related bone and mineral disease |
| HPTH | Hyperparathyroidism |

## References


1. Tilici, D.M.; Paun, D.L.; Arnautu, A.M.; Mirica, A.; Duta, C.; Costea, M.; Guja, C. The Intricate Relationship Between Thyroid Disorders and Type 2 Diabetes—A Narrative Review. *Diabetology* **2025**, *6*, 41. [CrossRef]
2. Shaw, J.E.; Sicree, R.A.; Zimmet, P.Z. Global estimates of the prevalence of diabetes for 2010 and 2030. *Diabetes Res. Clin. Pract.* **2010**, *87*, 4–14. [CrossRef]

3. Dagar, N.; Das, P.; Bisht, P.; Taraphdar, A.K.; Velayutham, R.; Arumugam, S. Diabetic nephropathy: A twisted thread to unravel. *Life Sci.* **2021**, *278*, 119635. [CrossRef]
4. Thipsawat, S. Early detection of diabetic nephropathy in patient with type 2 diabetes mellitus: A review of the literature. *Diabetes Vasc. Dis. Res.* **2021**, *18*, 14791641211058856. [CrossRef]
5. Kanwar, Y.S.; Sun, L.; Xie, P.; Liu, F.Y.; Chen, S. A glimpse of various pathogenetic mechanisms of diabetic nephropathy. *Annu. Rev. Pathol. Mech. Dis.* **2011**, *6*, 395–423. [CrossRef] [PubMed]
6. Soltani-Fard, E.; Taghvimi, S.; Karimi, F.; Vahedi, F.; Khatami, S.H.; Behrooj, H.; Hayati, M.D.; Movahedpour, A.; Ghasemi, H. Urinary biomarkers in diabetic nephropathy. *Clin. Chim. Acta* **2024**, *561*, 119762. [CrossRef]
7. Dwivedi, S.; Sikarwar, M.S. Diabetic Nephropathy: Pathogenesis, Mechanisms, and Therapeutic Strategies. *Horm. Metab. Res.* **2024**, *57*, 7–17. [CrossRef] [PubMed]
8. Drüeke, T.B. Hyperparathyroidism in Chronic Kidney Disease. Endotext. February 2025. Available online: https://www.ncbi.nlm.nih.gov/books/NBK278975/ (accessed on 8 July 2025).
9. Koizumi, M.; Komaba, H.; Fukagawa, M. Parathyroid function in chronic kidney disease: Role of FGF23-klotho axis. *Contrib. Nephrol.* **2013**, *180*, 110–123. [CrossRef] [PubMed]
10. Gupta, K.; Kumar, C.; Puri, G. The Panorama of Hyperparathyroidism. *Indian J. Endocr. Surg. Res.* **2022**, *17*, 40–51. [CrossRef]
11. Xu, Y.; Evans, M.; Carrero, J.J.; Soro, M.; Barany, P. Secondary hyperparathyroidism and adverse health outcomes in adults with chronic kidney disease. *Clin. Kidney J.* **2021**, *14*, 2213–2220. [CrossRef]
12. Lemoine, S.; Figueres, L.; Bacchetta, J.; Frey, S.; Dubourg, L. Calcium homeostasis and hyperparathyroidism: Nephrologic and endocrinologic points of view. *Ann. Endocrinol.* **2022**, *83*, 237–243. [CrossRef]
13. Pasieka, J.L.; Parsons, L.L. A prospective surgical outcome study assessing the impact of parathyroidectomy on symptoms in patients with secondary and tertiary hyperparathyroidism. *Surgery* **2000**, *128*, 531–539. [CrossRef]
14. Leeds, J.; Doyle, A. Tertiary Hyperparathyroidism. In *Medical and Surgical Treatment of Parathyroid Diseases*; Springer: Cham, Switzerland, 2025; pp. 93–100. [CrossRef]
15. Hu, L.; Napoletano, A.; Provenzano, M.; Garofalo, C.; Bini, C.; Comai, G.; La Manna, G. Mineral Bone Disorders in Kidney Disease Patients: The Ever-Current Topic. *Int. J. Mol. Sci.* **2022**, *23*, 12223. [CrossRef]
16. Palumbo, V.D.; Damiano, G.; Messina, M.; Fazzotta, S.; Monte, G.L.; Monte, A.I.L. Tertiary hyperparathyroidism: A review. *Clin. Ther.* **2021**, *172*, 241–246. [CrossRef]
17. Sutton, W.; Chen, X.; Patel, P.; Karzai, S.; Prescott, J.D.; Segev, D.L.; McAdams-DeMarco, M.; Mathur, A. Prevalence and risk factors for tertiary hyperparathyroidism in kidney transplant recipients. *Surgery* **2022**, *171*, 69–76. [CrossRef] [PubMed]
18. Dulfer, R.R.; Franssen, G.J.H.; Hesselink, D.A.; Hoorn, E.J.; van Eijck, C.H.J.; van Ginhoven, T.M. Systematic review of surgical and medical treatment for tertiary hyperparathyroidism. *Br. J. Surg.* **2017**, *104*, 804–813. [CrossRef] [PubMed]
19. Özdemir, F.N.; Afsar, B.; Akgul, A.; Usluoğullari, C.; Akçay, A.; Haberal, M. Persistent hypercalcemia is a significant risk factor for graft dysfunction in renal transplantation recipients. *Transplant. Proc.* **2006**, *38*, 480–482. [CrossRef]
20. Egbuna, O.I.; Taylor, J.G.; Bushinsky, D.A.; Zand, M.S. Elevated calcium phosphate product after renal transplantation is a risk factor for graft failure. *Clin. Transplant.* **2007**, *21*, 558–566. [CrossRef]
21. Hryciuk, M.; Heleniak, Z.; Ślizień, A.D. Management of chronic kidney disease–mineral and bone disorder. *Ren. Dis. Transplant. Forum* **2023**, *16*, 65–80. [CrossRef]
22. van der Plas, W.Y.; Noltes, M.E.; van Ginhoven, T.M.; Kruijff, S. Secondary and Tertiary Hyperparathyroidism: A Narrative Review. *Scand. J. Surg.* **2020**, *109*, 271–278. [CrossRef]
23. Aaltonen, L.; Koivuviita, N.; Seppänen, M.; Burton, I.S.; Kröger, H.; Löyttyniemi, E.; Metsärinne, K. Bone Histomorphometry and 18F-Sodium Fluoride Positron Emission Tomography Imaging: Comparison Between only Bone Turnover-based and Unified TMV-based Classification of Renal Osteodystrophy. *Calcif. Tissue Int.* **2021**, *109*, 605–614. [CrossRef]
24. Aguilar, A.; Gifre, L.; Ureña-Torres, P.; Carrillo-López, N.; Rodriguez-García, M.; Massó, E.; da Silva, I.; López-Báez, V.; Sánchez-Bayá, M.; Prior-Español, Á.; et al. Pathophysiology of bone disease in chronic kidney disease: From basics to renal osteodystrophy and osteoporosis. *Front. Physiol.* **2023**, *14*, 1177829. [CrossRef] [PubMed]
25. Pazianas, M.; Miller, P.D. Osteoporosis and Chronic Kidney Disease–Mineral and Bone Disorder (CKD-MBD): Back to Basics. *Am. J. Kidney Dis.* **2021**, *78*, 582–589. [CrossRef] [PubMed]
26. Cannata-Andía, J.B.; García, M.R.; Alonso, C.G. Osteoporosis and adynamic bone in chronic kidney disease. *J. Nephrol.* **2013**, *26*, 73–80. [CrossRef] [PubMed]
27. Cannata-Andía, J.B.; Martín-Carro, B.; Martín-Vírgala, J.; Rodríguez-Carrio, J.; Bande-Fernández, J.J.; Alonso-Montes, C.; Carrillo-López, N. Chronic Kidney Disease—Mineral and Bone Disorders: Pathogenesis and Management. *Calcif. Tissue Int.* **2021**, *108*, 410–422. [CrossRef]
28. Kazama, J.J. Chronic kidney disease and fragility fracture. *Clin. Exp. Nephrol.* **2017**, *21* (Suppl. S1), 46–52. [CrossRef]
29. Shah, A.; Hashmi, M.F.; Aeddula, N.R. Chronic Kidney Disease-Mineral Bone Disorder (CKD-MBD). StatPearls. April 2024. Available online: https://www.ncbi.nlm.nih.gov/books/NBK560742/ (accessed on 19 August 2025).

30. Evenepoel, P.; Stenvinkel, P.; Shanahan, C.; Pacifici, R. Inflammation and gut dysbiosis as drivers of CKD–MBD. *Nat. Rev. Nephrol.* **2023**, *19*, 646–657. [CrossRef]
31. Zoccali, C.; Mallamaci, F.; Adamczak, M.; de Oliveira, R.B.; A Massy, Z.; Sarafidis, P.; Agarwal, R.; Mark, P.B.; Kotanko, P.; Ferro, C.J.; et al. Cardiovascular complications in chronic kidney disease: A review from the European Renal and Cardiovascular Medicine Working Group of the European Renal Association. *Cardiovasc. Res.* **2023**, *119*, 2017–2032. [CrossRef]
32. Marando, M.; Tamburello, A.; Salera, D.; Di Lullo, L.; Bellasi, A. Phosphorous metabolism and manipulation in chronic kidney disease. *Nephrology* **2024**, *29*, 791–800. [CrossRef]
33. Cernaro, V.; Santoro, D.; Lacquaniti, A.; Costantino, G.; Visconti, L.; Buemi, A.; Buemi, M. Phosphate binders for the treatment of chronic kidney disease: Role of iron oxyhydroxide. *Int. J. Nephrol. Renovasc. Dis.* **2016**, *9*, 11–19. [CrossRef]
34. Barna, M.M.; Kapoian, T.; O'MAra, N.B. Sevelamer Carbonate. *Ann. Pharmacother.* **2010**, *44*, 127–134. [CrossRef]
35. Perry, C.M.; Plosker, G.L. Sevelamer Carbonate: A Review in Hyperphosphataemia in Adults with Chronic Kidney Disease. *Drugs* **2014**, *74*, 771–792. [CrossRef]
36. Kim, J.S.; Hwang, H.S. Vascular calcification in chronic kidney disease: Distinct features of pathogenesis and clinical implication. *Korean Circ. J.* **2021**, *51*, 961–982. [CrossRef]
37. Caravaca, F.; Caravaca-Fontán, F.; Azevedo, L.; Luna, E. Changes in renal function after discontinuation of vitamin D analogues in advanced chronic kidney disease. *Nefrología* **2018**, *38*, 179–189. [CrossRef] [PubMed]
38. Tsai, S.H.; Kan, W.-C.; Jhen, R.-N.; Chang, Y.-M.; Kao, J.-L.; Lai, H.-Y.; Liou, H.-H.; Shiao, C.-C. Secondary hyperparathyroidism in chronic kidney disease: A narrative review focus on therapeutic strategy. *Clin. Med. J. R. Coll. Physicians Lond.* **2024**, *24*, 100238. [CrossRef] [PubMed]
39. Patel, R.; Nandini; Kharkwal, H.; Saha, M.; Sankaranarayanan, M.; Sharma, S.; Chander, S. Recent Advancements Towards the Use of Vitamin D Isoforms and the Development of Their Synthetic Analogues as New Therapeutics. *Biomedicines* **2025**, *13*, 1002. [CrossRef] [PubMed]
40. Dou, D.; Yang, B.; Gan, H.; Xie, D.; Lei, H.; Ye, N. Vitamin D supplementation for the improvement of vascular function in patients with chronic kidney disease: A meta-analysis of randomized controlled trials. *Int. Urol. Nephrol.* **2019**, *51*, 851–858. [CrossRef]
41. Franchi, M.; Gunnarsson, J.; Gonzales-Parra, E.; Ferreira, A.; Ström, O.; Corrao, G. Paricalcitol and Extended-Release Calcifediol for Treatment of Secondary Hyperparathyroidism in Non-Dialysis Chronic Kidney Disease: Results From a Network Meta-Analysis. *J. Clin. Endocrinol. Metab.* **2023**, *108*, E1424–E1432. [CrossRef]
42. Pan, S.; Yang, K.; Shang, Y.; Yu, R.; Liu, L.; Jin, J.; He, Q. Effect of regulated vitamin D increase on vascular markers in patients with chronic kidney disease: A systematic review and meta-analysis of randomized controlled trials. *Nutr. Metab. Cardiovasc. Dis.* **2024**, *34*, 33–44. [CrossRef]
43. Nakai, K.; Kono, K.; Yamada, S.; Taniguchi, M.; Hamano, T.; Fukagawa, M. Calcimimetics treatment strategy for serum calcium and phosphate management in patients with secondary hyperparathyroidism undergoing dialysis: A systematic review and meta-analysis of randomized studies. *Ther. Apher. Dial.* **2024**, *28*, 557–571. [CrossRef]
44. Li, X.; Ding, W.; Zhang, H. Cinacalcet use in secondary hyperparathyroidism: A machine learning-based systematic review. *Front. Endocrinol.* **2023**, *14*, 1146955. [CrossRef] [PubMed]
45. Bernardor, J.; De Mul, A.; Bacchetta, J.; Schmitt, C.P. Impact of Cinacalcet and Etelcalcetide on Bone Mineral and Cardiovascular Disease in Dialysis Patients. *Curr. Osteoporos. Rep.* **2023**, *21*, 193–204. [CrossRef] [PubMed]
46. Ni, Z.; Liang, X.; Jin, K.; Kim, Y.-L.; Lu, K.-C.; Chan, T.M.; Fukagawa, M.; Kinoshita, J.; Nagai, C.; Kojima, M.; et al. Comparison of the Oral Calcimimetics Evocalcet and Cinacalcet in East Asian Patients on Hemodialysis with Secondary Hyperparathyroidism. *Kidney Int. Rep.* **2023**, *8*, 2294. [CrossRef]
47. Pereira, L.; Meng, C.; Marques, D.; Frazão, J.M. Old and new calcimimetics for treatment of secondary hyperparathyroidism: Impact on biochemical and relevant clinical outcomes. *Clin. Kidney J.* **2018**, *11*, 80–88. [CrossRef] [PubMed]
48. Freitas, P.; Pereira, L. Etelcalcetide: What we know eight years since its approval. *Nefrologia* **2025**, *45*, 116–134. [CrossRef]
49. Dörr, K.; Kainz, A.; Oberbauer, R. Lessons from effect of etelcalcetide on left ventricular hypertrophy in patients with end-stage kidney disease. *Curr. Opin. Nephrol. Hypertens.* **2022**, *31*, 339–343. [CrossRef]
50. Hoy, S.M. Upacicalcet: First Approval. *Drugs* **2021**, *81*, 1593–1596. [CrossRef]
51. Shigematsu, T.; Koiwa, F.; Isaka, Y.; Fukagawa, M.; Hagita, K.; Watanabe, Y.S.; Honda, D.; Akizawa, T. Efficacy and Safety of Upacicalcet in Hemodialysis Patients with Secondary Hyperparathyroidism A Randomized Placebo-Controlled Trial. *Clin. J. Am. Soc. Nephrol.* **2023**, *18*, 1300–1309. [CrossRef]
52. Casella, C.; Guarneri, C.; Campanile, M.; Adhoute, X.; Gelera, P.P.; Morandi, R. Surgical treatment of tertiary hyperparathyroidism: Does one fit for all? *Front. Endocrinol.* **2023**, *14*, 1226917. [CrossRef]
53. Bollerslev, J.; Rejnmark, L.; Zahn, A.; Heck, A.; Appelman-Dijkstra, N.M.; Cardoso, L.; Hannan, F.M.; Cetani, F.; Sikjaer, T.; Formenti, A.M.; et al. European Expert Consensus on Practical Management of Specific Aspects of Parathyroid Disorders in Adults and in Pregnancy: Recommendations of the ESE Educational Program of Parathyroid Disorders. *Eur. J. Endocrinol.* **2022**, *186*, R33–R63. [CrossRef]

54. Oltmann, S.C.; Madkhali, T.M.; Sippel, R.S.; Chen, H.; Schneider, D.F. Kidney Disease Improving Global Outcomes guidelines and parathyroidectomy for renal hyperparathyroidism. *J. Surg. Res.* **2015**, *199*, 115–120. [CrossRef] [PubMed]
55. Finnegan, K.P.J.; Dowling, G.P.; El Wahab, S.A.; Hill, A.D.K. Intraoperative parathyroid hormone monitoring in parathyroidectomy: Predicting cure and recurrence in tertiary hyperparathyroidism—Systematic review. *Am. J. Otolaryngol. Head Neck Med. Surg.* **2025**, *46*, 104609. [CrossRef]
56. Zhang, L.X.; Zhang, B.; Liu, X.-Y.; Wang, Z.-M.; Qi, P.; Zhang, T.-Y.; Zhang, Q. Advances in the treatment of secondary and tertiary hyperparathyroidism. *Front. Endocrinol.* **2022**, *13*, 1059828. [CrossRef]
57. Wang, R.; Abraham, P.; Lindeman, B.; Chen, H.; Fazendin, J. Is preoperative parathyroid localization necessary for tertiary hyperparathyroidism? *Am. J. Surg.* **2022**, *224*, 918–922. [CrossRef]
58. Rao, S.S.; Rao, H.; Moinuddin, Z.; Rozario, A.P.; Augustine, T. Preservation of parathyroid glands during thyroid and neck surgery. *Front. Endocrinol.* **2023**, *14*, 1173950. [CrossRef]
59. Latus, J.; Roesel, M.; Fritz, P.; Braun, N.; Ulmer, C.; Steurer, W.; Biegger, D.; Alscher, D.; Kimmel, M.; Alscher, M.D. Incidence of and risk factors for hungry bone syndrome in 84 patients with secondary hyperparathyroidism. *Int. J. Nephrol. Renovasc. Dis.* **2013**, *6*, 131. [CrossRef]
60. Tan, Z.K.; Looi, W.L.; Chen, F.; Yeo, S.C.; Bairy, M. Determinants of Severe Hypocalcemia After Parathyroidectomy in Patients with End-Stage Kidney Disease and Renal Hyperparathyroidism: A Retrospective Cohort Study. *J. Clin. Med.* **2025**, *14*, 379. [CrossRef] [PubMed]
61. Alqahtani, S.M. Post-thyroidectomy hypoparathyroidism: A clinical surgical dilemma. *Saudi Med. J.* **2024**, *45*, 1305–1311. [CrossRef] [PubMed]

*biomedicines*

MDPI

*Review*

# Next-Generation SGLT2 Inhibitors: Innovations and Clinical Perspectives

**Dana Movila [1,2], Daniel Duda Seiman [1,2,*] and Simona Ruxanda Dragan [1,2]**

[1] Department VI—Cardiology, University Clinic of Internal Medicine and Ambulatory Care, Prevention and Cardiovascular Recovery, "Victor Babes" University of Medicine and Pharmacy, 300041 Timisoara, Romania; dana.movila@umft.ro (D.M.); simona.dragan@umft.ro (S.R.D.)

[2] Research Centre of Timisoara Institute of Cardiovascular Diseases, "Victor Babes" University of Medicine and Pharmacy, 300041 Timisoara, Romania

* Correspondence: daniel.duda-seiman@umft.ro; Tel.: +40-723219416

**Abstract**

Sodium–glucose cotransporter 2 (SGLT2) inhibitors have substantially reshaped the management of type 2 diabetes mellitus (T2DM), owing not only to their glucose-lowering properties but also to their consistent cardiovascular and renal protective effects. Beyond their initial metabolic indication, these agents have emerged as disease-modifying therapies across a broad spectrum of cardiometabolic and renal conditions. Building on the clinical success of first-generation SGLT2 inhibitors, such as empagliflozin and dapagliflozin, next-generation SGLT2-based therapies have been developed with the aim of refining pharmacological selectivity, optimizing pharmacokinetic profiles, and expanding therapeutic applicability beyond diabetes. These innovations include dual SGLT1/SGLT2 inhibition, alternative dosing strategies, and molecular designs tailored to specific clinical phenotypes, such as heart failure with preserved ejection fraction (HFpEF) and chronic kidney disease (CKD). This narrative review critically evaluates the evolving landscape of next-generation SGLT2 inhibitors, with a focus on structural and pharmacokinetic innovations, transporter selectivity, glucose-independent mechanisms, and emerging clinical implications. A comprehensive literature search was conducted using PubMed/MEDLINE, Scopus, and Web of Science, encompassing publications from inception to March 2025. Eligible sources included randomized clinical trials, observational studies, meta-analyses, and authoritative reviews published in English. Available evidence indicates that, while conventional SGLT2 inhibitors confer robust and reproducible cardiorenal benefits, newer agents may further extend therapeutic potential through incretin-related effects, modulation of extra-renal pathways, and disease-specific cardiac and renal mechanisms. Nevertheless, evidence supporting incremental clinical benefit beyond established SGLT2 inhibitors remains limited and heterogeneous, particularly for recently developed compounds. Overall safety profiles appear broadly consistent within the class, although long-term data for next-generation agents are still evolving. Key limitations of the current evidence base include reliance on emerging or indirect mechanistic data, heterogeneity in study populations and clinical endpoints, and the relative scarcity of large, outcome-driven trials for newer SGLT2-based therapies. Future research should prioritize mechanism-driven clinical trials, precision-oriented patient stratification, and head-to-head comparative studies to more clearly define the role of next-generation SGLT2 inhibitors in cardiovascular, renal, and metabolic disease management.

 https://doi.org/10.3390/biomedicines14010081

**Keywords:** SGLT2 inhibitors; dual inhibitors; heart failure; metabolic disease; pharmacotherapy; renal protection

## 1. Introduction

Sodium–glucose cotransporter 2 inhibitors (SGLT2i) exert their principal pharmacological effect by selectively inhibiting sodium–glucose cotransporter 2 proteins expressed in the S1 segment of the renal proximal tubule, which under physiological conditions account for approximately 80–90% of filtered glucose reabsorption [1–3]. Pharmacological blockade of SGLT2 induces glycosuria and lowers plasma glucose concentrations through an insulin-independent, glycemia-dependent mechanism, resulting in an average reduction in glycated hemoglobin (HbA1c) of approximately 0.6–1.0% [1,2].

Overall, SGLT2 inhibitors are well tolerated. Nevertheless, class-related adverse effects have been reported, including genitourinary infections, volume depletion–related hypotension, and rare cases of diabetic ketoacidosis, with incidence influenced by patient characteristics, comorbidities, and clinical context [2].

Although initially developed as glucose-lowering agents for the treatment of type 2 diabetes mellitus (T2DM), SGLT2 inhibitors were subsequently shown to confer substantial cardiovascular and renal benefits that extend beyond glycemic control. Large cardiovascular and renal outcome trials consistently demonstrated significant reductions in heart failure hospitalization and slowing of chronic kidney disease progression, effects that could not be fully explained by glucose-lowering alone. These findings prompted a fundamental shift in the therapeutic positioning of this drug class [3].

From a regulatory perspective, several SGLT2 inhibitors have received approvals for indications extending well beyond diabetes. Empagliflozin is approved for T2DM, reduction in cardiovascular mortality in adults with T2DM, and for heart failure across both reduced and preserved ejection fraction phenotypes. Canagliflozin is indicated for T2DM and diabetic kidney disease and has demonstrated reductions in cardiovascular events and heart failure hospitalization. Dapagliflozin is approved for T2DM, heart failure with reduced ejection fraction (HFrEF), and chronic kidney disease (CKD), irrespective of diabetes status, whereas ertugliflozin remains approved for the treatment of T2DM [4].

Given that individuals with T2DM frequently present with multiple comorbidities—including age-related cardiovascular and renal disease—SGLT2 inhibitors have assumed an increasingly prominent role across multiple clinical disciplines, extending well beyond traditional endocrinology [5].

In this evolving context, the term "next-generation SGLT2 inhibitors" has emerged to describe contemporary SGLT2-targeting therapies that move beyond glucose lowering to deliver robust cardiovascular and renal protection in both diabetic and non-diabetic populations. These agents are characterized by enhanced cardiorenal benefits, improved pharmacologic selectivity, optimized pharmacokinetic and safety profiles, and expanded therapeutic indications, particularly in heart failure and chronic kidney disease [6].

Importantly, next-generation SGLT2 inhibitors are defined not solely by molecular novelty, but by their capacity to modulate cardiorenal and metabolic pathways beyond renal glucose excretion. Through mechanisms such as natriuresis, hemodynamic modulation, metabolic reprogramming, and anti-inflammatory effects, often independent of glycemic control, these therapies contribute to reductions in heart failure hospitalization, attenuation of CKD progression, and improved clinical outcomes [7].

Accordingly, this review moves beyond a descriptive overview of SGLT2 inhibition to critically examine the pharmacological innovations, mechanistic diversity, and emerging

https://doi.org/10.3390/biomedicines14010081

clinical implications of next-generation SGLT2-based therapies, with particular emphasis on cardiovascular and renal disease modification [8].

### *1.1. Next-Generation SGLT2-Based Inhibitors: Developmental Status and Clinical Evidence*

The term next-generation SGLT2 inhibitors encompasses a heterogeneous group of agents designed to extend or refine the therapeutic profile of conventional SGLT2 inhibitors. These compounds differ with respect to transporter selectivity (including dual SGLT1/SGLT2 inhibition), molecular structure, pharmacokinetics, and intended clinical applications, and are currently at varying stages of clinical development. Importantly, not all next-generation agents have achieved regulatory approval, and many remain investigational or indication-specific [9].

### *1.2. Sotagliflozin (Dual SGLT1/SGLT2 Inhibitor)*

Sotagliflozin is the most clinically advanced next-generation agent, combining renal SGLT2 inhibition with partial intestinal SGLT1 blockade. Regulatory approval has been granted in selected regions for specific indications, while evaluation continues elsewhere. Clinical evidence derives primarily from SOLOIST-WHF and SCORED, which demonstrated reductions in cardiovascular death and heart failure events in high-risk populations. However, comparative efficacy relative to selective SGLT2 inhibitors cannot be established due to the absence of head-to-head trials, and proposed advantages related to incretin-mediated effects remain hypothesis-generating [10].

### *1.3. Licogliflozin (Dual SGLT1/SGLT2 Inhibitor)*

Licogliflozin exhibits a more balanced SGLT1/SGLT2 inhibitory profile. Its clinical development has focused largely on metabolic and inflammatory conditions rather than cardiovascular outcomes. Available evidence is limited to small phase II trials assessing biomarkers and short-term metabolic endpoints, with no large outcome-driven cardiovascular or renal studies completed to date. Gastrointestinal adverse effects associated with intestinal SGLT1 inhibition have constrained further development, and its comparative clinical value remains undetermined [11].

### *1.4. Emerging and Experimental Approaches*

Additional next-generation strategies include ultra-selective SGLT2 inhibitors, hybrid molecules designed to optimize tissue distribution or pharmacokinetic stability, and combination approaches integrating SGLT2 inhibition with incretin-based or metabolic modulation. Most of these agents remain in early-phase development, with limited publicly available efficacy data [12].

### *1.5. Comparative Efficacy: Current Limitations*

At present, comparative efficacy among next-generation and conventional SGLT2 inhibitors cannot be conclusively established, owing to the lack of head-to-head randomized trials, heterogeneity in study designs and endpoints, and the limited availability of outcome-driven evidence for newer compounds (Table 1). Accordingly, claims of superiority should be avoided, and next-generation agents should be viewed as potentially complementary rather than definitively superior to established therapies [13].

https://doi.org/10.3390/biomedicines14010081

**Table 1.** Next-Generation SGLT2-Based Inhibitors: Regulatory Status and Clinical Development.

| Agent | Transporter Profile | Regulatory Status | Clinical Phase | Key Trials | Primary Focus |
|---|---|---|---|---|---|
| Sotagliflozin | SGLT1/SGLT2 | Approved (region/ indication-specific) | Phase III completed | SOLOIST-WHF (~1200); SCORED (~10,500) | HF, CKD, CV risk |
| Licogliflozin | SGLT1/SGLT2 | Not approved | Phase II | Multiple small RCTs (<300) | Metabolic/HFpEF |
| Emerging agents | SGLT2-based | Not approved | Phase I–II | Early studies | Precision therapy |

## 2. Effects of SGLT2 Inhibitors on Cardiovascular Outcomes

Sodium–glucose cotransporter 2 inhibitors (SGLT2i) have become integral components of contemporary therapy for type 2 diabetes (T2D), heart failure (HF), and chronic kidney disease (CKD). Although originally developed as glucose-lowering agents, large randomized clinical trials have demonstrated robust cardiovascular benefits, including reductions in heart failure hospitalization and cardiovascular mortality, across diverse patient populations, including individuals without diabetes [14].

### 2.1. Cardiovascular Outcome Evidence

Large cardiovascular outcome trials have firmly established SGLT2 inhibitors as disease-modifying therapies in cardiovascular medicine. Empagliflozin and dapagliflozin demonstrated consistent reductions in heart failure hospitalization and cardiovascular mortality in EMPA-REG OUTCOME, DAPA-HF, EMPEROR-Reduced, and EMPEROR-Preserved, while canagliflozin showed cardiovascular and renal benefits in CANVAS and CREDENCE [1–5]. The dual SGLT1/SGLT2 inhibitor sotagliflozin further expanded this evidence base through SOLOIST-WHF and SCORED, particularly in patients with recent worsening heart failure or advanced cardiometabolic risk [15].

These consistent outcome benefits, observed across heart failure phenotypes and independent of glycemic status, indicate that the cardioprotective effects of SGLT2 inhibitors cannot be attributed solely to glucose lowering.

### 2.2. Direct Myocardial SGLT2 Inhibition: Evidence and Controversy

One proposed mechanism for the cardiovascular benefits of SGLT2 inhibitors is direct inhibition of SGLT2 within myocardial tissue. However, the existence and functional relevance of myocardial SGLT2 expression in humans remain highly controversial [16].

Several experimental and translational studies have reported detectable SGLT2 expression in cardiomyocytes under specific pathological conditions. For example, Marfella et al. identified SGLT2 expression in cardiomyocytes from patients with end-stage heart failure, with higher levels observed in individuals with diabetes, and demonstrated glucose-dependent upregulation in vitro [17]. Similar findings were reported by Scisciola et al. in patients with low-flow, low-gradient aortic stenosis, where SGLT2 overexpression was associated with myocardial metabolic remodeling and impaired energetic efficiency. In addition, studies using human induced pluripotent stem cell–derived cardiomyocytes exposed to high glucose demonstrated increased SGLT2 expression alongside hypertrophic and contractile abnormalities, which were partially reversed by empagliflozin [3,18,19].

In contrast, other investigations have failed to detect SGLT2 expression in adult human myocardial tissue. Notably, analyses of atrial and ventricular cardiomyocytes obtained from patients undergoing cardiac surgery demonstrated no detectable SGLT2 mRNA, identifying only SGLT1, GLUT1, and GLUT4 transporters [3]. Larger transcriptomic and proteomic datasets, including bulk and single-cell RNA sequencing of human myocardium,

https://doi.org/10.3390/biomedicines14010081

have similarly reported minimal or absent SGLT2 expression when compared with the kidney [20].

*2.3. Methodological Sources of Heterogeneity*

Discrepancies among studies likely reflect substantial methodological heterogeneity, including differences in tissue source (atrial vs. ventricular myocardium), disease state, species, antibody specificity, detection thresholds, and analytical techniques. Many positive findings derive from small, single-center studies, non-human models, neonatal cardiomyocytes, or diseased myocardium, where cross-reactivity with other glucose transporters cannot be excluded [21].

Importantly, reported myocardial SGLT2 expression levels are typically orders of magnitude lower than renal expression and may fall below the threshold required for meaningful inhibition at clinically achieved drug concentrations [22].

*2.4. Indirect and Glucose-Independent Mechanisms of Cardioprotection*

Given these limitations, the weight of evidence favors indirect and glucose-independent mechanisms as the primary drivers of cardiovascular benefit (Figure 1) [23]. These include the following:

- Osmotic diuresis and natriuresis leading to hemodynamic unloading;
- Improved myocardial energetics through increased ketone body availability;
- Modulation of intracellular sodium and calcium handling, potentially via effects on the $Na^+/H^+$ exchanger;
- Attenuation of inflammation, oxidative stress, and myocardial fibrosis;
- Secondary effects mediated through renal–cardiac and gut–cardiac axes, including reductions in hyperuricemia, improved endothelial function, and changes in erythropoietin signaling [24].

Crucially, many experimentally observed cardiac effects of SGLT2 inhibitors—such as modulation of sodium homeostasis or mitochondrial efficiency—do not require direct myocardial SGLT2 expression and may reflect downstream or off-target signaling pathways [25].

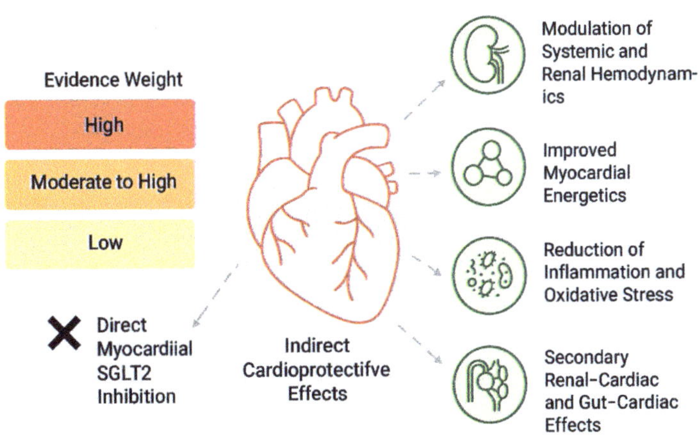


**Figure 1.** Cardioprotective SGLT2 Mechanism Diagram.

*2.5. Interpretative Framework*

Taken together, although experimental studies exploring myocardial SGLT2 expression have generated important hypotheses, current evidence does not support direct cardiac

https://doi.org/10.3390/biomedicines14010081

SGLT2 inhibition as a dominant mechanism in humans. Instead, the consistent cardiovascular benefits observed across large-scale randomized trials are best explained by an integrated network of systemic, hemodynamic, metabolic, and renal-mediated mechanisms, which operate independently of direct myocardial SGLT2 expression (Table 2) [25].

**Table 2.** Methodological approaches and interpretative strength of studies assessing myocardial SGLT2 expression.

| Study Type/ Reference | Species/Tissue | Sample Size | Methodology | Key Findings | Major Limitations | Interpretative Weight |
|---|---|---|---|---|---|---|
| Immunohistochemistry studies (early reports) | Rodent myocardium/neonatal cardiomyocytes | Small ($n < 10$) | IHC using commercial antibodies | Detectable SGLT2 signal in cardiomyocytes | Antibody cross-reactivity; non-human tissue; neonatal cells | Hypothesis-generating only |
| Western blot analyses | Rodent or diseased human myocardium | Small ($n < 15$) | Western blot | Low-level SGLT2 protein bands | Lack of validated controls; possible non-specific binding | Low |
| RT-PCR–based studies | Diseased human myocardial samples | Small ($n < 20$) | RT-PCR | Low or variable SGLT2 mRNA expression | High Ct values; sampling bias; atrial vs. ventricular tissue | Low |
| Bulk RNA sequencing | Adult human left ventricular myocardium | Moderate–large | Transcriptomic profiling | Minimal or absent SGLT2 transcripts | Detection threshold limitations | Moderate–high (negative evidence) |
| Single-cell RNA sequencing | Adult human cardiomyocytes | Large datasets | scRNA-seq | No consistent SGLT2 expression in cardiomyocytes | Dropout effects inherent to scRNA-seq | High |
| Proteomic analyses | Adult human myocardium | Moderate | Mass spectrometry | No detectable SGLT2 protein | Sensitivity limits for low-abundance proteins | High |
| Comparative renal–cardiac expression studies | Human kidney vs. myocardium | Moderate | Multi-tissue profiling | Marked SGLT2 expression in kidney, absent in heart | None significant | High |

This interpretation provides a more robust and clinically coherent framework for understanding the cardiovascular efficacy of SGLT2 inhibitors and aligns mechanistic hypotheses with outcome-driven clinical evidence [26].

This comparison highlights that positive reports of myocardial SGLT2 expression are predominantly derived from small, single-center, preclinical or antibody-dependent studies, whereas larger human transcriptomic and proteomic analyses consistently fail to demonstrate meaningful SGLT2 expression in adult cardiomyocytes. The cumulative weight of evidence therefore argues against direct myocardial SGLT2 inhibition as a dominant mechanism of cardioprotection, favoring indirect, systemic, or downstream cellular effects [26].

### 2.6. Heart Failure Outcomes: Effect Size, Timing, and Mechanistic Interpretation

Large randomized controlled trials have demonstrated consistent reductions in heart failure (HF)–related outcomes with SGLT2 inhibitors across a broad spectrum of ejection fraction phenotypes. In DAPA-HF, dapagliflozin reduced the composite of worsening HF or cardiovascular death by 26% (hazard ratio [HR] 0.74; 95% CI 0.65–0.85), while EMPEROR-Reduced reported a 25% relative risk reduction with empagliflozin (HR 0.75; 95% CI 0.65–0.86) in patients with HFrEF. In HFpEF populations, EMPEROR-Preserved and DELIVER demonstrated more modest but significant benefits, with HRs ranging from 0.79 to 0.82, accompanied by wider confidence intervals reflecting greater phenotypic heterogeneity [27–31].

https://doi.org/10.3390/biomedicines14010081

### 2.7. Trial-Level Limitations and Interpretation

Despite robust outcome signals, these trials share important limitations. Patients enrolled were highly selected, often younger and with fewer comorbidities than real-world HF populations, introducing potential selection bias. Moreover, baseline background therapy varied substantially, particularly with respect to mineralocorticoid receptor antagonists, angiotensin receptor–neprilysin inhibitors, and device therapy, complicating attribution of effect size to SGLT2 inhibition alone [24–27].

Event-driven designs and composite endpoints further limit mechanistic inference, underscoring the need for cautious interpretation when extrapolating trial results to broader clinical settings.

### 2.8. Early Initiation and Interaction with Diuretic Therapy

Recent studies suggest that initiation of SGLT2 inhibitors during or shortly after hospitalization for acute or worsening HF is associated with early clinical benefit. However, this strategy requires careful consideration of volume status and concomitant diuretic therapy. SGLT2 inhibitors induce osmotic diuresis and natriuresis, which may augment the effects of loop diuretics, potentially increasing the risk of volume depletion or hypotension in susceptible patients [32].

Although most trials did not mandate systematic diuretic dose adjustments, real-world implementation of early initiation strategies should consider individualized diuretic down-titration and close monitoring, particularly in elderly patients or those with marginal renal perfusion [33].

### 2.9. Mechanistic Hypotheses: Evidence and Caution

Multiple mechanisms have been proposed to explain the HF benefits of SGLT2 inhibitors, including improved myocardial energetics, reduction in inflammation and fibrosis, and modulation of intracellular sodium and calcium homeostasis. Among these, inhibition of the cardiac sodium–hydrogen exchanger (NHE) has received considerable attention [34].

However, evidence supporting clinically relevant NHE inhibition remains largely preclinical, derived from in vitro and animal models. To date, no direct clinical biomarker or imaging evidence has confirmed NHE modulation as a dominant mechanism in humans treated with SGLT2 inhibitors. Accordingly, such mechanisms should be viewed as hypothesis-generating rather than definitively established, and interpreted within the broader context of systemic and hemodynamic effects [35].

### 2.10. Integrative Perspective

Taken together, the benefits of SGLT2 inhibitors in HF are supported by consistent clinical outcome data but should be interpreted with awareness of trial-specific limitations, background therapy heterogeneity, and mechanistic uncertainty. Future studies integrating mechanistic endpoints, real-world populations, and standardized background therapy will be essential to refine understanding of both effect magnitude and biological pathways [35].

## 3. SGLT2 Inhibitors in Acute and Chronic Heart Failure

### 3.1. Clinical Evidence Across the Heart Failure Spectrum

Large-scale randomized clinical trials—including DAPA-HF, EMPEROR-Reduced, and DECLARE–TIMI 58—have consistently demonstrated the efficacy of SGLT2 inhibitors in patients with heart failure with reduced ejection fraction (HFrEF). Across these studies, SGLT2 inhibitor therapy was associated with relative reductions of approximately 30–40% in heart failure hospitalizations and ~20% in cardiovascular mortality [33]. In

https://doi.org/10.3390/biomedicines14010081

DECLARE–TIMI 58, dapagliflozin significantly reduced cardiovascular death, heart failure hospitalization, and all-cause mortality specifically in the HFrEF subgroup [28,36,37].

Subsequent analyses and dedicated trials extended these benefits to patients with mildly reduced and preserved ejection fraction (EF $\geq$ 40%), demonstrating improvements in heart failure–related outcomes across a broader range of ventricular function. Beyond hard clinical endpoints, both dapagliflozin and empagliflozin have been shown to improve symptoms, exercise capacity, and health-related quality of life, underscoring their clinical relevance beyond event reduction [38].

### 3.2. Early Initiation in Acute Heart Failure

An expanding body of evidence supports the early initiation of SGLT2 inhibitors in acute heart failure (AHF). In the EMPULSE trial, empagliflozin initiated during hospitalization for acute decompensated HF improved overall clinical status and reduced mortality at 90 days, irrespective of diabetes status [39]. Complementary observational data, including the study by Carvalho et al., reported reductions in all-cause mortality and HF readmissions among acutely decompensated patients treated with SGLT2 inhibitors, without adverse effects on renal function [31].

While these findings support feasibility and short-term benefit, the heterogeneity of AHF populations and relatively short follow-up durations necessitate cautious interpretation [31].

### 3.3. Mechanistic Basis of Heart Failure Benefit

The cardiovascular benefits of SGLT2 inhibitors appear to arise from a convergence of hemodynamic, metabolic, and cellular mechanisms, rather than a single dominant pathway. Osmotic diuresis and natriuresis promote effective decongestion, reducing preload and afterload while minimizing counter-regulatory neurohormonal activation, and may reduce reliance on loop diuretics [31].

From a metabolic perspective, SGLT2 inhibition promotes a shift toward ketone body utilization, providing a more energetically efficient substrate for the failing myocardium. At the cellular level, experimental studies suggest indirect modulation of the cardiac $Na^+/H^+$ exchanger (NHE1), leading to reduced intracellular sodium accumulation, improved calcium handling, and attenuation of oxidative stress—mechanisms collectively described within the "sodium-interactome" hypothesis. However, these effects are supported predominantly by preclinical data [40].

In parallel, preclinical models demonstrate attenuation of pro-inflammatory signaling, including reductions in IL-6, TNF-α, and NLRP3 inflammasome activation, along with mitigation of myocardial fibrosis and adverse remodeling. The extent to which these pathways translate into clinically meaningful benefit in humans remains an area of active investigation [41].

### 3.4. Post–Myocardial Infarction States

Interest has also grown in the potential role of SGLT2 inhibitors in mitigating post–myocardial infarction (MI) left ventricular dysfunction. Although underlying mechanisms are incompletely defined, metabolic modulation—particularly increased myocardial availability of β-hydroxybutyrate—has been proposed. β-Hydroxybutyrate may inhibit NLRP3 inflammasome activation, thereby reducing post-ischemic myocardial inflammation, an effect associated with reductions in NT-proBNP levels and improvements in ventricular function in clinical studies [42].

Observational data from a large Swedish registry suggested that early post-MI initiation of SGLT2 inhibitors was associated with reduced heart failure hospitalization and all-cause mortality. These findings were supported by randomized trials. In DAPA-MI,

https://doi.org/10.3390/biomedicines14010081

early dapagliflozin therapy in patients without prior diabetes or heart failure improved markers of ventricular recovery and reduced HF hospitalizations. Similarly, Hernandez et al. reported that empagliflozin reduced both first and recurrent HF hospitalizations in patients with recent MI complicated by left ventricular dysfunction or congestion. Notably, patients with chronic kidney disease or frailty appeared to derive particular benefit [43,44].

### 3.5. Differential Pathophysiology of Acute and Chronic Heart Failure

Heart failure represents a heterogeneous clinical syndrome encompassing acute heart failure (AHF) and chronic heart failure (CHF), each characterized by distinct yet overlapping pathophysiological processes. AHF is dominated by abrupt hemodynamic decompensation, intense neurohormonal activation, sodium and calcium dysregulation, endothelial dysfunction, and inflammatory signaling. In contrast, CHF reflects a sustained disease state driven by chronic neurohormonal activation, adverse ventricular remodeling, myocardial fibrosis, metabolic inflexibility, and progressive mitochondrial dysfunction [32,45,46].

### 3.6. Shared and Condition-Specific Mechanisms of SGLT2 Inhibition

Across both acute and chronic HF, SGLT2 inhibitors target shared mechanisms, including effective decongestion, reduction in preload and afterload, improved myocardial energetics, and attenuation of maladaptive neurohormonal signaling. These common pathways likely underlie the consistent reductions in HF hospitalization observed across HF phenotypes in large outcome trials [47].

In acute HF, benefits appear to be driven predominantly by early hemodynamic effects, plasma volume redistribution, and modulation of intracellular sodium handling during decompensation. In chronic HF, sustained SGLT2 inhibition engages longer-term disease-modifying mechanisms, including suppression of adverse remodeling, reduction in myocardial fibrosis and inflammation, stabilization of the cardiorenal axis, and improvement in metabolic efficiency [5].

### 3.7. Integrative Perspective

Post-MI states occupy a transitional position within the HF continuum, sharing mechanistic features with both acute decompensation and chronic remodeling. By influencing inflammatory pathways, myocardial energetics, and renal–cardiac interactions, SGLT2 inhibitors may plausibly modify the trajectory from acute ischemic injury to chronic HF, although definitive causal evidence remains incomplete [48].

Collectively, the mechanistic actions of SGLT2 inhibitors span the entire HF spectrum, integrating early hemodynamic and metabolic effects with long-term remodeling and renal benefits. This time-dependent, multi-level mechanistic profile provides a biologically plausible framework for their consistent efficacy across acute HF, chronic HF, and post–myocardial infarction settings [49].

## 4. Effects of SGLT2 Inhibitors on Renal Outcomes

Large-scale randomized trials have established that SGLT2 inhibitors (SGLT2i) provide clinically meaningful renoprotection in patients at risk of chronic kidney disease (CKD), initially observed as secondary outcomes in cardiovascular outcome trials (CVOTs). Across EMPA-REG OUTCOME, CANVAS, and DECLARE–TIMI 58, SGLT2i therapy was associated with a slower decline in estimated glomerular filtration rate (eGFR) and reduced progression of albuminuria, supporting a reproducible signal of kidney protection beyond glycemic control [27–30,36,50,51].

The CREDENCE trial was the first dedicated renal outcomes study in patients with type 2 diabetes (T2D) and diabetic kidney disease receiving maximally tolerated renin–angiotensin–aldosterone system (RAAS) blockade. Canagliflozin reduced the risk

https://doi.org/10.3390/biomedicines14010081

of the composite renal endpoint by approximately 30%, including sustained eGFR decline to very low levels, doubling of serum creatinine, kidney replacement therapy, or renal/cardiovascular death, with consistent benefit across baseline eGFR strata and a clinically relevant reduction in albuminuria [29,52].

Subsequently, DAPA-CKD extended these findings to a broader CKD population, demonstrating that dapagliflozin reduced the risk of sustained eGFR decline (≥50%), end-stage kidney disease, and renal or cardiovascular death irrespective of diabetes status [30]. EMPA-KIDNEY further strengthened the evidence base by enrolling patients with advanced CKD (eGFR 20–45 mL/min/1.73 $m^2$) or those with higher eGFR accompanied by significant albuminuria, with Ref. [53] showing that empagliflozin reduced the risk of kidney disease progression or cardiovascular death [31]. Together, these trials confirm that renal benefit persists even at lower eGFR values, where glycosuric effects are attenuated—supporting a largely glucose-independent mechanism [54].

### 4.1. Mechanistic Basis of Renoprotection: A Hierarchical Framework

Although the mechanistic basis of renoprotection is multifactorial, evidence is strongest for hemodynamic and tubuloglomerular pathways. By inhibiting proximal tubular sodium and glucose reabsorption, SGLT2i increase sodium delivery to the macula densa, restoring tubuloglomerular feedback and promoting adenosine-mediated afferent arteriolar vasoconstriction. This reduces intraglomerular pressure and single-nephron hyperfiltration, thereby limiting long-term nephron injury and slowing CKD progression [32,33]. Concurrent osmotic diuresis and natriuresis also contribute to reductions in blood pressure and intravascular volume [33].

Beyond these primary mechanisms, translational studies suggest additional contributions from improved endothelial function, reductions in oxidative stress and inflammatory signaling, and favorable effects on intrarenal hemodynamics—changes that may be reflected by reductions in the renal resistive index (RI) [34]. SGLT2i also consistently reduces albuminuria, plausibly through combined effects on intraglomerular pressure and mitigation of tubular toxicity associated with filtered albumin [37].

Several proposed molecular pathways—such as inhibition of epithelial-to-mesenchymal transition (EMT), modulation of mTORC1 signaling, and direct podocyte protection—are supported predominantly by experimental models and indirect human evidence [35,36,40,41]. Accordingly, these mechanisms should be considered hypothesis-generating, rather than definitive drivers of clinical outcome benefits, until validated by robust human studies.

### 4.2. Next-Generation SGLT2-Based Therapies and Renal Implications

Next-generation SGLT2-targeting therapies aim to extend therapeutic scope beyond conventional gliflozins by optimizing selectivity, pharmacokinetic profiles, tolerability, and potential tissue-specific effects (Tables 3–5) [55]. In some regions (e.g., Japan) [56], agents such as ipragliflozin, tofogliflozin, and luseogliflozin have been approved, while others—including bexagliflozin and dual-acting agents—remain under clinical development, often with broader metabolic or cardio-renal aims [1].

**Table 3.** Structural and Pharmacological Comparison.

| Feature | Conventional SGLT2i | Next-Generation SGLT2i |
|---|---|---|
| Chemical backbone | C-aryl glucoside | Modified glucoside/hybrid |
| SGLT2 selectivity | High | Ultra-high or dual |
| SGLT1 inhibition | Minimal | Partial (agent-specific) |
| PK profile | Standard | Optimized/prolonged |
| Tissue distribution | Predominantly renal | Multi-organ |

https://doi.org/10.3390/biomedicines14010081

**Table 4.** Clinical Evidence and Emerging Indications.

| Agent | Target | Key Trials | Novel Indications |
|---|---|---|---|
| Dapagliflozin | SGLT2 | DAPA-HF, DAPA-CKD | HF, CKD |
| Empagliflozin | SGLT2 | EMPEROR, EMPA-KIDNEY | HFpEF |
| Sotagliflozin | SGLT1/2 | SOLOIST-WHF | Acute HF |
| Emerging agents | SGLT2-based | Ongoing | Precision therapy |

**Table 5.** Structural and Pharmacological Differences between Conventional and Next-Generation SGLT2 Inhibitors.

| Feature | Conventional SGLT2 Inhibitors | Next-Generation SGLT2-Based Agents |
|---|---|---|
| Chemical backbone | C-aryl glucoside | Modified glucoside/hybrid structures |
| Molecular stability | High | Enhanced stability and binding kinetics |
| SGLT2 selectivity | High | Ultra-high or deliberately balanced |
| SGLT1 inhibition | Minimal or absent | Partial (agent-specific) |
| Intestinal effects | Limited | Delayed glucose absorption |
| Incretin activation | Minimal | Increased GLP-1 and GIP secretion |
| Pharmacokinetics | Standard half-life | Optimized exposure and duration |
| Tissue distribution | Predominantly renal | Multi-organ (renal, cardiac, vascular) |
| Glucose-independent effects | Secondary | Prominent and targeted |

A major innovation within this landscape is the development of dual SGLT1/SGLT2 inhibitors, which combine renal SGLT2 blockade with intestinal SGLT1 inhibition. While renal effects promote glycosuria and natriuresis, intestinal SGLT1 inhibition delays glucose absorption, blunts postprandial excursions, and enhances incretin signaling (GLP-1, GIP), potentially providing additive metabolic and cardiometabolic benefits [1].

Sotagliflozin is the most extensively studied dual inhibitor. In SOLOIST-WHF and SCORED, sotagliflozin was associated with early and sustained reductions in heart failure events and cardiovascular outcomes, including when initiated shortly after acute decompensation [42]. These data support the concept that dual transporter inhibition may be particularly relevant in high-risk clinical contexts, although direct comparative superiority over selective SGLT2 inhibitors remains uncertain in the absence of head-to-head trials.

Licogliflozin, another dual SGLT1/SGLT2 inhibitor evaluated primarily in phase 2 programs, in Ref. [57] has demonstrated favorable effects on weight and postprandial glucose handling, with signals of improvement in selected cardiometabolic biomarkers (e.g., NT-proBNP). However, available evidence remains limited by small sample sizes, short follow-up, and the absence of outcome-driven renal or cardiovascular endpoints; further studies are required to clarify long-term efficacy and tolerability, including potential gastrointestinal effects related to intestinal SGLT1 inhibition [2,57].

### *4.3. Expanded Indications and Ongoing Research*

The clinical positioning of SGLT2 inhibitors has expanded markedly beyond diabetes, supported by outcome trials in heart failure and CKD—including in non-diabetic

https://doi.org/10.3390/biomedicines14010081

populations. Trials such as DAPA-HF [28], DELIVER [58], EMPEROR-Reduced [59], EMPEROR-Preserved [60], DAPA-CKD, and EMPA-KIDNEY have established consistent cardio-renal benefits and influenced guideline recommendations in both cardiology and nephrology [3,4,61]. In acute settings, EMPULSE [62] demonstrated that initiating empagliflozin during hospitalization for acute decompensated heart failure is feasible and associated with improved short-term clinical outcomes, supporting early initiation strategies in selected patients [5].

### *4.4. Renal Outcomes: Clinical Effect Size, eGFR Contextualization, and Mechanistic Hierarchy*

Large randomized controlled trials have consistently demonstrated that SGLT2 inhibitors slow the progression of chronic kidney disease (CKD) across diabetic and non-diabetic populations. In DAPA-CKD, dapagliflozin reduced the primary composite renal endpoint by 39% (HR 0.61; 95% CI 0.51–0.72), corresponding to an absolute risk reduction of approximately 5.3% over 2.4 years and a number-needed-to-treat (NNT) of ~19. Similarly, EMPA-KIDNEY reported a relative risk reduction of 28% (HR 0.72; 95% CI 0.64–0.82), translating into an absolute risk reduction of ~3.8%, reflecting the broader and lower-risk population enrolled (Table 6) [63,64].

**Table 6.** Clinical Evidence Distinguishing Conventional and Next-Generation SGLT2 Inhibitors.

| Agent | Transporter Profile | Key Trials | Population | Distinct Clinical Features |
|---|---|---|---|---|
| Dapagliflozin | SGLT2 | DAPA-HF, DAPA-CKD | Chronic HF, CKD | Strong chronic cardiorenal protection |
| Empagliflozin | SGLT2 | EMPEROR-Reduced, EMPEROR-Preserved, EMPA-KIDNEY | HF spectrum, CKD | Robust HFpEF evidence |
| Canagliflozin | SGLT2 | CREDENCE | Diabetic CKD | Early renal outcome data |
| Sotagliflozin | SGLT1/SGLT2 | SOLOIST-WHF, SCORED | Recent worsening HF, CKD | Incretin-mediated and acute HF benefits |
| Emerging agents | SGLT2-based | Ongoing | Non-diabetic HF, post-MI | Precision-medicine strategies |

These absolute measures provide essential clinical context, highlighting that the magnitude of benefit varies according to baseline renal risk and disease stage.

### *4.5. Clinical Interpretation of eGFR Thresholds*

The efficacy of SGLT2 inhibitors has been demonstrated across a wide range of baseline eGFR values, including patients with eGFR as low as 20 mL/min/1.73 m$^2$. From a clinical standpoint, lower eGFR thresholds identify populations at higher absolute risk, in whom relative risk reductions yield greater absolute benefit [65].

However, reduced glycosuric efficacy at lower eGFR values underscores that renal protection is largely glucose-independent. Importantly, early eGFR "dips" observed after treatment initiation reflect hemodynamic adaptation rather than true nephron loss and should be interpreted cautiously in clinical practice [66].

### *4.6. Hierarchical Organization of Renoprotective Mechanisms*

To avoid overgeneralization, renoprotective mechanisms of SGLT2 inhibitors are best conceptualized within a hierarchical framework based on strength of evidence [67].

https://doi.org/10.3390/biomedicines14010081

### 4.7. Primary Mechanisms (Strong Human Clinical Evidence)

- Restoration of tubuloglomerular feedback and reduction in intraglomerular pressure;
- Hemodynamic stabilization of glomerular filtration;
- Reduction in albuminuria, strongly correlated with long-term renal outcomes.

### 4.8. Secondary Mechanisms (Supported by Translational and Indirect Clinical Data)

- Improved renal oxygenation through reduced proximal tubular workload;
- Attenuation of renal inflammation and oxidative stress;
- Modulation of renal hemodynamics within the cardiorenal axis.

### 4.9. Exploratory Mechanisms (Predominantly Preclinical Evidence)

Mechanisms such as epithelial–mesenchymal transition (EMT) inhibition, mTORC1 signaling modulation, and direct podocyte protection have been reported primarily in animal models or cellular systems. To date, direct human clinical evidence supporting these pathways remains limited, and their contribution to observed renal outcomes should be regarded as hypothesis-generating rather than definitive [67].

### 4.10. Expanded Indications of SGLT2 Inhibitors: Evidence, Guidelines, and Implementation Challenges

The clinical indications for SGLT2 inhibitors have expanded substantially over the past decade, evolving from glucose-lowering therapy in type 2 diabetes mellitus to agents with established roles in heart failure and chronic kidney disease. While this expansion is sometimes described as a "paradigm shift", such terminology should be reserved for contexts in which practice-changing evidence has been incorporated into international guidelines [68].

### 4.11. Guideline-Supported Expansion of Indications

Current guidelines from major professional societies now recommend SGLT2 inhibitors as foundational therapy in multiple non-glycemic indications. The 2021 and 2023 ESC Heart Failure Guidelines endorse SGLT2 inhibitors for patients with heart failure with reduced ejection fraction (HFrEF), and more recently for heart failure with preserved ejection fraction (HFpEF), independent of diabetes status. Similarly, AHA/ACC/HFSA guidelines include SGLT2 inhibitors as core components of guideline-directed medical therapy in heart failure.

In nephrology, KDIGO 2022 guidelines recommend SGLT2 inhibitors for patients with chronic kidney disease at risk of progression, including non-diabetic populations, based on robust outcome data from DAPA-CKD and EMPA-KIDNEY. These guideline endorsements provide a clear justification for considering SGLT2 inhibitors as disease-modifying therapies beyond glycemic control.

### 4.12. Acute and Early Use: EMPULSE Trial in Context

The EMPULSE trial evaluated the initiation of empagliflozin during hospitalization for acute heart failure and demonstrated an early net clinical benefit across a composite hierarchical endpoint. However, interpretation of these findings requires acknowledgment of several limitations. EMPULSE was a moderate-sized trial with a relatively short follow-up period, employed a composite endpoint combining clinical and patient-reported outcomes, and excluded patients with severe hypotension or advanced renal dysfunction. Consequently, while the results support the feasibility and short-term safety of early initiation, they do not establish definitive long-term outcome benefits or universal applicability across all acute heart failure phenotypes [69].

https://doi.org/10.3390/biomedicines14010081

### 4.13. Terminology and Evidence Thresholds

Accordingly, the expanded use of SGLT2 inhibitors can be considered a practice-changing development in selected conditions where guideline-level endorsement exists. In other emerging indications—such as post–myocardial infarction states, valvular heart disease, or acute decompensated settings—the evidence remains incomplete, and claims of paradigm change should be avoided until supported by adequately powered randomized trials [70].

### 4.14. Cost-Effectiveness and Access Considerations

Despite strong clinical evidence, real-world implementation of SGLT2 inhibitors is influenced by cost, reimbursement policies, and healthcare system constraints. Cost-effectiveness analyses generally support their use in heart failure and CKD, particularly in high-risk populations where absolute risk reduction is greatest. However, access disparities persist, especially in low- and middle-income settings, potentially limiting the population-level impact of these therapies [71].

Future research and policy efforts should address health economic considerations, optimize patient selection to maximize value, and ensure equitable access to these agents as indications continue to expand.

### 4.15. Timing of SGLT2 Inhibitor Initiation in Relation to TAVI: Hemodynamics, Risk, and Remodeling

Transcatheter aortic valve implantation (TAVI) induces a profound and abrupt alteration in cardiovascular hemodynamics, transitioning patients from a state of chronic pressure overload and fixed outflow obstruction to one of immediate afterload reduction. The timing of SGLT2 inhibitor initiation relative to TAVI, therefore, warrants careful consideration, as physiological priorities differ across the pre-procedural, peri-procedural, and post-procedural phases [72].

### 4.16. Pre-TAVI Initiation: Hemodynamic Vulnerability

Before TAVI, patients with severe aortic stenosis often exhibit preload dependence, limited cardiac output reserve, and impaired baroreflex responses. In this setting, the natriuretic and osmotic diuretic effects of SGLT2 inhibitors may theoretically increase the risk of hypotension, renal hypoperfusion, or syncope, particularly in low-flow or advanced disease phenotypes [72].

While pre-TAVI initiation could confer metabolic or renal benefits in selected patients treated for approved indications (e.g., heart failure or diabetes), routine initiation specifically for AS-related indications cannot be justified in the absence of prospective safety data [72].

### 4.17. Peri-Procedural Considerations: Procedural Risk and Renal Protection

The peri-TAVI period is characterized by hemodynamic instability, contrast exposure, inflammatory activation, and transient renal stress. Although SGLT2 inhibitors have been associated with renal protection in other clinical contexts, their initiation immediately before or during the peri-procedural window raises concerns regarding volume depletion, hypotension, and acute kidney injury, especially when combined with procedural fasting and contrast load.

Accordingly, available evidence does not support routine peri-procedural initiation, and most procedural protocols currently favor temporary withholding of agents with diuretic or hemodynamic effects [73].

https://doi.org/10.3390/biomedicines14010081

### *4.18. Post-TAVI Initiation: Reverse Remodeling and Stabilization*

Following successful TAVI, relief of valvular obstruction leads to rapid hemodynamic stabilization, improved forward flow, and initiation of reverse left ventricular remodeling, including regression of hypertrophy and gradual improvement in diastolic function. This phase may represent a more physiologically favorable window for SGLT2 inhibitor initiation [74].

In the post-TAVI setting, SGLT2 inhibitors could theoretically support:

- Optimization of heart failure therapy;
- Reduction in residual congestion;
- Renal protection during remodeling;
- Metabolic efficiency during myocardial recovery [74].

However, these potential benefits remain hypothesis-generating, as no randomized trials have evaluated the impact of SGLT2 inhibitor initiation on post-TAVI remodeling, outcomes, or safety.

### *4.19. Integrative Perspective*

Overall, renal outcome benefits of SGLT2 inhibitors are robustly supported by clinical trial data, particularly when interpreted through absolute risk metrics and patient-specific baseline risk. Mechanistic explanations should prioritize pathways validated in humans, while acknowledging that several molecular effects remain incompletely translated to clinical settings [75].

In summary, expansion of SGLT2 inhibitor indications is best viewed as a graduated evolution grounded in guideline-supported evidence, rather than a uniform paradigm shift across all clinical contexts. Careful alignment of terminology with evidence strength, acknowledgment of trial limitations, and consideration of cost-effectiveness are essential for responsible translation into clinical practice.

Taken together, the post-TAVI period, once hemodynamic stability has been achieved, appears to be the most rational phase for consideration of SGLT2 inhibitor initiation when clinically indicated. Pre- or peri-procedural initiation should be approached with caution and individualized assessment. Importantly, any discussion of timing must acknowledge the absence of randomized evidence, and clinical decisions should prioritize established indications and patient-specific risk profiles [76].

## 5. Conclusions

### *5.1. Pleiotropic and Emerging Biological Effects of SGLT2 Inhibitors*

Beyond their established hemodynamic and metabolic actions, SGLT2 inhibitors are increasingly recognized to exert a range of pleiotropic biological effects that may contribute to cardiovascular, renal, and systemic protection. These include anti-inflammatory, antifibrotic, and antioxidative actions, which have been implicated in improved vascular and myocardial remodeling, attenuation of interstitial fibrosis, and modulation of maladaptive neurohormonal activation [77].

Emerging experimental and early clinical evidence also suggests potential roles for SGLT2 inhibition in neuroprotection and in the management of metabolic liver disease, including non-alcoholic fatty liver disease (NAFLD) and non-alcoholic steatohepatitis (NASH). In these contexts, SGLT2 inhibitors may improve hepatic steatosis, insulin sensitivity, and systemic inflammatory burden, although robust outcome data remain limited [78].

Collectively, these expanding biological effects reinforce a shift in how SGLT2 inhibitors are conceptualized—not merely as glucose-lowering agents, but as multisystem modulators acting across the heart, kidney, vasculature, and metabolic networks. Ongoing translational and clinical research will be essential to clarify the clinical relevance of these

https://doi.org/10.3390/biomedicines14010081

non-glycemic mechanisms and to define optimal therapeutic combinations and timing strategies [79].

*5.2. Conclusions and Future Perspectives*

This review highlights the evolving pharmacological landscape of SGLT2-based therapies, reflecting a transition from conventional glucose-lowering drugs toward next-generation disease-modifying interventions in cardiovascular and renal medicine. While first-generation SGLT2 inhibitors have firmly established robust benefits in heart failure and chronic kidney disease across diabetic and non-diabetic populations, emerging strategies suggest that newer SGLT2-based approaches may engage additional and qualitatively distinct mechanisms.

Next-generation SGLT2 inhibitors are characterized by structural and pharmacokinetic innovations, refined or deliberately balanced SGLT2/SGLT1 selectivity profiles, and enhanced engagement of extra-renal and glucose-independent pathways, including myocardial energetics, inflammatory modulation, vascular function, and gut–heart signaling. Dual SGLT1/SGLT2 inhibition further broadens therapeutic scope by incorporating incretin-mediated mechanisms and intestinal glucose modulation, which may be particularly relevant in high-risk or acute clinical settings.

Importantly, accumulating evidence indicates that the renal benefits of SGLT2 inhibition are disease-specific rather than uniform, with differential mechanistic relevance across proteinuric glomerular diseases, tubulointerstitial CKD, hereditary nephropathies, and renal anemia. Similarly, in heart failure, SGLT2-based therapies appear to target both acute hemodynamic decompensation and chronic myocardial remodeling, supporting their use across the heart failure continuum, including post–myocardial infarction states.

At the same time, important uncertainties persist. The extent to which next-generation agents confer incremental clinical benefit beyond established SGLT2 inhibitors remains unclear, owing to the lack of head-to-head randomized trials and the limited availability of outcome-driven evidence for newer compounds. Many proposed mechanisms—particularly those involving direct myocardial effects, intracellular signaling modulation, or podocyte protection—are supported predominantly by preclinical or indirect data, and their relevance in human disease requires further validation.

Key knowledge gaps include the optimal clinical positioning of dual SGLT1/SGLT2 inhibition; safety and efficacy in underrepresented populations such as patients with severe valvular heart disease, advanced CKD, or acute hemodynamic instability; and the long-term implications of early initiation strategies in acute heart failure. Additional challenges relate to implementation, including cost-effectiveness, access, and integration into guideline-directed therapy without unintended adverse interactions.

Finally, a balanced appraisal of negative, neutral, and conflicting evidence underscores that the benefits of SGLT2-based therapies are context-dependent rather than universal. Recognition of these limitations is essential to avoid overgeneralization, guide appropriate patient selection, and frame future research priorities.

In conclusion, next-generation SGLT2-based therapies represent a promising yet still evolving chapter in cardiometabolic and renal medicine. While current evidence supports their role as foundational disease-modifying agents in selected indications, future progress will depend on rigorously designed randomized trials, human-validated mechanistic studies, and precision-oriented clinical strategies to determine where pharmacological innovation translates into meaningful incremental benefit.

**Author Contributions:** Conceptualization, D.M. and S.R.D.; methodology, D.D.S.; software, D.M.; resources, D.M.; data curation, D.D.S.; writing—original draft preparation, D.M.; writing—review

https://doi.org/10.3390/biomedicines14010081

and editing, D.M. and D.D.S.; visualization, S.R.D.; supervision, S.R.D. All authors have read and agreed to the published version of the manuscript.

**Funding:** This research received no external funding.

**Institutional Review Board Statement:** Not applicable.

**Informed Consent Statement:** Not applicable.

**Data Availability Statement:** The original contributions presented in this study are included in the article. Further inquiries can be directed to the corresponding author.

**Acknowledgments:** We would like to acknowledge Victor Babes University of Medicine and Pharmacy, Timisoara, for their support in covering the costs of publication of this research paper.

**Conflicts of Interest:** The authors declare no conflicts of interest.

## References

1. Bonner, C.; Kerr-Conte, J.; Gmyr, V.; Queniat, G.; Moerman, E.; Thévenet, J.; Beaucamps, C.; Delalleau, N.; Popescu, I.; Malaisse, W.J.; et al. Inhibition of the glucose transporter SGLT2 with dapagliflozin in pancreatic alpha cells triggers glucagon secretion. *Nat. Med.* **2015**, *21*, 512–517. [CrossRef] [PubMed]
2. Usman, M.S.; Siddiqi, T.J.; Anker, S.D.; Bakris, G.L.; Bhatt, D.L.; Filippatos, G.; Fonarow, G.C.; Greene, S.J.; Januzzi, J.L., Jr.; Khan, M.S.; et al. Effect of SGLT2 Inhibitors on Cardiovascular Outcomes Across Various Patient Populations. *J. Am. Coll. Cardiol.* **2023**, *81*, 2377–2387. [CrossRef] [PubMed]
3. Scisciola, L.; Paolisso, P.; Belmonte, M.; Gallinoro, E.; Delrue, L.; Taktaz, F.; Fontanella, R.A.; Degrieck, I.; Pesapane, A.; Casselman, F.; et al. Myocardial sodium-glucose cotransporter 2 expression and cardiac remodelling in patients with severe aortic stenosis: The BIO-AS study. *Eur. J. Heart Fail.* **2024**, *26*, 471–482. [CrossRef] [PubMed]
4. Madaan, T.; Akhtar, M.; Najmi, A.K. Sodium glucose CoTransporter 2 (SGLT2) inhibitors: Current status and future perspective. *Eur. J. Pharm. Sci.* **2016**, *93*, 244–252. [CrossRef]
5. Bianco, M.; Di Nora, C.; De Maria, R.; Mousavi, A.H.; Carigi, S.; De Gennaro, L.; Manca, P.; Matassini, M.V.; Rizzello, V.; Tinti, M.D.; et al. SGLT2-is in Acute Heart Failure. *J. Clin. Med.* **2025**, *14*, 7799. [CrossRef]
6. Brown, E.; Wilding, J.P.H.; Alam, U.; Barber, T.M.; Karalliedde, J.; Cuthbertson, D.J. The expanding role of SGLT2 inhibitors beyond glucose-lowering to cardiorenal protection. *Ann. Med.* **2021**, *53*, 2072–2089. [CrossRef]
7. Hou, Y.C.; Zheng, C.M.; Yen, T.H.; Lu, K.C. Molecular Mechanisms of SGLT2 Inhibitor on Cardiorenal Protection. *Int. J. Mol. Sci.* **2020**, *21*, 7833. [CrossRef]
8. Vallon, V. State-of-the-Art-Review: Mechanisms of Action of SGLT2 Inhibitors and Clinical Implications. *Am. J. Hypertens.* **2024**, *37*, 841–852. [CrossRef]
9. Vaduganathan, M.; Docherty, K.F.; Claggett, B.L.; Jhund, P.S.; de Boer, R.A.; Hernandez, A.F.; Inzucchi, S.E.; Kosiborod, M.N.; Lam, C.S.P.; Martinez, F.; et al. SGLT-2 inhibitors in patients with heart failure: A comprehensive meta-analysis of five randomised controlled trials. *Lancet* **2022**, *400*, 757–767, Erratum in *Lancet* **2023**, *401*, 104.. [CrossRef]
10. Aggarwal, R.; Bhatt, D.L.; Szarek, M.; Cannon, C.P.; Leiter, L.A.; Inzucchi, S.E.; Lopes, R.D.; McGuire, D.K.; Lewis, J.B.; Riddle, M.C.; et al. Effect of sotagliflozin on major adverse cardiovascular events: A prespecified secondary analysis of the SCORED randomised trial. *Lancet Diabetes Endocrinol.* **2025**, *13*, 321–332. [CrossRef]
11. Kamrul-Hasan, A.; Mondal, S.; Zahura Aalpona, F.T.; Nagendra, L.; Dutta, D. Role of Sodium-Glucose Cotransporter-2 Inhibitors in Managing Polycystic Ovary Syndrome: A Systematic Review. *touchREV Endocrinol.* **2025**, *21*, 32–41. [CrossRef] [PubMed]
12. Lau, K.T.K.; Wong, C.K.H.; Au, I.C.H.; Lau, W.C.Y.; Man, K.K.C.; Chui, C.S.L.; Wong, I.C.K. Switching to Versus Addition of Incretin-Based Drugs Among Patients with Type 2 Diabetes Taking Sodium-Glucose Cotransporter-2 Inhibitors. *J. Am. Heart Assoc.* **2022**, *11*, e023489. [CrossRef] [PubMed]
13. Gorboulev, V.; Schürmann, A.; Vallon, V.; Kipp, H.; Jaschke, A.; Klessen, D.; Friedrich, A.; Scherneck, S.; Rieg, T.; Cunard, R.; et al. $Na^{+}$-D-glucose cotransporter SGLT1 is pivotal for intestinal glucose absorption and glucose-dependent incretin secretion. *Diabetes* **2012**, *61*, 187–196. [PubMed]
14. Inzucchi, S.E.; Zinman, B.; Fitchett, D.; Wanner, C.; Ferrannini, E.; Schumacher, M.; Schmoor, C.; Ohneberg, K.; Johansen, O.E.; George, J.T.; et al. How Does Empagliflozin Reduce Cardiovascular Mortality? Insights From a Mediation Analysis of the EMPA-REG OUTCOME Trial. *Diabetes Care* **2018**, *41*, 356–363.
15. Fitchett, D.; Inzucchi, S.E.; Zinman, B.; Wanner, C.; Schumacher, M.; Schmoor, C.; Ohneberg, K.; Ofstad, A.P.; Salsali, A.; George, J.T.; et al. Mediators of the improvement in heart failure outcomes with empagliflozin in the EMPA-REG OUTCOME trial. *ESC Heart Fail.* **2021**, *8*, 4517–4527.

https://doi.org/10.3390/biomedicines14010081

16. Hu, S.; Tang, T.; Yu, Q.; Tong, X.; You, Y.; Zhang, S.; Chen, C.; Tang, J.; Wang, C.; Wang, H.; et al. Cardiovascular Outcome of the SGLT2 Inhibitor in Acute Myocardial Infarction: A Meta-Analysis. *Rev. Cardiovasc. Med.* **2025**, *26*, 26136. [CrossRef]
17. Marfella, R.; Scisciola, L.; D'Onofrio, N.; Maiello, C.; Trotta, M.C.; Sardu, C.; Panarese, I.; Ferraraccio, F.; Capuano, A.; Barbieri, M.; et al. Sodium-glucose cotransporter-2 (SGLT2) expression in diabetic and non-diabetic failing human cardiomyocytes. *Pharmacol. Res.* **2022**, *184*, 106448.
18. Paolisso, P.; Belmonte, M.; Gallinoro, E.; Scarsini, R.; Bergamaschi, L.; Portolan, L.; Armillotta, M.; Esposito, G.; Moscarella, E.; Benfari, G.; et al. SGLT2-inhibitors in diabetic patients with severe aortic stenosis and cardiac damage undergoing transcatheter aortic valve implantation (TAVI). *Cardiovasc. Diabetol.* **2024**, *23*, 420, Erratum in *Cardiovasc. Diabetol.* **2025**, *24*, 14.. [CrossRef]
19. Raposeiras-Roubin, S.; Amat-Santos, I.J.; Rossello, X.; González Ferreiro, R.; González Bermúdez, I.; Lopez Otero, D.; Nombela-Franco, L.; Gheorghe, L.; Diez, J.L.; Baladrón Zorita, C.; et al. Dapagliflozin in Patients Undergoing Transcatheter Aortic-Valve Implantation. *N. Engl. J. Med.* **2025**, *392*, 1396–1405.
20. Shah, T.; Zhang, Z.; Shah, H.; Fanaroff, A.C.; Nathan, A.S.; Parise, H.; Lutz, J.; Sugeng, L.; Bellumkonda, L.; Redfors, B.; et al. Effect of Sodium-Glucose Cotransporter-2 Inhibitors on the Progression of Aortic Stenosis. *JACC Cardiovasc. Interv.* **2025**, *18*, 738–748.
21. Baptiste, T.M.G.; Rodero, C.; Sillett, C.P.; Strocchi, M.; Lanyon, C.W.; Augustin, C.M.; Lee, A.W.C.; Solís-Lemus, J.A.; Roney, C.H.; Ennis, D.B.; et al. Regional heterogeneity in left atrial stiffness impacts passive deformation in a cohort of patient-specific models. *PLoS Comput. Biol.* **2025**, *21*, e1013656. [CrossRef]
22. Matthews, J.; Herat, L.; Schlaich, M.P.; Matthews, V. The Impact of SGLT2 Inhibitors in the Heart and Kidneys Regardless of Diabetes Status. *Int. J. Mol. Sci.* **2023**, *24*, 14243. [CrossRef] [PubMed]
23. Ndumele, C.E.; Rangaswami, J.; Chow, S.L.; Neeland, I.J.; Tuttle, K.R.; Khan, S.S.; Coresh, J.; Mathew, R.O.; Baker-Smith, C.M.; Carnethon, M.R.; et al. Cardiovascular-Kidney-Metabolic Health: A Presidential Advisory From the American Heart Association. *Circulation* **2023**, *148*, 1606–1635, Erratum in *Circulation* **2024**, *149*, e1023.. [CrossRef] [PubMed]
24. Curaj, A.; Vanholder, R.; Loscalzo, J.; Quach, K.; Wu, Z.; Jankowski, V.; Jankowski, J. Cardiovascular Consequences of Uremic Metabolites: An Overview of the Involved Signaling Pathways. *Circ. Res.* **2024**, *134*, 592–613. [CrossRef] [PubMed]
25. Chen, S.; Coronel, R.; Hollmann, M.W.; Weber, N.C.; Zuurbier, C.J. Direct cardiac effects of SGLT2 inhibitors. *Cardiovasc. Diabetol.* **2022**, *21*, 45. [CrossRef]
26. Dai, H.; Lee, Y.A.; Bian, J.; Guo, J. Heterogeneous Effects of Sodium-Glucose Cotransporter-2 Inhibitors on Acute Kidney Injury: A Causal Learning Approach. *medRxiv* **2025**. [CrossRef]
27. Mohebi, R.; Liu, Y.; Felker, G.M.; Prescott, M.F.; Ward, J.H.; PiÑA, I.L.; Butler, J.; Solomon, S.D.; Januzzi, J.L. Heart Failure Duration and Mechanistic Efficacy of Sacubitril/Valsartan in Heart Failure with Reduced Ejection Fraction. *J. Card. Fail.* **2022**, *28*, 1673–1682. [CrossRef]
28. McMurray, J.J.V.; Solomon, S.D.; Inzucchi, S.E.; Køber, L.; Kosiborod, M.N.; Martinez, F.A.; Ponikowski, P.; Sabatine, M.S.; Anand, I.S.; Bělohlávek, J.; et al. Dapagliflozin in Patients with Heart Failure and Reduced Ejection Fraction. *N. Engl. J. Med.* **2019**, *381*, 1995–2008. [CrossRef]
29. Packer, M.; Januzzi, J.L.; Ferreira, J.P.; Anker, S.D.; Butler, J.; Filippatos, G.; Pocock, S.J.; Brueckmann, M.; Jamal, W.; Cotton, D.; et al. Concentration-dependent clinical and prognostic importance of high-sensitivity cardiac troponin T in heart failure and a reduced ejection fraction and the influence of empagliflozin: The EMPEROR-Reduced trial. *Eur. J. Heart Fail.* **2021**, *23*, 1529–1538. [CrossRef]
30. Solomon, S.D.; de Boer, R.A.; DeMets, D.; Hernandez, A.F.; Inzucchi, S.E.; Kosiborod, M.N.; Lam, C.S.P.; Martinez, F.; Shah, S.J.; Lindholm, D.; et al. Dapagliflozin in heart failure with preserved and mildly reduced ejection fraction: Rationale and design of the DELIVER trial. *Eur. J. Heart Fail.* **2021**, *23*, 1217–1225. [CrossRef]
31. Voors, A.A.; Angermann, C.E.; Teerlink, J.R.; Collins, S.P.; Kosiborod, M.; Biegus, J.; Ferreira, J.P.; Nassif, M.E.; Psotka, M.A.; Tromp, J.; et al. The SGLT2 inhibitor empagliflozin in patients hospitalized for acute heart failure: A multinational randomized trial. *Nat. Med.* **2022**, *28*, 568–574. [CrossRef] [PubMed]
32. Raffaello, W.; Henrina, J.; Huang, I.; Lim, M.; Suciadi, L.; Siswanto, B.; Pranata, R. Clinical Characteristics of De Novo Heart Failure and Acute Decompensated Chronic Heart Failure: Are They Distinctive Phenotypes That Contribute to Different Outcomes? *Card. Fail. Rev.* **2021**, *7*, e02. [CrossRef] [PubMed]
33. Mazzotta, R.; Garofalo, M.; Salvi, S.; Orlandi, M.; Marcaccini, G.; Susini, P.; Checchi, L.; Palazzuoli, A.; Di Mario, C.; Pieroni, M.; et al. Early administration of SGLT2 inhibitors in hospitalized patients: A practical guidance from the current evidence. *ESC Heart Fail.* **2025**, *12*, 2631–2642. [CrossRef] [PubMed]
34. McLean, P.; Bennett, J.; Trey Woods, E.; Chandrasekhar, S.; Newman, N.; Mohammad, Y.; Khawaja, M.; Rizwan, A.; Siddiqui, R.; Birnbaum, Y.; et al. SGLT2 inhibitors across various patient populations in the era of precision medicine: The multidisciplinary team approach. *npj Metab. Health Dis.* **2025**, *3*, 29. [CrossRef]
35. Preda, A.; Montecucco, F.; Carbone, F.; Camici, G.G.; Lüscher, T.F.; Kraler, S.; Liberale, L. SGLT2 inhibitors: From glucose-lowering to cardiovascular benefits. *Cardiovasc. Res.* **2024**, *120*, 443–460. [CrossRef]

36. Wiviott, S.D.; Raz, I.; Bonaca, M.P.; Mosenzon, O.; Kato, E.T.; Cahn, A.; Silverman, M.G.; Zelniker, T.A.; Kuder, J.F.; Murphy, S.A.; et al. Dapagliflozin and Cardiovascular Outcomes in Type 2 Diabetes. *N. Engl. J. Med.* **2019**, *380*, 347–357. [CrossRef]
37. Gao, H.; Wei, Q.; Zou, A.; Yu, K.; Song, D.; Li, J.; Han, H.; Liu, A. Evaluation of three mechanisms of action (SGLT2 inhibitors, GLP-1 receptor agonists, and sulfonylureas) in treating type 2 diabetes with heart failure: A systematic review and network meta-analysis of RCTs. *Front. Endocrinol.* **2025**, *16*, 1562815. [CrossRef]
38. Desai, N.; Olewinska, E.; Famulska, A.; Remuzat, C.; Francois, C.; Folkerts, K. Heart failure with mildly reduced and preserved ejection fraction: A review of disease burden and remaining unmet medical needs within a new treatment landscape. *Heart Fail. Rev.* **2024**, *29*, 631–662. [CrossRef]
39. Cherbi, M.; Lairez, O.; Baudry, G.; Gautier, P.; Roubille, F.; Delmas, C. Early Initiation of Sodium–Glucose Cotransporter 2 Inhibitors in Acute Heart Failure: A Systematic Review and Meta-Analysis. *J. Am. Heart Assoc.* **2025**, *14*, e039105. [CrossRef]
40. Saucedo-Orozco, H.; Voorrips, S.N.; Yurista, S.R.; de Boer, R.A.; Westenbrink, B.D. SGLT2 Inhibitors and Ketone Metabolism in Heart Failure. *J. Lipid Atheroscler.* **2022**, *11*, 1–19. [CrossRef]
41. Xu, W.; Huang, Y.; Zhou, R. NLRP3 inflammasome in neuroinflammation and central nervous system diseases. *Cell. Mol. Immunol.* **2025**, *22*, 341–355. [CrossRef] [PubMed]
42. Xu, M.L.; Wang, H.; Ouyang, D.; Qi, H.; Li, X.H. The impact of SGLT2 inhibitors on cardiac remodeling after myocardial infarction: An updated meta-analysis of randomized controlled trials. *Front. Pharmacol.* **2025**, *16*, 1699066. [CrossRef] [PubMed]
43. Svanström, H.; Mkoma, G.F.; Hviid, A.; Pasternak, B. SGLT-2 inhibitors and mortality among patients with heart failure with reduced ejection fraction: Linked database study. *BMJ* **2024**, *387*, e080925. [CrossRef] [PubMed]
44. Semirani-Nezhad, D.; Soleimani, H.; Taebi, M.; Roozbehi, K.; Jahangiri, S.; Sattartabar, B.; Takaloo, F.; Parastooei, B.; Asfa, E.; Salabat, D.; et al. Early initiation of SGLT2 inhibitors in acute myocardial infarction and cardiovascular outcomes, an updated systematic review and meta-analysis. *BMC Cardiovasc. Disord.* **2025**, *25*, 527. [CrossRef]
45. Arrigo, M.; Jessup, M.; Mullens, W.; Reza, N.; Shah, A.M.; Sliwa, K.; Mebazaa, A. Acute heart failure. *Nat. Rev. Dis. Primers* **2020**, *6*, 16. [CrossRef]
46. Schwinger, R.H.G. Pathophysiology of heart failure. *Cardiovasc. Diagn. Ther.* **2021**, *11*, 263–276. [CrossRef]
47. Crispino, S.P.; Segreti, A.; Nafisio, V.; Valente, D.; Crisci, F.; Ferro, A.; Cavallari, I.; Nusca, A.; Ussia, G.P.; Grigioni, F. The Role of SGLT2-Inhibitors Across All Stages of Heart Failure and Mechanisms of Early Clinical Benefit: From Prevention to Advanced Heart Failure. *Biomedicines* **2025**, *13*, 608. [CrossRef]
48. Jiang, H.; Fang, T.; Cheng, Z. Mechanism of heart failure after myocardial infarction. *J. Int. Med. Res.* **2023**, *51*, 3000605231202573. [CrossRef]
49. Patel, R.B. Mechanistic Insights of SGLT2 Inhibition in Heart Failure Through Proteomics: Appreciating the Cardiovascular-Kidney Connection. *J. Am. Coll. Cardiol.* **2024**, *84*, 1995–1998. [CrossRef]
50. Soliman, A.R.; Elkhatib, M.; El-Khashab, S.; Darwish, R.A.; Fayed, A.; Abdelaziz, T.S.; Hammad, H.; Ahmed, R.M.; Maamoun, H.A. Dual-faced guardians: SGLT2 inhibitors' kidney protection and health challenges: A position statement by Kasralainy nephrology group (KANG). *Diabetol. Metab. Syndr.* **2025**, *17*, 214. [CrossRef]
51. Mark, P.B.; Sarafidis, P.; Ekart, R.; Ferro, C.J.; Balafa, O.; Fernandez-Fernandez, B.; Herrington, W.G.; Rossignol, P.; Del Vecchio, L.; Valdivielso, J.M.; et al. SGLT2i for evidence-based cardiorenal protection in diabetic and non-diabetic chronic kidney disease: A comprehensive review by EURECA-m and ERBP working groups of ERA. *Nephrol. Dial. Transplant.* **2023**, *38*, 2444–2455. [CrossRef] [PubMed]
52. Perkovic, V.; Jardine, M.J.; Neal, B.; Bompoint, S.; Heerspink, H.J.L.; Charytan, D.M.; Edwards, R.; Agarwal, R.; Bakris, G.; Bull, S.; et al. Canagliflozin and Renal Outcomes in Type 2 Diabetes and Nephropathy. *N. Engl. J. Med.* **2019**, *380*, 2295–2306. [CrossRef] [PubMed]
53. The EMPA-KIDNEY Collaborative Group. Empagliflozin in Patients with Chronic Kidney Disease. *N. Engl. J. Med.* **2023**, *388*, 117–127. [CrossRef] [PubMed]
54. Heerspink, H.J.L.; Stefánsson, B.V.; Correa-Rotter, R.; Chertow, G.M.; Greene, T.; Hou, F.-F.; Mann, J.F.E.; McMurray, J.J.V.; Lindberg, M.; Rossing, P.; et al. Dapagliflozin in Patients with Chronic Kidney Disease. *N. Engl. J. Med.* **2020**, *383*, 1436–1446. [CrossRef]
55. Checa-Ros, A.; Okojie, O.J.; D'Marco, L. SGLT2 Inhibitors: Multifaceted Therapeutic Agents in Cardiometabolic and Renal Diseases. *Metabolites* **2025**, *15*, 536. [CrossRef]
56. Poole, R.M.; Dungo, R.T. Ipragliflozin: First Global Approval. *Drugs* **2014**, *74*, 611–617. [CrossRef]
57. Bays, H.E.; Kozlovski, P.; Shao, Q.; Proot, P.; Keefe, D. Licogliflozin, a Novel SGLT1 and 2 Inhibitor: Body Weight Effects in a Randomized Trial in Adults with Overweight or Obesity. *Obesity* **2020**, *28*, 870–881. [CrossRef]
58. Solomon, S.D.; McMurray, J.J.V.; Claggett, B.; Boer, R.A.d.; DeMets, D.; Hernandez, A.F.; Inzucchi, S.E.; Kosiborod, M.N.; Lam, C.S.P.; Martinez, F.; et al. Dapagliflozin in Heart Failure with Mildly Reduced or Preserved Ejection Fraction. *N. Engl. J. Med.* **2022**, *387*, 1089–1098. [CrossRef]

https://doi.org/10.3390/biomedicines14010081

59. Packer, M.; Anker, S.D.; Butler, J.; Filippatos, G.; Pocock, S.J.; Carson, P.; Januzzi, J.; Verma, S.; Tsutsui, H.; Brueckmann, M.; et al. Cardiovascular and Renal Outcomes with Empagliflozin in Heart Failure. *N. Engl. J. Med.* **2020**, *383*, 1413–1424. [CrossRef]
60. Anker, S.D.; Butler, J.; Filippatos, G.; Ferreira, J.P.; Bocchi, E.; Böhm, M.; Rocca, H.-P.B.L.; Choi, D.-J.; Chopra, V.; Chuquiure-Valenzuela, E.; et al. Empagliflozin in Heart Failure with a Preserved Ejection Fraction. *N. Engl. J. Med.* **2021**, *385*, 1451–1461. [CrossRef]
61. Song, S.; Guo, Y.; Lin, Y.; Gao, H.; Ren, H. Impact of Empagliflozin on Cardiovascular Outcomes and Renal Function in Patients with Obesity and Type 2 Diabetes: A Retrospective Cohort Study. *Diabetes Ther.* **2025**, *16*, 1451–1463. [CrossRef] [PubMed]
62. Kosiborod, M.N.; Angermann, C.E.; Collins, S.P.; Teerlink, J.R.; Ponikowski, P.; Biegus, J.; Comin-Colet, J.; Ferreira, J.P.; Mentz, R.J.; Nassif, M.E.; et al. Effects of Empagliflozin on Symptoms, Physical Limitations, and Quality of Life in Patients Hospitalized for Acute Heart Failure: Results From the EMPULSE Trial. *Circulation* **2022**, *146*, 279–288. [CrossRef]
63. Podestà, M.A.; Sabiu, G.; Galassi, A.; Ciceri, P.; Cozzolino, M. SGLT2 Inhibitors in Diabetic and Non-Diabetic Chronic Kidney Disease. *Biomedicines* **2023**, *11*, 279. [CrossRef]
64. Cherney, D.Z.I.; Verma, S. DAPA-CKD: The Beginning of a New Era in Renal Protection. *JACC Basic. Transl. Sci.* **2021**, *6*, 74–77. [CrossRef]
65. Dong, Z.; Mo, W.; Ling, Z.; Hou, L.; Deng, T. Efficacy of SGLT2 inhibitors on acute kidney injury in patients with chronic kidney disease, cardiovascular disease, and type 2 diabetes: A meta-analysis. *J. Natl. Med. Assoc.* **2025**, *117*, 458–469. [CrossRef]
66. Cherney David, Z.; Odutayo, A.; Aronson, R.; Ezekowitz, J.; Parker John, D. Sodium Glucose Cotransporter-2 Inhibition and Cardiorenal Protection. *JACC* **2019**, *74*, 2511–2524. [CrossRef]
67. Nishiyama, A.; Kitada, K. Possible renoprotective mechanisms of SGLT2 inhibitors. *Front. Med.* **2023**, *10*, 1115413. [CrossRef]
68. Wong, E.; Nguyen, T.V. Expanding the Role of SGLT2 Inhibitors Beyond Diabetes: A Case-Based Approach. *Sr. Care Pharm.* **2023**, *38*, 233–238. [CrossRef]
69. Biegus, J.; Voors, A.A.; Collins, S.P.; Kosiborod, M.N.; Teerlink, J.R.; Angermann, C.E.; Tromp, J.; Ferreira, J.P.; Nassif, M.E.; Psotka, M.A.; et al. Impact of empagliflozin on decongestion in acute heart failure: The EMPULSE trial. *Eur. Heart J.* **2023**, *44*, 41–50. [CrossRef]
70. Cuttone, A.; Cannavò, V.; Abdullah, R.M.S.; Fugazzotto, P.; Arena, G.; Brancati, S.; Muscarà, A.; Morace, C.; Quartarone, C.; Ruggeri, D.; et al. Expanding the Use of SGLT2 Inhibitors in T2D Patients Across Clinical Settings. *Cells* **2025**, *14*, 668. [CrossRef]
71. Li, X.; Hoogenveen, R.; El Alili, M.; Knies, S.; Wang, J.; Beulens, J.W.J.; Elders, P.J.M.; Nijpels, G.; van Giessen, A.; Feenstra, T.L. Cost-Effectiveness of SGLT2 Inhibitors in a Real-World Population: A MICADO Model-Based Analysis Using Routine Data from a GP Registry. *Pharmacoeconomics* **2023**, *41*, 1249–1262. [CrossRef] [PubMed]
72. Wu, C.Y.; Shah, B.R.; Sharma, A.; Sheng, Y.; Liu, P.P.; Kopp, A.; Saskin, R.; Edwards, J.D.; Swardfager, W. Timing of sodium-glucose cotransporter 2 inhibitor initiation and post-discharge outcomes in acute heart failure with diabetes: A population-based cohort study. *Eur. J. Heart Fail.* **2025**, *27*, 307–316. [CrossRef] [PubMed]
73. Heyman, S.N.; Aronson, D.; Abassi, Z. SGLT2 Inhibitors and the Risk of Contrast-Associated Nephropathy Following Angiographic Intervention: Contradictory Concepts and Clinical Outcomes. *Int. J. Mol. Sci.* **2024**, *25*, 10759. [CrossRef] [PubMed]
74. An, J.; Wang, H.; Liao, H.; Yan, F.; Shi, F.; Liu, S. Transcatheter aortic valve implantation improves left ventricular function and hemodynamics and reduces severe ventricular arrhythmias in patients with severe aortic stenosis. *BMC Cardiovasc. Disord.* **2025**, *25*, 778. [CrossRef]
75. Lee, Y.S.; Park, G.; Lee, K.; Jang, H.R.; Lee, J.E.; Huh, W.; Jeon, J. SGLT2 inhibitor use and renal outcomes in low-risk population with diabetes mellitus and normal or low body mass index. *BMJ Open Diabetes Res. Care* **2025**, *13*, e004876. [CrossRef]
76. Trimaille, A.; Marchandot, B.; Morel, O. SGLT2 Inhibition After Aortic Valve Replacement: The Role of the Native Valve. *JACC Adv.* **2025**, *4*, 102136. [CrossRef]
77. Salvatore, T.; Galiero, R.; Caturano, A.; Rinaldi, L.; Di Martino, A.; Albanese, G.; Di Salvo, J.; Epifani, R.; Marfella, R.; Docimo, G.; et al. An Overview of the Cardiorenal Protective Mechanisms of SGLT2 Inhibitors. *Int. J. Mol. Sci.* **2022**, *23*, 3651. [CrossRef]
78. Abbas, M.S.; Dandamudi, M.; Rehman, T.; Faizan, M.A.; Zahra, I.; Mojica, J.C.; Ahmed, M.R.; Bai, K.; Shakeel, I.; Shahzad, Z.; et al. "Efficacy of SGLT2 inhibitors in non-diabetic non-alcoholic fatty liver disease: A systematic review and meta-analysis". *J. Diabetes Metab. Disord.* **2025**, *24*, 285. [CrossRef]
79. Wu, Q.; Zhang, J.; Zhang, F.; Li, D. SGLT2 inhibitors as metabolic modulators: Beyond glycemic control in type 2 diabetes. *Front. Endocrinol.* **2025**, *16*, 1601633. [CrossRef]




https://doi.org/10.3390/biomedicines14010081

*biomedicines*

MDPI

*Review*

# Statins, Vitamin D, and Cardiovascular Health: A Comprehensive Review

**Dragos Cozma [1,2,3], Cristina Tudoran [4,5,6,*] and Cristina Văcărescu [1,2,3]**

1 Institute of Cardiovascular Diseases Timisoara, 300310 Timisoara, Romania; dragos.cozma@umft.ro (D.C.); cristina.vacarescu@umft.ro (C.V.)

2 Cardiology Department, Victor Babeș University of Medicine and Pharmacy, 300041 Timisoara, Romania

3 Research Center of the Institute of Cardiovascular Diseases Timisoara, 300310 Timisoara, Romania

4 Department VII, Internal Medicine II, Discipline of Cardiology, Victor Babeș University of Medicine and Pharmacy, E. Murgu Square, Nr. 2, 300041 Timisoara, Romania

5 County Emergency Hospital "Pius Brinzeu", L. Rebreanu, Nr. 156, 300723 Timisoara, Romania

6 Center of Molecular Research in Nephrology and Vascular Disease, Faculty of the Victor Babeș University of Medicine and Pharmacy, E. Murgu Square, Nr. 2, 300041 Timisoara, Romania

* Correspondence: tudoran.cristina@umft.ro

**Abstract**

Statins are widely used lipid-lowering agents that significantly reduce cardiovascular morbidity and mortality by lowering LDL-cholesterol. Vitamin D, traditionally known for its skeletal role, is increasingly recognized for its cardiovascular relevance. This study aims to focus on the complex relationship between statins, vitamin D, and their impact on cardiovascular outcomes. Both molecules intersect metabolically at 7-dehydrocholesterol, raising interest in their potential interactions. While theoretical concerns exist about statins impairing vitamin D synthesis, clinical studies suggest a neutral or modestly positive effect on circulating 25(OH)D levels. Statins may increase vitamin D levels by inhibiting its catabolism (via CYP3A4) and enhancing absorption. Observational data also suggest synergy between statins and vitamin D in reducing inflammation, oxidative stress, endothelial dysfunction, and atherogenesis. Though large trials showed no benefit of vitamin D supplementation in cardiovascular event reduction among vitamin D–replete individuals, select subgroups (those deficient or with statin-induced myalgia) may benefit from targeted supplementation. Optimizing vitamin D status could improve statin tolerability and adherence, especially in high-risk populations such as the elderly or those with metabolic syndrome. This review highlights the complex interplay between statins and vitamin D and supports a personalized approach to supplementation in statin-treated patients, aiming to enhance cardiovascular protection without overtreatment.

**Keywords:** statin-vitamin D interaction; cardiometabolic synergy; endothelial inflammation modulation

## 1. Introduction

Statins are among the most commonly prescribed lipid-lowering agents worldwide and have been pivotal in the prevention and management of cardiovascular disease (CVD). By inhibiting HMG-CoA reductase, the rate-limiting enzyme in the cholesterol biosynthesis pathway, statins effectively reduce low-density lipoprotein cholesterol (LDL-C), thereby lowering the risk of myocardial infarction, stroke, and cardiovascular mortality. Their widespread use and proven efficacy have made them a cornerstone in both primary and secondary cardiovascular prevention strategies [1,2].

Vitamin D, a fat-soluble secosteroid hormone traditionally associated with bone health and calcium-phosphorus metabolism, has recently gained substantial attention for its broader biological effects, particularly in cardiovascular physiology and pathology. Observational studies have suggested associations between vitamin D deficiency (VDD) (serum 25-hydroxyvitamin D (25(OH)D) under 20 nmol/L), and an increased risk of hypertension, heart failure, atherosclerosis, and coronary artery disease (CAD) [3–6]. These findings have sparked interest in understanding the cardiovascular implications of vitamin D beyond its classical endocrine functions.

Interestingly, the biosynthetic pathways of cholesterol and vitamin D intersect at a common precursor, 7-dehydrocholesterol. This biochemical connection raises important questions about the potential impact of statin therapy on vitamin D metabolism. Since statins inhibit cholesterol synthesis, concerns have been raised that they might also inadvertently affect endogenous vitamin D production, particularly in populations already at risk for deficiency, such as the elderly or those with limited sun exposure [7,8].

Cardiometabolic synergy refers to the complementary effects of interventions that target overlapping pathways involved in both metabolic dysfunction and cardiovascular disease. Statins reduce LDL cholesterol and exert anti-inflammatory and endothelial-stabilizing effects, while vitamin D modulates similar pathways through RAAS suppression, cytokine inhibition, and endothelial function enhancement. Mechanistically, both agents may interact through shared enzymes such as CYP3A4 and transport proteins like SR-BI and NPC1L1, influencing vitamin D metabolism and absorption. Together, their combined use may enhance lipid control, attenuate inflammation, and improve vascular health—demonstrating a form of cardiometabolic synergy that supports integrated risk reduction, particularly in patients with coexisting dyslipidemia, insulin resistance, or vitamin D deficiency.

This review explores the complex and evolving relationship between statin therapy and vitamin D status. We will examine whether statin use significantly influences serum vitamin D concentrations and whether supplementation is warranted in statin-treated individuals [9,10]. Furthermore, we will delve into the role of vitamin D in cardiovascular disease, with a specific focus on its involvement in endothelial function, inflammation, atherosclerosis, and myocardial health [11]. Drawing upon recent studies on pathophysiological mechanisms, clinical trials, and large-scale meta-analyses, this review seeks to provide a comprehensive and evidence-based perspective on the interplay between statins, vitamin D, and cardiovascular outcomes.

## 2. Statins and Vitamin D: Mechanistic Interactions

### 2.1. *Shared Precursors and Metabolic Pathways*

Cholesterol and 25(OH)D, the main circulating form of vitamin D, are both synthesized through pathways that involve a common biochemical precursor—7-dehydrocholesterol (7-DHC). This molecule is a pivotal substrate located in the skin, where it undergoes photochemical conversion to previtamin D3 upon exposure to ultraviolet B (UVB) radiation, eventually forming vitamin D3. At the same time, 7-DHC is also a precursor for endogenous cholesterol synthesis through a series of enzymatic transformations. Statins exert their therapeutic effect by inhibiting HMG-CoA reductase, the rate-limiting enzyme in the mevalonate pathway responsible for cholesterol biosynthesis. By doing so, they reduce the production of cholesterol and its upstream intermediates, theoretically including 7-DHC. Based on this biochemical intersection, early theoretical concerns arose that statin therapy could inadvertently reduce the availability of 7-DHC for vitamin D synthesis, thereby impairing cutaneous production of vitamin D3 [7,12].

However, subsequent research and clinical evidence have shown that this interaction is far more complex than initially assumed. While the inhibition of upstream cholesterol synthesis could potentially reduce 7-DHC levels, compensatory mechanisms in the skin and liver, along with dietary and supplemental sources of vitamin D, appear to mitigate any significant clinical impact. Indeed, no well-conducted clinical trials to date have demonstrated a consistent decrease in serum 25(OH)D concentrations as a direct consequence of statin use [12,13].

### *2.2. Vitamin D Receptor Signaling and Cardiovascular Implications*

Vitamin D exerts its physiological functions primarily through binding to the vitamin D receptor (VDR), a nuclear hormone receptor expressed in a wide range of tissues, including vascular endothelial cells, cardiomyocytes, smooth muscle cells, and immune cells. Upon activation by 1,25-dihydroxyvitamin D [1,25$(OH)_2$D], VDR heterodimerizes with the retinoid X receptor (RXR) and binds to vitamin D response elements (VDREs) in the promoter regions of target genes, thereby regulating transcription [14,15]. This genomic signaling modulates key processes such as inhibition of pro-inflammatory cytokine production (e.g., TNF-$\alpha$, IL-6), suppression of RAAS (renin–angiotensin–aldosterone system), enhancement of endothelial nitric oxide synthase (eNOS) expression, and inhibition of vascular smooth muscle proliferation and fibrosis [16,17]. Non-genomic actions have also been described, involving rapid activation of signaling pathways such as PI3K/Akt, MAPK, and intracellular calcium fluxes, which may contribute to vasodilatory and cardioprotective effects. These diverse actions of VDR signaling provide a mechanistic basis for the observed associations between vitamin D status and cardiovascular outcomes, supporting its role as a modifiable factor in atherosclerosis and heart failure [18].

### *2.3. Molecular Mediators and Transport Proteins Involved in Statin–Vitamin D Interactions*

To further elucidate the biochemical intersection between statin therapy and vitamin D metabolism, it is critical to consider the specific enzymes and transport proteins that mediate the synthesis, activation, and catabolism of both molecules [19–21]. Statins exert their primary action through the inhibition of HMG-CoA reductase, the rate-limiting enzyme in the mevalonate pathway, thereby reducing downstream production of 7-dehydrocholesterol (7-DHC) and cholesterol. However, 7-DHC also serves as the substrate for UVB-mediated photoconversion in the skin to previtamin D3, which is subsequently isomerized into vitamin D3 (cholecalciferol). Vitamin D3 is then hydroxylated in the liver by cytochrome P450 2R1 (CYP2R1) to form 25-hydroxyvitamin D [25(OH)D], and further activated in the kidney via CYP27B1 into 1,25-dihydroxyvitamin D [1,25$(OH)_2$D], the hormonally active form.

In parallel, CYP3A4, a major hepatic enzyme, is involved in the catabolism of both statins (particularly lipophilic agents like simvastatin and atorvastatin) and vitamin D metabolites. Competitive inhibition at CYP3A4 by statins may reduce the degradation of vitamin D, thereby increasing its circulating levels. In the gastrointestinal tract, vitamin D absorption is facilitated by sterol transporters, including Niemann–Pick C1-Like 1 (NPC1L1), Scavenger Receptor Class B type 1 (SR-B1), and Cluster of Differentiation 36 (CD36). Preclinical studies suggest that statins may modulate the expression or activity of these proteins, potentially enhancing vitamin D absorption from dietary or supplemental sources. Finally, CYP24A1, which catalyzes the inactivation of vitamin D metabolites, may also be indirectly influenced by statins through feedback mechanisms or modulation of inflammatory pathways, though this remains under investigation. These complex, overlapping molecular pathways underscore the need for further mechanistic studies to determine whether statins exert direct or tissue-specific regulatory effects on vitamin D homeostasis [20].

### 2.4. Influence on Vitamin D Synthesis and Metabolism

Several plausible mechanisms have been proposed to explain how statins might affect the metabolism and bioavailability of vitamin D, suggesting not only a neutral but potentially favorable effect on circulating vitamin D levels.

- CYP3A4 Competition: Many statins, particularly the lipophilic ones like atorvastatin and simvastatin, are metabolized in the liver via the cytochrome P450 3A4 (CYP3A4) enzyme system—the same enzymatic pathway responsible for catabolizing vitamin D into inactive metabolites. Competitive inhibition at this enzymatic site could slow the breakdown of vitamin D, thereby leading to higher circulating levels of 25(OH)D. This mechanism offers one explanation for why some studies have observed increased vitamin D levels in statin-treated individuals [22–24].
- Upregulation of Intestinal Absorption: Vitamin D is absorbed in the small intestine, a process facilitated by cholesterol transporters such as SR-BI (Scavenger Receptor Class B type I), CD36, and NPC1L1 (Niemann–Pick C1-like 1). These transporters are known to play a role in the absorption of lipid-soluble substances, including vitamin D. Statins may enhance the expression of these transporters, especially under experimental or high-dose conditions, potentially improving the efficiency of dietary vitamin D absorption [20].
- Alteration of 7-DHC Availability in the Skin: Despite reducing cholesterol synthesis overall, statins may cause a buildup of certain precursors, including 7-DHC, in specific tissues such as the skin. This could theoretically increase the substrate pool available for vitamin D3 production under UVB exposure, possibly explaining why some statin users exhibit improved bone health markers, such as increased bone mineral density. However, the evidence in human studies for this mechanism is still limited and mostly inferential [19,22].
- Vitamin D Receptor Activation: A more speculative but intriguing theory is that statins might directly or indirectly activate the vitamin D receptor (VDR). Since both statins and vitamin D exert overlapping cardioprotective effects—such as reducing inflammation, modulating immune responses, and improving endothelial function—it has been proposed by researchers like Bhattacharyya et al. that statins could mimic some of the actions of vitamin D by acting as partial agonists of VDR. Although compelling in concept, this idea remains largely hypothetical and has not yet been substantiated through direct clinical evidence [25].

Although cholesterol and vitamin D share a common metabolic origin in 7-DHC, and statins modulate this biosynthetic pathway, the net effect of statin therapy on vitamin D status appears to be neutral or mildly beneficial, rather than detrimental. Statins do not induce VDD at a population level, and in some cases, may even lead to modestly elevated 25(OH)D levels due to mechanisms like reduced catabolism and enhanced absorption [10,26].

Figure 1 illustrates the biochemical and physiological interactions between statin therapy and vitamin D metabolism, emphasizing their overlapping roles in cholesterol synthesis, inflammation modulation, endothelial function, and cardiovascular protection.

However, these effects likely vary depending on several factors, including: the specific statin used (lipophilic vs. hydrophilic), the dose and duration of therapy, individual genetic polymorphisms affecting metabolism, baseline vitamin D status, and concurrent dietary or environmental influences (e.g., sun exposure, nutritional intake) [27].

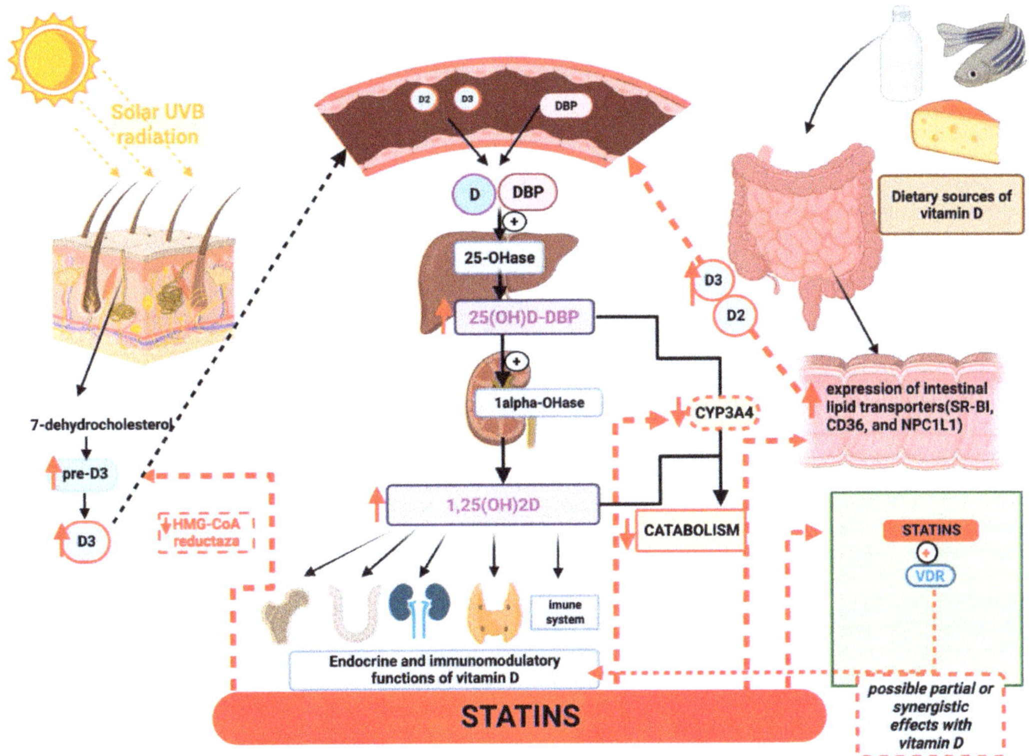


**Figure 1.** Statins and vitamin D: molecular links and shared pathways. $D_3$—cholecalciferol (vitamin $D_3$); $D_2$—ergocalciferol (vitamin $D_2$); DBP—vitamin D-binding protein; 25(OH)D—25-hydroxyvitamin D (calcidiol); $1,25(OH)_2D$—1,25-dihydroxyvitamin D (calcitriol); 25-OHase—hepatic 25-hydroxylase; 1α-OHase—renal 1α-hydroxylase; CYP3A4—cytochrome P450 3A4; HMG-CoA reductase—3-hydroxy-3-methylglutaryl-CoA reductase; SR-BI—scavenger receptor class B type I; CD36—cluster of differentiation 36; NPC1L1—niemann–pick C1-like 1 protein; VDR—vitamin D receptor.

Understanding these nuanced interactions can help clinicians make more informed decisions about vitamin D monitoring and supplementation in patients undergoing long-term statin therapy.

In summary, while statins and vitamin D share metabolic pathways, statin therapy does not appear to induce VDD on a population level. On the contrary, pathophysiological pathways suggest statins might modestly increase vitamin D levels via reduced catabolism and enhanced absorption. The net effect likely varies among different statin drugs and patient contexts.

### *2.5. Combined Role of Statins and Vitamin D in Cardiovascular Risk Reduction: Synergy or Redundancy?*

Overlap in Mechanisms of Action:

- Anti-Inflammatory Effects

Both statins and vitamin D exhibit potent anti-inflammatory properties, which are central to mitigating the pathogenesis of atherosclerosis. Statins reduce the production of pro-inflammatory cytokines such as IL-6, CRP, and TNF-α, and suppress macrophage activity within atherosclerotic plaques [28]. Vitamin D, through activation of the VDR, downregulates pro-inflammatory gene expression and modulates immune responses by promoting a more efficient anti-inflammatory cytokine response (e.g., increased IL-10) [29]. This dual anti-inflammatory action could theoretically stabilize atherosclerotic plaques more effectively than either agent alone, reducing the risk of plaque rupture and thrombosis.

Platelets may play a pivotal role in the pro-inflammatory and pro-aggregatory response which is one of the main pathophysiological mechanisms involved in atherosclerosis and atherogenesis. In his study, Ahmadi et al. [30] demonstrated that patients with CAD had significantly elevated platelet TGF-β1 levels, both latent and mature, compared to healthy subjects. Soluble TGF-β1 was also increased in patients with clinical atherosclerosis. This study determined significant correlations between mature/active TGF-β1 and platelet pro-inflammatory markers (P-selectin and CD40L) as well as common indicators of inflammation (CRP and ESR). Several studies highlighted an interaction between statins and platelets, with the implication of multiple mechanisms [31]. A recent study demonstrated that atorvastatin prevented TGF-β1-induced fibrogenesis in human ventricular fibroblasts, at least partially by inhibiting the Smad3 and MAPK signaling pathways, and therefore, this statin can be a promising drug for the treatment of myocardial fibrosis [27]. Through the activation of the VDR, vitamin D demonstrated antithrombotic effects in in vivo studies by upregulating platelet aggregation and by increasing endothelial NO synthesis. Other pathophysiological pathways consist of enhanced gene expression of antithrombin in the liver and of thrombomodulin in the liver, as well as in the aorta and kidney and in reducing tissue factor mRNA expression [32].

- Endothelial Protection

Endothelial dysfunction is a hallmark of cardiovascular disease and a target for both statins and vitamin D. Statins enhance endothelial nitric oxide synthase (eNOS) activity, improving nitric oxide (NO) bioavailability, which leads to vasodilation and vascular homeostasis [16,33]. Vitamin D also improves endothelial function by reducing oxidative stress, lowering endothelin-1 levels, and modulating NO production [34]. The concurrent use of statins and vitamin D might result in additive improvement in endothelial responsiveness, particularly in patients with diabetes or hypertension, where endothelial dysfunction is pronounced.

- Plaque Stabilization and Atheroprotection

Both agents have shown plaque-stabilizing effects in animal models and clinical studies. Statins promote fibrous cap thickening, reduce lipid core size, and inhibit matrix metalloproteinase activity, thereby reducing plaque vulnerability [35]. Vitamin D's regulatory effects on macrophage activity, smooth muscle cell function, and vascular calcification may complement statins by slowing plaque progression and improving plaque composition [36].

- Modulation of Thrombogenesis

A pro-thrombotic state contributes to acute coronary syndromes. Statins reduce platelet aggregation and improve fibrinolysis [37]. VDD has been associated with a hypercoagulable state, including increased tissue factor expression and platelet activation. Supplementation may normalize hemostatic balance, especially in vitamin D-deficient individuals [5].

- Oxidative Stress Reduction

Oxidative stress promotes endothelial dysfunction, lipid peroxidation, and plaque instability. Statins decrease oxidative stress by downregulating NADPH oxidase and enhancing antioxidant enzymes [38]. Vitamin D exerts similar effects by reducing reactive oxygen species (ROS) and enhancing antioxidant defenses. Together, these agents may synergistically reduce oxidative burden, contributing to improved vascular integrity and myocardial protection [39]. Several studies indicate that the inhibition of ROS generation can have benefic effects on platelets [40,41].

Table 1 provides a summary of overlapping mechanisms by which statins and vitamin D exert cardiovascular protective effects, including anti-inflammatory actions, endothelial protection, plaque stabilization, modulation of thrombogenesis, and reduction in oxidative stress. Most references focus on human clinical or observational data, with a few incorporating animal or in vitro mechanistic findings (notably references [26,29]).

**Table 1.** Overlap of Statins and Vitamin D Mechanisms.

| Mechanism/Pathway | Statin Mechanism | Vitamin D Mechanism | Representative Evidence (Study Type/Population) |
|---|---|---|---|
| Anti-inflammatory/immuno-modulation | ↓ IL-6, CRP, TNF-α; suppress macrophage activation in plaques | VDR activation; suppress pro-inflammatory genes; ↑ IL-10 | Review of clinical and experimental studies on cardiovascular inflammation [1] |
| Endothelial protection/vasodilation | ↑ eNOS activity, improved NO bioavailability | ↓ oxidative stress; ↓ endothelin-1; modulate NO synthesis | Experimental endothelial cell studies; small vascular trials in humans [1] |
| Plaque stabilization/atheroprotection | Promote fibrous cap thickening; reduce lipid core; inhibit metalloproteinases | Regulate macrophages, smooth muscle cells; modulate vascular calcification | Histopathological plaque studies; cardiovascular biology reviews [1] |
| Modulation of thrombogenesis/coagulation | Reduce platelet aggregation; improve fibrinolysis | Vitamin D deficiency linked to hypercoagulability: ↑ tissue factor, ↑ platelet activation | Reviews on coagulation and vitamin D deficiency in CVD [1] |
| Oxidative stress reduction/redox balance | ↓ NADPH oxidases; ↑ antioxidant enzymes (SOD, catalase) | Reduce ROS, enhance antioxidant defenses | In vivo and cross-sectional studies in hypertension and general populations [42] |
| Statin-associated muscle symptoms (SAMS) | — | Observational link: low 25(OH)D associated with increased SAMS risk | Meta-analysis of observational studies ($n \approx 2400$) [43]; RCT (VITAL sub-study, $n = 2083$) found no difference [43] |
| Adherence/persistence with statin therapy | — | Monthly vitamin D improved persistence (HR = 1.15, $p = 0.02$) but not adherence | Secondary analysis of long-term statin users in randomized vitamin D supplementation trial [44] |
| Cardiovascular outcomes (MACE) | Robust reduction in major adverse cardiovascular events across many RCTs | No consistent reduction in MACE; large RCTs and meta-analyses generally neutral | Meta-analysis of 14 RCTs ($n \approx 80{,}500$, age 50–74) showed no significant benefit [21,45] DAYLIGHT trial no biomarker improvement [46] |

↑—increased value, ↓—decreased value.

## 3. Changes in Vitamin D Levels in Statin Users: Clinical Evidence

Multiple clinical studies have explored how statin therapy impacts serum 25(OH)D concentrations, with mixed results:

- Rosuvastatin's Notable Effect: Initial reports by Yavuz et al. observed a striking rise in 25(OH)D after initiating rosuvastatin. In an 8-week observational study, mean 25(OH)D increased from ~14 ng/mL to 36 ng/mL on rosuvastatin [7]. A subsequent randomized trial (STATIN-D) comparing rosuvastatin (10 mg) to fluvastatin found a similar tripling of 25(OH)D with rosuvastatin (11.8 to 35.2 ng/mL over 8 weeks), whereas fluvastatin (80 mg) showed no significant change. These findings led to the hypothesis of a rosuvastatin-specific effect. However, a critical re-analysis questioned the magnitude of this increase, suggesting that factors like seasonal sunlight variation might have contributed [19].
- Other Statins (Atorvastatin, Simvastatin, Lovastatin): Several studies suggest moderate increases in vitamin D with other statins. For example, an analysis by Pérez-Castrillón et al. in patients with CAD and baseline VDD found that 12 months of atorvastatin

(10–80 mg/day) raised mean 25(OH)D from 16.4 to 18.8 ng/mL ($p$ = 0.003) [47]. Similarly, small studies reported that lovastatin (20–80 mg) for 3 months led to higher 25(OH)D levels, and simvastatin therapy increased 25(OH)D and even active 1,25-dihydroxyvitamin D_3 levels [48]. Sathyapalan et al. observed a ~47% increase in 25(OH)D with 20 mg atorvastatin in patients with polycystic ovary syndrome [10]. On the other hand, Rejnmark et al. reported no significant change in vitamin D metabolites with simvastatin 40 mg over 8 weeks [49].

- Pravastatin and Fluvastatin: Hydrophilic statins like pravastatin have generally shown no effect on vitamin D levels. Trials of pravastatin (10–80 mg daily) demonstrated no difference in 25(OH)D between treated and control groups during a follow-up of 2 to 6 months [7]. As noted above, fluvastatin (another hydrophilic statin) did not alter vitamin D in the STATIN-D trial [7].
- Cross-Sectional Analyses: In a retrospective cross-sectional study of 384 patients, those on statins (mostly atorvastatin) had slightly higher serum vitamin D on average than matched non-users [22]. Short-term and medium-term statin users in that study had significantly higher 25(OH)D levels than non-users [34]. Notably, no group of statin-treated patients showed a decline in vitamin D compared to controls.

Meta-Analyses: When pooling data from various trials, the overall impact of statins on vitamin D appears modest. A meta-analysis of seven studies (including five RCTs) found no significant overall effect of statin therapy on plasma 25(OH)D levels [23]. Some discrepancies across studies may relate to differences in statin potency, lipophilicity, duration of therapy, baseline vitamin D status, and seasonal timing of measurements.

Table 2 summarizes the effects of various statins on serum 25(OH)D levels, presenting findings from different study designs—including observational studies, randomized controlled trials, and meta-analyses—while highlighting the type of statin used and reported changes in vitamin D concentrations.

**Table 2.** Effects of stations on Vitamin D levels.

| Statin(s) Studied | Study Design | Effect on Vitamin D Levels | Remarks/Limitations |
|---|---|---|---|
| Rosuvastatin | Observational study. [34] | Substantial increase (~14 to 36 ng/mL over 8 weeks) | Possible confounding, small sample |
| Rosuvastatin vs. Fluvastatin | Randomized controlled trial (STATIN-D) [35] | Rosuvastatin tripled 25(OH)D levels; Fluvastatin showed no change | Head-to-head comparison, limited duration |
| Atorvastatin | Longitudinal study in CAD patients [37] | Moderate increase (from 16.4 to 18.8 ng/mL over 12 months) | Slow change; possible seasonal/lifestyle influences |
| Lovastatin | Small clinical study [38] | Reported increase after 3 months | Very small sample, uncontrolled |
| Simvastatin | Small studies and one trial [40] | Reported increase in 25(OH)D and 1,25-dihydroxyvitamin $D_3$; some studies show no change | Heterogeneous designs and populations |
| Atorvastatin (Cross-sectional) | Retrospective cross-sectional study (citation [43]) | Slightly higher vitamin D levels in statin users vs. non-users | Association only; cannot prove causation |
| Pravastatin | Multiple clinical trials [41] | No significant difference compared to control over 2–6 months | Negative/null results in short-term trials |
| Fluvastatin | Randomized trial (STATIN-D) [42] | No significant change | Similarly to control arm in STATIN-D trial |
| Various (Meta-analysis) | Meta-analysis of 7 studies including 5 RCTs [36] | Overall modest or no significant effect | |

Key Point: Current evidence does not support the notion that statins induce VDD. Statin therapy is often associated with stable or modestly increased vitamin D levels. For instance, rosuvastatin has been shown to increase serum 25(OH)D concentrations from approximately 14 ng/mL to 36 ng/mL over 8 weeks [7], while atorvastatin therapy led to an increase from 16.4 to 18.8 ng/mL over 12 months in patients with coronary artery disease [47].

## 4. Vitamin D Supplementation in Statin-Treated Patients

Given that statins do not typically cause vitamin D deficiency, is there a need for vitamin D supplementation in patients on statins? The answer depends on the clinical context:

- General Statin Users: For the average patient on statins, vitamin D supplementation should be guided by the same principles as in the general population—namely, treating vitamin D insufficiency or deficiency if present. There is no evidence that statin users benefit from extra vitamin D beyond the recommended dietary intake and sun exposure for bone health, unless they are deficient [5,7].
- Statin-Associated Muscle Symptoms: An important scenario is statin-induced myalgia or myopathy, which has been linked with low vitamin D status. Vitamin D deficiency can cause myalgias and muscle weakness on its own, and studies have observed that patients experiencing statin-associated muscle symptoms often have lower 25(OH)D levels than those who tolerate statins. In such cases, correcting the deficiency may improve symptoms. Notably, some patients with statin-associated myalgia and low vitamin D levels find relief after vitamin D repletion [43,50]. For example, one analysis reported that over 88% of previously statin-intolerant patients (with baseline 25(OH)D < 32 ng/mL) were able to tolerate statin rechallenge without myalgia after vitamin D supplementation corrected their deficiency. Another cohort found that 95% of statin-intolerant patients became free of muscle symptoms after 24 months of vitamin D repletion and statin retry [50,51]. While these are not randomized trials, they suggest that ensuring adequate vitamin D can mitigate statin myopathic side effects in susceptible individuals.
- Guidelines and Clinical Practice: Formal guidelines do not mandate vitamin D testing for all statin users, but many clinicians assess 25(OH)D in patients who develop muscle symptoms on statins. Ensuring a sufficient vitamin D status (e.g., >30 ng/mL) is a reasonable strategy to potentially reduce myopathic side effects and improve adherence to statin therapy. Vitamin D supplementation is inexpensive and safe in moderate doses, making it a practical consideration for patients at risk of deficiency or those with muscle complaints [52,53].

In summary, vitamin D supplementation is not universally required for every patient on a statin. However, identifying and treating VDD can be beneficial, particularly in those with statin-associated muscle complaints or other risk factors for low vitamin D. This targeted approach may enhance the tolerability and continued use of statins, indirectly supporting better cardiovascular outcomes.

### *Guideline Recommendations for Vitamin D Testing and Supplementation*

While formal guidelines do not currently mandate routine vitamin D screening for all statin users, several authoritative bodies do provide recommendations for vitamin D assessment in broader populations. For instance, the Endocrine Society advises screening individuals at high risk of deficiency, including older adults, those with osteoporosis, malabsorption syndromes, chronic kidney disease, or individuals presenting with unexplained musculoskeletal pain or weakness. Similarly, the National Osteoporosis Foundation and the Institute of Medicine recommend vitamin D supplementation for individuals with documented deficiency (<20 ng/mL), with target serum 25(OH)D levels generally set between 20 and 30 ng/mL depending on comorbidities and clinical context. In clinical practice, vitamin D testing is frequently considered in statin-treated patients who report myalgias or other muscle symptoms, as repletion may improve tolerance and reduce discontinuation risk. However, there is no current recommendation for universal screening or supplementation in statin users without specific risk factors or symptoms [54].

## 5. Vitamin D, Atherosclerosis, and Coronary Artery Disease

### 5.1. Vitamin D Levels and Cardiovascular Risk

A growing body of research has explored the link between vitamin D status and atherosclerotic cardiovascular disease. Observational studies consistently report that low 25(OH)D levels are associated with greater prevalence of cardiovascular risk factors (hypertension, diabetes, metabolic syndrome) and higher incidence of events. For instance, among male health professionals, those with VDD (<15 ng/mL) had approximately twice the risk of myocardial infarction compared to men with sufficient levels [4,55]. Similarly, in an observational cohort, each ~20 ng/mL increment in 25(OH)D was associated with a 33% reduction in the risk of fatal stroke [56]. A meta-analysis of more than 10 prospective studies (nearly 66,000 participants) found that individuals in the lowest vitamin D category had a 52% higher risk of total cardiovascular events compared to those with the highest vitamin D levels. This analysis showed a generally linear inverse relationship between 25(OH)D and cardiovascular risk down to about 60 nmol/L (~24 ng/mL), suggesting that deficiency is linked to elevated risk [57]. VDD has also been correlated with subclinical markers of atherosclerosis, such as increased carotid intima–media thickness and arterial stiffness, in some studies [58].

### 5.2. Pathophysiological Role of Vitamin D in Atherosclerosis

Biological plausibility for these associations is strong. VDRs are expressed in vascular endothelial cells, cardiomyocytes, vascular smooth muscle cells, and immune cells. Active vitamin D (1,25-dihydroxyvitamin D) exerts anti-inflammatory and immunomodulatory effects that could attenuate the chronic inflammation of atherosclerosis. Vitamin D also promotes endothelial function (by enhancing eNO availability and reducing oxidative stress) and may inhibit macrophage cholesterol uptake and foam cell formation, potentially stabilizing atherosclerotic plaques. Furthermore, vitamin D influences other cardiovascular factors: deficiency is linked to activation of the renin–angiotensin system (raising blood pressure), increased arterial stiffness, insulin resistance, and a pro-thrombotic profile—all of which can exacerbate atherosclerosis. Thus, adequate vitamin D levels might protect against atherosclerosis and its clinical consequences [11,59,60].

### 5.3. Vitamin D Supplementation and Cardiovascular Outcomes

Despite compelling observational data, randomized controlled trials have yet to conclusively demonstrate that vitamin D supplementation reduces cardiovascular events. Two large trials—VITAL (Vitamin D and Omega-3 Trial) and ViDA (Vitamin D Assessment Study)—tested moderate-to-high dose vitamin D in generally healthy, vitamin D–D-replete adults. Both reported no significant difference in major cardiovascular outcomes between the vitamin D and placebo groups [21,61]. In VITAL, for example, daily 2000 IU of cholecalciferol did not lower the incidence of myocardial infarction, stroke, or cardiovascular death over five years compared to placebo. These null results align with several prior smaller trials and meta-analyses, which collectively show no clear benefit of vitamin D supplementation for preventing heart attacks or strokes in unselected populations. It is important to note that the trial populations were largely vitamin D sufficient at baseline (mean 25(OH)D ~77 nmol/L in VITAL), leaving open the question of whether truly deficient patients might benefit [61]. Some subgroup analyses have hinted that participants with very low baseline vitamin D could have modest risk reductions with supplementation, but results have been inconsistent and not definitive.

### 5.4. Current Consensus

Based on available evidence, maintaining adequate vitamin D status is considered part of overall health optimization, but vitamin D alone is not a proven therapy to reduce established atherosclerosis or prevent CAD events. Current public health guidelines restrict vitamin D supplementation recommendations to bone health and fracture prevention, not cardiovascular protection. Nonetheless, given vitamin D's low cost and safety, ongoing research is examining if certain high-risk groups (e.g., patients with both VDD and cardiovascular risk factors, or those with chronic kidney disease) might derive cardiovascular benefit from targeted vitamin D supplementation. For now, treating vitamin D deficiency in patients with atherosclerosis is reasonable for general health, while recognized preventive measures (statins, blood pressure control, diet, etc.) remain paramount for reducing cardiovascular risk [45,54].

### 5.5. Identifying Target Populations for Combined Therapy

Given that the benefits of vitamin D are more pronounced in deficient individuals, not all patients on statins will necessarily benefit from routine co-supplementation. Therefore, a targeted approach is key:

- Patients with confirmed vitamin D deficiency (<20 ng/mL): These individuals are most likely to benefit from supplementation, both for skeletal and possible cardiovascular protection. When treated with statins, addressing the deficiency could improve muscle tolerability and overall metabolic balance [18,21,43].
- Elderly or institutionalized patients: These populations frequently exhibit both low vitamin D levels and high cardiovascular risk. They are prime candidates for dual intervention [62].
- Patients with chronic inflammatory conditions (e.g., rheumatoid arthritis, metabolic syndrome): These conditions are associated with persistent low-grade inflammation, endothelial dysfunction, and oxidative stress—areas where vitamin D and statins may have complementary actions [63,64].

### 5.6. Optimizing Dosage and Monitoring

Statin Therapy: Prescribed according to current cardiovascular risk stratification guidelines: Consider using lipophilic statins (e.g., atorvastatin, simvastatin) in patients also receiving vitamin D, due to potential shared metabolic pathways (e.g., CYP3A4), but be cautious of interactions [23].

Vitamin D Supplementation: Follow standard dosing protocols for deficiency (e.g., 1000–2000 IU/day or weekly loading doses in severe cases). Monitor 25(OH)D levels periodically, especially in those with comorbidities such as chronic kidney disease or malabsorption and aim for a target level of >30 ng/mL, especially in patients with muscular symptoms or inflammation [54,65].

### 5.7. Preventive Cardiovascular Strategies

Incorporating both agents into multifactorial cardiovascular risk reduction plans may be especially useful in secondary prevention of myocardial infarction or stroke, especially in patients with low vitamin D level, in patients undergoing cardiac rehabilitation, where improvements in endothelial function and muscle performance may be enhanced by vitamin D status optimization and for diabetic or metabolic syndrome patients, where vitamin D may assist in insulin sensitivity and vascular health alongside statin therapy [66–68].

### 5.8. Multidisciplinary Integration

To implement this strategy effectively, primary care providers, cardiologists, endocrinologists, and clinical pharmacists should collaborate in identifying patients who may benefit from combination therapy. Integration into electronic health records (EHRs) could prompt clinicians to assess vitamin D status when prescribing statins, particularly for at-risk individuals, also patient education on lifestyle (diet, sun exposure) and medication adherence is critical to maximizing therapeutic outcomes.

## 6. Summary of Key Findings and Future Therapeutic Directions

### 6.1. Summary of Key Findings

- Statins and Vitamin D Levels: Contrary to initial expectations, statins do not appear to induce VDD. Some statins—notably rosuvastatin and atorvastatin—have even been associated with increased 25(OH)D levels in clinical studies. Proposed mechanisms include reduced vitamin D catabolism via CYP3A4 inhibition and enhanced intestinal vitamin D absorption via upregulated cholesterol transporters. Overall, no significant drop in vitamin D has been documented with statin use, and meta-analyses show a neutral net effect on 25(OH)D levels.
- Vitamin D Supplementation in Statin Users: Routine vitamin D supplementation solely because a patient is on statin therapy is not necessary if the patient is vitamin D replete. However, vitamin D status should be optimized in all individuals. In statin users who are vitamin D-deficient, repletion is advised—especially if they experience statin-associated myalgias. Evidence suggests correcting low vitamin D can alleviate statin-related muscle symptoms and improve tolerance to therapy. Ensuring adequate vitamin D may help patients continue statins, maximizing cardiovascular benefits.
- Vitamin D and Cardiovascular Disease: Low vitamin D levels have been consistently linked to a higher risk of atherosclerosis and adverse cardiovascular outcomes in observational studies. Vitamin D's anti-inflammatory, endothelial, and metabolic effects provide a plausible explanation for these associations. However, randomized trials (in mostly vitamin D-sufficient populations) have not shown significant reductions in heart attacks or strokes with vitamin D supplementation. Therefore, while maintaining adequate vitamin D is important for overall health, it should complement—not replace—established cardiovascular prevention strategies. Further research is needed to determine if targeted vitamin D therapy in deficient, high-risk groups can improve cardiovascular outcomes.

### 6.2. Future Therapeutic Directions

- Fixed-dose combination therapy: Although not yet commercially available, the idea of combining statins with vitamin D in a single pill could be explored for specific populations.
- Personalized medicine approaches using genetic profiling (e.g., CYP3A4 polymorphisms, VDR gene variants) could help predict which patients will benefit most from combined therapy.

Adjunctive trials: Future randomized controlled trials should evaluate whether vitamin D supplementation in deficient statin users improves hard cardiovascular endpoints (e.g., MI, stroke, mortality), beyond symptom relief. Looking ahead, the integration of genetic and molecular profiling into clinical decision-making may enable more personalized strategies for combining statins and vitamin D. Specific genetic polymorphisms such as CYP3A4 variants, which influence the hepatic metabolism of lipophilic statins, and polymorphisms in the vitamin D receptor (VDR) gene have been implicated in interindividual variability in vitamin D responsiveness and could serve as predictors of treatment efficacy or adverse effects. Future clinical trials should investigate whether these genetic markers can stratify patients likely to derive enhanced cardiovascular or metabolic benefit from combined therapy. Additionally, randomized controlled trials should be designed to evaluate robust cardiovascular endpoints (e.g., myocardial infarction, stroke, cardiovascular death), alongside functional biomarkers such as endothelial reactivity, serum inflammatory mediators (e.g., IL-6, CRP), and statin tolerability in vitamin D-deficient individuals. These studies could clarify whether targeted co-supplementation improves clinical outcomes beyond those achievable by statins alone.

## 7. Limitations of the Study

An important methodological limitation of this review lies in the interpretation of subgroup analyses derived from the primary studies discussed. While subgroup findings can offer valuable preliminary insights into potential differential effects—such as those influenced by baseline vitamin D status, specific statin pharmacokinetics, or comorbid conditions—they are frequently exploratory in nature and lack adequate statistical power to support definitive conclusions. Moreover, many subgroup observations do not meet the threshold for statistical interaction testing and are not independently replicated across trials. As a result, they should be regarded as hypothesis-generating rather than confirmatory.

In the context of this manuscript, differential responses to statins or vitamin D supplementation across subgroups (e.g., vitamin D-deficient individuals or users of specific statin types) are reported primarily as observational trends. While mechanistically plausible, these findings must be interpreted with caution until validated by prospectively designed, adequately powered randomized controlled trials specifically addressing these interactions.

Also, it is important to acknowledge the potential for publication bias in the available literature. Trials or observational studies reporting favorable effects of vitamin D supplementation or statin–vitamin D interactions may be more likely to be published than studies showing neutral or null findings. This asymmetry could influence pooled or narrative assessments of the evidence, resulting in an overestimation of benefit or mechanistic plausibility.

Collectively, these limitations highlight the need for cautious interpretation of subgroup findings and published results, as well as the importance of future high-quality, prospective studies—including those with rigorous publication of null or negative outcomes—to more definitively characterize the interplay between statins, vitamin D, and cardiovascular risk.

## 8. Conclusions

Statin therapy does not appear to induce vitamin D deficiency and may modestly increase serum 25(OH)D concentrations, particularly with lipophilic statins such as rosuvastatin and atorvastatin. Although routine vitamin D supplementation is not indicated in all statin-treated individuals, targeted repletion in those with confirmed deficiency—especially in the context of statin-associated muscle symptoms—may enhance tolerability and therapeutic adherence. While low vitamin D is linked to higher cardiovascular risk, supplementation has not shown consistent benefit in reducing major cardiovascular events, highlighting the need for targeted research in high-risk, vitamin D-deficient populations.

**Author Contributions:** Conceptualization, D.C. and C.V.; methodology, D.C.; software, D.C.; validation, C.T. and C.V.; formal analysis, C.V.; investigation, D.C.; resources, D.C.; data curation, C.V.; writing—original draft preparation, D.C.; writing—review and editing, C.V.; visualization, C.T.; supervision, C.T.; project administration, D.C.; funding acquisition, C.V. All authors have read and agreed to the published version of the manuscript.

**Funding:** This research received no external funding. Internal funding: We would like to acknowledge Victor Babes University of Medicine and Pharmacy Timisoara for their support in covering the publication costs for this research paper.

**Institutional Review Board Statement:** Not applicable.

**Informed Consent Statement:** Not applicable.

**Data Availability Statement:** No new data were created or analyzed in this study. Data sharing is not applicable to this article.

**Conflicts of Interest:** The authors declare no conflicts of interest.

## Abbreviations

The following abbreviations are used in this manuscript:

| | |
|---|---|
| CVD | Cardiovascular Disease |
| LDL-C | Low-Density Lipoprotein Cholesterol |
| CAD | Coronary Artery Disease |
| HMG-CoA | 3-Hydroxy-3-Methylglutaryl-Coenzyme A |
| 25(OH)D | 25-Hydroxyvitamin D |
| 7-DHC | 7-Dehydrocholesterol |
| UVB | Ultraviolet B |
| VDR | Vitamin D Receptor |
| CYP3A4 | Cytochrome P450 3A4 |
| SR-BI | Scavenger Receptor Class B type I |
| NPC1L1 | Niemann–Pick C1-like 1 |
| IL-6 | Interleukin 6 |
| CRP | C-Reactive Protein |
| TNF-Î± | Tumor Necrosis Factor Alpha |
| IL-10 | Interleukin 10 |
| eNOS | Endothelial Nitric Oxide Synthase |
| NO | Nitric Oxide |
| ROS | Reactive Oxygen Species |
| RCT | Randomized Controlled Trial |
| IU | International Units |
| MI | Myocardial Infarction |
| EHR | Electronic Health Record |

## References

1. Haider, F.; Ghafoor, H.; Hassan, O.F.; Farooqui, K.; Bel Khair, A.O.M.; Shoaib, F. Vitamin D and Cardiovascular Diseases: An Update. *Cureus* **2023**, *15*, e49734. [CrossRef]
2. Cholesterol Treatment Trialists' (CTT) Collaboration; Baigent, C.; Blackwell, L.; Emberson, J.; Holland, L.E.; Reith, C.; Bhala, N.; Peto, R.; Barnes, E.H.; Keech, A.; et al. Efficacy and Safety of More Intensive Lowering of LDL Cholesterol: A Meta-Analysis of Data from 170,000 Participants in 26 Randomised Trials. *Lancet* **2010**, *376*, 1670–1681. [CrossRef]
3. Zittermann, A.; Schleithoff, S.S.; Koerfer, R. Vitamin D and Vascular Calcification. *Curr. Opin. Lipidol.* **2007**, *18*, 41–46. [CrossRef]
4. Wang, T.J.; Pencina, M.J.; Booth, S.L.; Jacques, P.F.; Ingelsson, E.; Lanier, K.; Benjamin, E.J.; D'Agostino, R.B.; Wolf, M.; Vasan, R.S. Vitamin D Deficiency and Risk of Cardiovascular Disease. *Circulation* **2008**, *117*, 503–511. [CrossRef]
5. Pilz, S.; Tomaschitz, A.; März, W.; Drechsler, C.; Ritz, E.; Zittermann, A.; Cavalier, E.; Pieber, T.R.; Lappe, J.M.; Grant, W.B.; et al. Vitamin D, Cardiovascular Disease and Mortality. *Clin. Endocrinol.* **2011**, *75*, 575–584. [CrossRef]
6. Vacek, J.L.; Vanga, S.R.; Good, M.; Lai, S.M.; Lakkireddy, D.; Howard, P.A. Vitamin D Deficiency and Supplementation and Relation to Cardiovascular Health. *Am. J. Cardiol.* **2012**, *109*, 359–363. [CrossRef] [PubMed]
7. Yavuz, B.; Ertugrul, D.T.; Cil, H.; Ata, N.; Akin, K.O.; Yalcin, A.A.; Kucukazman, M.; Dal, K.; Hokkaomeroglu, M.S.; Yavuz, B.B.; et al. Increased Levels of 25 Hydroxyvitamin D and 1,25-Dihydroxyvitamin D after Rosuvastatin Treatment: A Novel Pleiotropic Effect of Statins? *Cardiovasc. Drugs Ther.* **2009**, *23*, 295–299. [CrossRef] [PubMed]

8. Jaber, H.; Mohmmed, K.M.; Flayyih, A.A.; Rasheed, A. Effect of Atorvastatin on Vitamin D Levels in Type 2 Diabetic Patients with Hypercholesterolemia. *J. Garmian Univ.* **2018**, *5*, 84–89. [CrossRef]
9. Gröber, U.; Spitz, J.; Reichrath, J.; Kisters, K.; Holick, M.F. Vitamin D. *Dermatoendocrinol* **2013**, *5*, 331–347. [CrossRef]
10. Sathyapalan, T.; Shepherd, J.; Arnett, C.; Coady, A.-M.; Kilpatrick, E.S.; Atkin, S.L. Atorvastatin Increases 25-Hydroxy Vitamin D Concentrations in Patients with Polycystic Ovary Syndrome. *Clin. Chem.* **2010**, *56*, 1696–1700. [CrossRef]
11. Talmor, Y.; Golan, E.; Benchetrit, S.; Bernheim, J.; Klein, O.; Green, J.; Rashid, G. Calcitriol Blunts the Deleterious Impact of Advanced Glycation End Products on Endothelial Cells. *Am. J. Physiol. Renal Physiol.* **2008**, *294*, F1059–F1064. [CrossRef]
12. Ott, C.; Raff, U.; Schneider, M.P.; Titze, S.I.; Schmieder, R.E. 25-Hydroxyvitamin D Insufficiency Is Associated with Impaired Renal Endothelial Function and Both Are Improved with Rosuvastatin Treatment. *Clin. Res. Cardiol.* **2013**, *102*, 299–304. [CrossRef] [PubMed]
13. Anagnostis, P.; Athyros, V.G.; Adamidou, F.; Florentin, M.; Karagiannis, A. Vitamin D and Cardiovascular Disease: A Novel Agent for Reducing Cardiovascular Risk? *Curr. Vasc. Pharmacol.* **2010**, *8*, 720–730. [CrossRef] [PubMed]
14. Marquina, C.; Mousa, A.; Scragg, R.; de Courten, B. Vitamin D and cardiometabolic disorders: A review of current evidence, genetic determinants and pathomechanisms. *Obes. Rev.* **2019**, *20*, 262–277. [CrossRef]
15. Parker, J.; Hashmi, O.; Dutton, D.; Mavrodaris, A.; Stranges, S.; Kandala, N.-B.; Clarke, A.; Franco, O.H. Levels of Vitamin D and Cardiometabolic Disorders: Systematic Review and Meta-Analysis. *Maturitas* **2010**, *65*, 225–236. [CrossRef] [PubMed]
16. Chen, W.-H.; Chen, C.-H.; Hsu, M.-C.; Chang, R.-W.; Wang, C.-H.; Lee, T.-S. Advances in the Molecular Mechanisms of Statins in Regulating Endothelial Nitric Oxide Bioavailability: Interlocking Biology between eNOS Activity and L-Arginine Metabolism. *Biomed. Pharmacother.* **2024**, *171*, 116192. [CrossRef]
17. Jain, S.S.; Liu, Q.; Raikhelkar, J.; Fried, J.; Elias, P.; Poterucha, T.J.; DeFilippis, E.M.; Rosenblum, H.; Wang, E.Y.; Redfors, B.; et al. Indications for and Findings on Transthoracic Echocardiography in COVID-19. *J. Am. Soc. Echocardiogr.* **2020**, *33*, 1278–1284. [CrossRef]
18. Hou, Q.; Pang, C.; Chen, Y. Association Between Vitamin D and Statin-Related Myopathy: A Meta-Analysis. *Am. J. Cardiovasc. Drugs* **2022**, *22*, 183–193. [CrossRef]
19. Glossmann, H.H.; Blumthaler, M. Does Rosuvastatin Increase Serum Levels of 25-Hydroxy-Vitamin D? *Derm. Endocrinol.* **2012**, *4*, 2–7. [CrossRef]
20. Robien, K.; Oppeneer, S.J.; Kelly, J.A.; Hamilton-Reeves, J.M. Drug-Vitamin D Interactions: A Systematic Review of the Literature. *Nutr. Clin. Pract.* **2013**, *28*, 194–208. [CrossRef]
21. Scragg, R.; Stewart, A.W.; Waayer, D.; Lawes, C.M.M.; Toop, L.; Sluyter, J.; Murphy, J.; Khaw, K.-T.; Camargo, C.A. Effect of Monthly High-Dose Vitamin D Supplementation on Cardiovascular Disease in the Vitamin D Assessment Study: A Randomized Clinical Trial. *JAMA Cardiol.* **2017**, *2*, 608–616. [CrossRef]
22. Erdal, G.S.; Caglar, N.T.; Karabulut, D.; Kocamaz, N.; Isiksacan, N.; Erdal, G.S.; Caglar, N.T.; Karabulut, D.; Kocamaz, N.; Isiksacan, N. Is Vitamin D Level Elevated in Patients on Statin Treatment? *Int. J. Cardiovasc. Acad.* **2019**, *5*, 159–162. [CrossRef]
23. Mazidi, M.; Rezaie, P.; Vatanparast, H.; Kengne, A.P. Effect of Statins on Serum Vitamin D Concentrations: A Systematic Review and Meta-Analysis. *Eur. J. Clin. Investig.* **2017**, *47*, 93–101. [CrossRef]
24. Kellick, K.A.; Bottorff, M.; Toth, P.P. The National Lipid Association's Safety Task Force, null A Clinician's Guide to Statin Drug-Drug Interactions. *J. Clin. Lipidol.* **2014**, *8*, S30–S46. [CrossRef] [PubMed]
25. Bhattacharyya, S.; Bhattacharyya, K.; Maitra, A. Possible Mechanisms of Interaction between Statins and Vitamin D. *QJM* **2012**, *105*, 487–491. [CrossRef]
26. Ferretti, G.; Bacchetti, T.; Sahebkar, A. Effect of Statin Therapy on Paraoxonase-1 Status: A Systematic Review and Meta-Analysis of 25 Clinical Trials. *Prog. Lipid Res.* **2015**, *60*, 50–73. [CrossRef] [PubMed]
27. Riche, K.D.; Arnall, J.; Rieser, K.; East, H.E.; Riche, D.M. Impact of Vitamin D Status on Statin-Induced Myopathy. *J. Clin. Transl. Endocrinol.* **2016**, *6*, 56–59. [CrossRef]
28. Jain, M.K.; Ridker, P.M. Anti-Inflammatory Effects of Statins: Clinical Evidence and Basic Mechanisms. *Nat. Rev. Drug Discov.* **2005**, *4*, 977–987. [CrossRef] [PubMed]
29. Bouillon, R.; Marcocci, C.; Carmeliet, G.; Bikle, D.; White, J.H.; Dawson-Hughes, B.; Lips, P.; Munns, C.F.; Lazaretti-Castro, M.; Giustina, A.; et al. Skeletal and Extraskeletal Actions of Vitamin D: Current Evidence and Outstanding Questions. *Endocr. Rev.* **2018**, *40*, 1109–1151. [CrossRef]
30. Ahmadi, J.; Hosseini, E.; Kargar, F.; Ghasemzadeh, M. Stable CAD Patients Show Higher Levels of Platelet-Borne TGF-B1 Associated with a Superior pro-Inflammatory State than the pro-Aggregatory Status; Evidence Highlighting the Importance of Platelet-Derived TGF-B1 in Atherosclerosis. *J. Thromb. Thrombolysis* **2023**, *55*, 102–115. [CrossRef]
31. Nenna, A.; Nappi, F.; Lusini, M.; Satriano, U.M.; Schilirò, D.; Spadaccio, C.; Chello, M. Effect of Statins on Platelet Activation and Function: From Molecular Pathways to Clinical Effects. *Biomed. Res. Int.* **2021**, *2021*, 6661847. [CrossRef] [PubMed]

32. Aihara, K.; Azuma, H.; Akaike, M.; Ikeda, Y.; Yamashita, M.; Sudo, T.; Hayashi, H.; Yamada, Y.; Endoh, F.; Fujimura, M.; et al. Disruption of Nuclear Vitamin D Receptor Gene Causes Enhanced Thrombogenicity in Mice. *J. Biol. Chem.* **2004**, *279*, 35798–35802. [CrossRef]
33. Tudoran, M.; Tudoran, C. Particularities of Endothelial Dysfunction in Hypothyroid Patients. *Kardiol. Pol.* **2015**, *73*, 337–343. [CrossRef]
34. Tarcin, O.; Yavuz, D.G.; Ozben, B.; Telli, A.; Ogunc, A.V.; Yuksel, M.; Toprak, A.; Yazici, D.; Sancak, S.; Deyneli, O.; et al. Effect of Vitamin D Deficiency and Replacement on Endothelial Function in Asymptomatic Subjects. *J. Clin. Endocrinol. Metab.* **2009**, *94*, 4023–4030. [CrossRef]
35. Crisby, M.; Nordin-Fredriksson, G.; Shah, P.K.; Yano, J.; Zhu, J.; Nilsson, J. Pravastatin Treatment Increases Collagen Content and Decreases Lipid Content, Inflammation, Metalloproteinases, and Cell Death in Human Carotid Plaques: Implications for Plaque Stabilization. *Circulation* **2001**, *103*, 926–933. [CrossRef]
36. Norman, A.W. From Vitamin D to Hormone D: Fundamentals of the Vitamin D Endocrine System Essential for Good Health. *Am. J. Clin. Nutr.* **2008**, *88*, 491S–499S. [CrossRef]
37. Undas, A.; Brummel-Ziedins, K.E.; Mann, K.G. Statins and Blood Coagulation. *Arterioscler. Thromb. Vasc. Biol.* **2005**, *25*, 287–294. [CrossRef]
38. Wassmann, S.; Laufs, U.; Bäumer, A.T.; Müller, K.; Ahlbory, K.; Linz, W.; Itter, G.; Rösen, R.; Böhm, M.; Nickenig, G. HMG-CoA Reductase Inhibitors Improve Endothelial Dysfunction in Normocholesterolemic Hypertension via Reduced Production of Reactive Oxygen Species. *Hypertension* **2001**, *37*, 1450–1457. [CrossRef]
39. Riccioni, G.; D'Orazio, N.; Palumbo, N.; Bucciarelli, V.; Di Ilio, E.; Bazzano, L.A.; Bucciarelli, T. Relationship between Plasma Antioxidant Concentrations and Carotid Intima-Media Thickness: The Asymptomatic Carotid Atherosclerotic Disease In Manfredonia Study. *Eur. J. Cardiovasc. Prev. Rehabil.* **2009**, *16*, 351–357. [CrossRef] [PubMed]
40. Hosseini, E.; Ghasemzadeh, M.; Atashibarg, M.; Haghshenas, M. ROS Scavenger, N-Acetyl-l-Cysteine and NOX Specific Inhibitor, VAS2870 Reduce Platelets Apoptosis While Enhancing Their Viability during Storage. *Transfusion* **2019**, *59*, 1333–1343. [CrossRef] [PubMed]
41. Hosseini, E.; Beyranvand, Z.; Schoenwaelder, S.M.; Farhid, F.; Ghasemzadeh, M. The Reactive Oxygen Species Scavenger N-Acetyl-L-Cysteine Reduces Storage-Dependent Decline in Integrin $\alpha$IIb$\beta$3-Mediated Platelet Function, Inhibiting Pre-Activation of Integrin and Its B3 Subunit Cleavage. *Oxidative Med. Cell. Longev.* **2025**, *2025*, 7499648. [CrossRef]
42. Grant, W.B.; Boucher, B.J.; Cheng, R.Z.; Pludowski, P.; Wimalawansa, S.J. Vitamin D and Cardiovascular Health: A Narrative Review of Risk Reduction Evidence. *Nutrients* **2025**, *17*, 2102. [CrossRef] [PubMed]
43. Fadah, K.; Mares, A.; Lange, R.A. Statin-Associated Muscle Symptoms and Vitamin D Supplementation. *Curr. Opin. Cardiol.* **2025**, *40*, 215–220. [CrossRef] [PubMed]
44. Wu, Z.; Camargo, C.A.; Khaw, K.-T.; Waayer, D.; Lawes, C.M.M.; Toop, L.; Scragg, R. Effects of Vitamin D Supplementation on Adherence to and Persistence with Long-Term Statin Therapy: Secondary Analysis from the Randomized, Double-Blind, Placebo-Controlled ViDA Study. *Atherosclerosis* **2018**, *273*, 59–66. [CrossRef] [PubMed]
45. Thompson, B.; Waterhouse, M.; English, D.R.; McLeod, D.S.; Armstrong, B.K.; Baxter, C.; Romero, B.D.; Ebeling, P.R.; Hartel, G.; Kimlin, M.G.; et al. Vitamin D Supplementation and Major Cardiovascular Events: D-Health Randomised Controlled Trial. *Br. Med. J.* **2023**, *381*, e075230. [CrossRef]
46. Miao, J.; Bachmann, K.N.; Huang, S.; Su, Y.R.; Dusek, J.; Newton-Cheh, C.; Arora, P.; Wang, T.J. Effects of vitamin D supplementation on cardiovascular and glycemic biomarkers. *J. Am. Heart Assoc.* **2021**, *10*, e017727. [CrossRef]
47. Pérez-Castrillón, J.L.; Vega, G.; Abad, L.; Sanz, A.; Chaves, J.; Hernandez, G.; Dueñas, A. Effects of Atorvastatin on Vitamin D Levels in Patients with Acute Ischemic Heart Disease. *Am. J. Cardiol.* **2007**, *99*, 903–905. [CrossRef]
48. Makariou, E.S.; Elisaf, M.; Kei, A.; Challa, A.; DiNicolantonio, J.J.; Liberopoulos, E. No Effect of Switching to High-Dose Rosuvastatin, Add-on Nicotinic Acid, or Add-on Fenofibrate on Serum Vitamin D Levels in Patients with Mixed Dyslipidemia. *Hippokratia* **2015**, *19*, 136–140.
49. Rejnmark, L.; Vestergaard, P.; Heickendorff, L.; Mosekilde, L. Simvastatin does not affect vitamin d status, but low vitamin d levels are associated with dyslipidemia: Results from a randomised, controlled trial. *Int. J. Endocrinol.* **2010**, *2010*, 957174. [CrossRef]
50. Glueck, C.J.; Budhani, S.B.; Masineni, S.S.; Abuchaibe, C.; Khan, N.; Wang, P.; Goldenberg, N. Vitamin D Deficiency, Myositis-Myalgia, and Reversible Statin Intolerance. *Curr. Med. Res. Opin.* **2011**, *27*, 1683–1690. [CrossRef]
51. Khayznikov, M.; Hemachrandra, K.; Pandit, R.; Kumar, A.; Wang, P.; Glueck, C.J. Statin Intolerance Because of Myalgia, Myositis, Myopathy, or Myonecrosis Can in Most Cases Be Safely Resolved by Vitamin D Supplementation. *N. Am. J. Med. Sci.* **2015**, *7*, 86–93. [CrossRef]
52. Backes, J.M.; Ruisinger, J.F.; Barnes, B.J.; Moriarty, P.M. Statin Intolerance and Vitamin D Supplementation: Sunny, but a Few Clouds Remain. *N. Am. J. Med. Sci.* **2015**, *7*, 337–338. [CrossRef] [PubMed]

53. Bitzur, R.; Cohen, H.; Kamari, Y.; Harats, D. Intolerance to Statins: Mechanisms and Management. *Diabetes Care* **2013**, *36* (Suppl. S2), S325–S330. [CrossRef]
54. Demay, M.B.; Pittas, A.G.; Bikle, D.D.; Diab, D.L.; Kiely, M.E.; Lazaretti-Castro, M.; Lips, P.; Mitchell, D.M.; Murad, M.H.; Powers, S.; et al. Vitamin D for the Prevention of Disease: An Endocrine Society Clinical Practice Guideline. *J. Clin. Endocrinol. Metab.* **2024**, *109*, 1907–1947. [CrossRef] [PubMed]
55. Giovannucci, E.; Liu, Y.; Hollis, B.W.; Rimm, E.B. 25-Hydroxyvitamin D and Risk of Myocardial Infarction in Men: A Prospective Study. *Arch. Intern. Med.* **2008**, *168*, 1174–1180. [CrossRef]
56. Brøndum-Jacobsen, P.; Benn, M.; Jensen, G.B.; Nordestgaard, B.G. 25-Hydroxyvitamin d Levels and Risk of Ischemic Heart Disease, Myocardial Infarction, and Early Death: Population-Based Study and Meta-Analyses of 18 and 17 Studies. *Arterioscler. Thromb. Vasc. Biol.* **2012**, *32*, 2794–2802. [CrossRef] [PubMed]
57. Chowdhury, R.; Kunutsor, S.; Vitezova, A.; Oliver-Williams, C.; Chowdhury, S.; Kiefte-de-Jong, J.C.; Khan, H.; Baena, C.P.; Prabhakaran, D.; Hoshen, M.B.; et al. Vitamin D and Risk of Cause Specific Death: Systematic Review and Meta-Analysis of Observational Cohort and Randomised Intervention Studies. *Br. Med. J.* **2014**, *348*, g1903. [CrossRef]
58. Liu, Y.; Pei, Y.; Zhang, S.; Du, Z.; Chen, L.; Yan, X.; Tian, J. Association of Serum 25-Hydroxyvitamin D3 Levels with Carotid Artery Intima-Media Thickness and Carotid Atherosclerotic Plaques in Smokers. *Vascul Pharmacol.* **2024**, *157*, 107416. [CrossRef]
59. Norman, P.E.; Powell, J.T. Vitamin D and Cardiovascular Disease. *Circ. Res.* **2014**, *114*, 379–393. [CrossRef]
60. Carrelli, A.; Bucovsky, M.; Horst, R.; Cremers, S.; Zhang, C.; Bessler, M.; Schrope, B.; Evanko, J.; Blanco, J.; Silverberg, S.J.; et al. Vitamin D Storage in Adipose Tissue of Obese and Normal Weight Women. *J. Bone Miner. Res.* **2017**, *32*, 237–242. [CrossRef]
61. Manson, J.E.; Cook, N.R.; Lee, I.-M.; Christen, W.; Bassuk, S.S.; Mora, S.; Gibson, H.; Gordon, D.; Copeland, T.; D'Agostino, D.; et al. Vitamin D Supplements and Prevention of Cancer and Cardiovascular Disease. *N. Engl. J. Med.* **2019**, *380*, 33–44. [CrossRef] [PubMed]
62. Anderson, J.L.; May, H.T.; Horne, B.D.; Bair, T.L.; Hall, N.L.; Carlquist, J.F.; Lappé, D.L.; Muhlestein, J.B.; Intermountain Heart Collaborative (IHC) Study Group. Relation of Vitamin D Deficiency to Cardiovascular Risk Factors, Disease Status, and Incident Events in a General Healthcare Population. *Am. J. Cardiol.* **2010**, *106*, 963–968. [CrossRef] [PubMed]
63. García-Gómez, C.; Bianchi, M.; de la Fuente, D.; Badimon, L.; Padró, T.; Corbella, E.; Pintó, X. Inflammation, Lipid Metabolism and Cardiovascular Risk in Rheumatoid Arthritis: A Qualitative Relationship? *World J. Orthop.* **2014**, *5*, 304–311. [CrossRef]
64. Avouac, J.; Meune, C.; Chenevier-Gobeaux, C.; Dieudé, P.; Borderie, D.; Lefevre, G.; Kahan, A.; Allanore, Y. Inflammation and Disease Activity Are Associated with High Circulating Cardiac Markers in Rheumatoid Arthritis Independently of Traditional Cardiovascular Risk Factors. *J. Rheumatol.* **2014**, *41*, 248–255. [CrossRef] [PubMed]
65. Milazzo, V.; De Metrio, M.; Cosentino, N.; Marenzi, G.; Tremoli, E. Vitamin D and Acute Myocardial Infarction. *World J. Cardiol.* **2017**, *9*, 14–20. [CrossRef]
66. Pilz, S.; Verheyen, N.; Grübler, M.R.; Tomaschitz, A.; März, W. Vitamin D and Cardiovascular Disease Prevention. *Nat. Rev. Cardiol.* **2016**, *13*, 404–417. [CrossRef]
67. Cipponeri, E.; Vitturi, N.; Mariano, V.; Boscari, F.; Galasso, S.; Crepaldi, C.; Fadini, G.P.; Vigili de Kreutzenberg, S.; Marescotti, M.C.; Iori, E.; et al. Vitamin D Status and Non-Alcoholic Fatty Liver Disease in Patients with Type 1 Diabetes. *J. Endocrinol. Investig.* **2019**, *42*, 1099–1107. [CrossRef]
68. Szymczak-Pajor, I.; Śliwińska, A. Analysis of Association between Vitamin D Deficiency and Insulin Resistance. *Nutrients* **2019**, *11*, 794. [CrossRef] [PubMed]

biomedicines

MDPI

*Article*

# Theophylline Attenuates the Release of Cardiovascular Disease-Related Triglyceride and Cholesterol by Inhibiting the Activity of Microsomal Triglyceride Transfer Protein in Rat Hepatocytes

**Min-Kyu Park [1,2,3], Hyeonha Jang [4], Jeong-Soo Bae [5], Jae-Ho Shin [1,2,3,*] and Hwa-Jin Park [6,*]**

1 Department of Applied Biosciences, Kyungpook National University, Daegu 41566, Republic of Korea; pmk601313@knu.ac.kr
2 KNU NGS Core Facility, Kyungpook National University, Daegu 41566, Republic of Korea
3 Microbalance Inc., Daegu 41566, Republic of Korea
4 College of Pharmacy, Kyungpook National University, Daegu 41566, Republic of Korea; jg.heana@gmail.com
5 Cardiovascular Laboratory, Medical Center of Dong-A University, 26 Daesingongwon-ro, Busan 49201, Republic of Korea; hamcholom@hanmail.net
6 Department of Biomedical Laboratory Science, College of Healthcare Medical Science and Engineering, Inje University, 197, Inje-ro, Gimhae 50834, Republic of Korea
* Correspondence: jhshin@knu.ac.kr (J.-H.S.); mlsjpark@inje.ac.kr (H.-J.P.)

**Abstract**

**Background/Objectives**: Cardiovascular diseases (CVD) remain the leading cause of diet-related mortality, with hepatic overproduction of very-low-density lipoprotein (VLDL) being a central driver of dyslipidemia. The microsomal triglyceride transfer protein (MTP) is essential for this process, and its activity is negatively regulated by cyclic adenosine monophosphate (cAMP). Theophylline, a methylxanthine found in tea, increases intracellular cAMP. This study aimed to evaluate whether physiologically relevant concentrations of theophylline could beneficially modulate lipoprotein secretion in an ex vivo model of diet-induced MTP activation. **Methods**: Primary hepatocytes were isolated from rats fed a high-fat, high-carbohydrate diet (HFCD). Cells were treated with 100 μM theophylline, and the secretion of triglyceride (TG), total cholesterol (TC), VLDL-cholesterol (VLDL-C), and HDL-cholesterol (HDL-C) was quantified. Hepatocellular MTP activity and atherogenic indices were also assessed. **Results**: Compared to untreated control cells, theophylline treatment significantly reduced the secretion of TG by 6% and TC by 24%. Specifically, VLDL-C secretion decreased by 6%, while HDL-C secretion increased substantially by 93%. These lipid-modulating effects were correlated with a 6.9% reduction in MTP activity. Consequently, significant improvements were observed in the atherogenic indices TG/HDL-C, TC/HDL-C, and the atherogenic index (AI) ($p < 0.01$). **Conclusions**: Our findings demonstrate that physiologically attainable concentrations of theophylline rebalance lipoprotein secretion by suppressing hepatic MTP activity, shifting the lipid profile toward an anti-atherogenic state. These results highlight the potential of theophylline as a functional dietary component for mitigating diet-induced dyslipidemia and reducing cardiovascular risk.

**Keywords:** theophylline; rat primary hepatocytes; hypertriglyceridemia; hypercholesterolemia; microsomal triglyceride transfer protein; high-density lipoprotein cholesterol; atherogenic index; cardiovascular disease

## 1. Introduction

Cardiovascular disease (CVD) remains the leading cause of diet-related mortality, accounting for over 18 million deaths annually [1]. Postprandial dyslipidemia, a hallmark of diet-induced CVD, is driven by the hepatic overproduction of very-low-density lipoprotein (VLDL). VLDL particles, which are rich in triglycerides (TG) and cholesterol esters, are rapidly converted in circulation to low-density lipoprotein (LDL)—the principal atherogenic lipoprotein [2–7]. Mounting evidence indicates that aberrant VLDL secretion arises from the upregulation of microsomal triglyceride transfer protein (MTP), a lipid-transfer chaperone located within the endoplasmic reticulum that is essential for the assembly of apolipoprotein B (ApoB) [8]. Nutritional and hormonal signals converge on MTP; for instance, high-fat, high-carbohydrate diets enhance hepatic TG synthesis and intracellular $Ca^{2+}$ levels, both of which proportionally elevate MTP activity and VLDL output, ultimately leading to hypertriglyceridemia and hypercholesterolemia [9,10]. Cyclic adenosine monophosphate (cAMP) functions as a negative regulator; elevated hepatocellular cAMP levels inhibit MTP-mediated TG transfer while concomitantly activating ATP-binding cassette transporter A1 (ABCA1), promoting nascent high-density lipoprotein (HDL) biogenesis [11,12]. These opposing effects position cAMP–MTP signaling as a nutritionally exploitable axis for rebalancing the release of atherogenic versus anti-atherogenic lipoproteins. Methylxanthines serve as key dietary modulators of cAMP, primarily through the non-selective inhibition of phosphodiesterases (PDEs). Theophylline (1,3-dimethylxanthine) is the most abundant methylxanthine found in young tea leaves (Camellia sinensis), with concentrations of 0.1–0.5 mg $g^{-1}$ dry weight; lower levels have been detected in cacao, coffee, and kola nut [13,14]. While synthetic theophylline is clinically used as a bronchodilator, its dietary counterpart, which can easily be extracted from green or partially fermented tea, represents a "clean-label" nutraceutical option. Notably, the dietary intake achievable via tea beverages (<10 μmol/day) is several orders of magnitude lower than the pharmacological doses that have yielded inconsistent lipid outcomes in asthmatic populations [13,15,16]. Whether such physiologically attainable concentrations can beneficially modulate MTP-driven lipoprotein secretion remains untested.

Here, we address this knowledge gap by utilizing primary hepatocytes isolated from rats fed a high-fat, high-carbohydrate diet (HFCD), a validated ex vivo model of diet-induced MTP activation. We hypothesized that tea-equivalent theophylline (i) inhibits MTP activity and the secretion of TG- and cholesterol-rich VLDL and (ii) promotes HDL release, thereby improving atherogenic indices. The present study delineates the direct hepatic actions of theophylline at nutritionally relevant concentrations and provides mechanistic insights regarding its potential application as a functional food component for the management of dyslipidemia.

## 2. Materials and Methods

### 2.1. Materials

Hanks' balanced salt solution (without calcium chloride, magnesium sulfate, or sodium bicarbonate) for collagenase perfusion and Williams' medium E (containing L-glutamine but excluding sodium bicarbonate) for hepatocyte culture were purchased from Sigma (St. Louis, MO, USA). Theophylline was also procured from Sigma (St. Louis, MO, USA). TG, total cholesterol (TC), and HDL cholesterol (HDL-C) assay kits were obtained from Nissui Pharmaceutical (Tokyo, Japan). MTP assay kits were acquired from Calbiochem® (an affiliate of Merck KGaA, Darmstadt, Germany). Collagenase (type IV) and additional chemical reagents were secured from Sigma (St. Louis, MO, USA).

### *2.2. Selection and Characteristics of Rat Diets for the Induction of TG and Cholesterol in Hepatocytes*

The production of triglycerides (TG) and total cholesterol (TC) is essential for assessing the effect of theophylline on their release by hepatocytes. Microsomal triglyceride transfer protein (MTP), which facilitates the secretion of VLDL containing TG and TC, shows increased activity in proportion to hepatic TG synthesis [7,10,17,18]. To establish this condition, male Sprague–Dawley rats (200–250 g; Samtako Bio, Osan, Republic of Korea) were fed a high-fat, high-carbohydrate diet (HFCD; Table 1) composed of 65.4% fat and carbohydrate (Sam Yang Oil & Fat Feed Co., Ltd., Seoul, Republic of Korea). The HFCD contained soybean oil and lard as fat sources and maltodextrin, cellulose, and sucrose as carbohydrate sources. These nutrients provide precursors for TG and TC synthesis in hepatocytes, thereby inducing obesity, fatty liver, and dyslipidemia, all of which are risk factors for cardiovascular disease (CVD) [10,15,19–22]. The HFCD (Table 1) was used as a normal control to induce obesity, a risk factor for CVD, in rats [23,24]. In this study, we administered HFCD to rats to increase the production of TG and TC, which are known to contribute to CVD, and investigated the effects of theophylline on lipid profiles. Therefore, the HFCD-administered group served as a control group for theophylline or 10% glucose plus HFCD groups.

**Table 1.** High-fats and carbohydrates diet composition (Sam Yang Oil & Fat Feed Co., Republic of Korea).

| **Ingredient** | | **Contents** | |
|---|---|---|---|
| | | **(g/kg)** | **(%)** |
| Fat and oil | Soybean oil | 25 | 65.4 |
| | Lard | 245 | |
| | Subtotal | 270 | |
| Carbohydrate | Maltodextrin | 125 | |
| | Sucrose | 68.8 | |
| | Cellulose | 50 | |
| | Subtotal | 243.8 | |
| Protein and amino acid | Casein | 200 | 25.8 |
| | L-Cystine | 3 | |
| Mineral and electrolytes | Mineral mix | 10 | 7.3 |
| | Dicalcium | 13 | |
| | Phosphate | 5.5 | |
| | Calcium | 16.5 | |
| | Carbonate | 10 | |
| | Potassium citrate | 2 | |
| Vitamins | Vitamin mix | 10 | 1.5 |
| | Choline bitartrate | 2 | |
| Total | | 785.8 | 100 |

Fat and oil, and carbohydrates representing HFCD are shaded.

To further amplify TG and TC production and thereby MTP activity, 10% glucose was supplemented to the HFCD [10]. Accordingly, hepatocytes isolated from rats fed either HFCD alone or HFCD plus glucose were used to evaluate whether theophylline modulates MTP activity and the secretion of TG and TC.

### *2.3. Experimental Design and Animal Administration*

As illustrated in Figure 1, male Sprague–Dawley rats (weight: 200–250 g, 6–7 weeks) secured from Samtako Bio (Osan, Republic of Korea) were categorized into two groups. The first group, designated as the HFCD group ($n$ = 5), was provided with HFCD and water *ad libitum*. The second group, referred to as the HFCD plus 10% glucose group ($n = 5$), received the HFCD along with 10% glucose *ad libitum*, with the glucose supplied in place of water. All rats were housed in an environment maintained at a temperature of 24 °C and humidity of 65% under normal light. The duration of rat care spanned 15 days to promote TG synthesis in the liver [25]. To prevent any alterations in TG levels and MTP activity, the rats were not fasted prior to post-mortem analysis [26]. Following the 15-day care period, the rats were anesthetized using 20% urethane before post-mortem procedures. All the animal experiments in this study were conducted in accordance with the Guidelines for Care and Use of Laboratory Animals issued by the Institutional Ethical Animal Care Committee of Kyungpook National University, Korea.

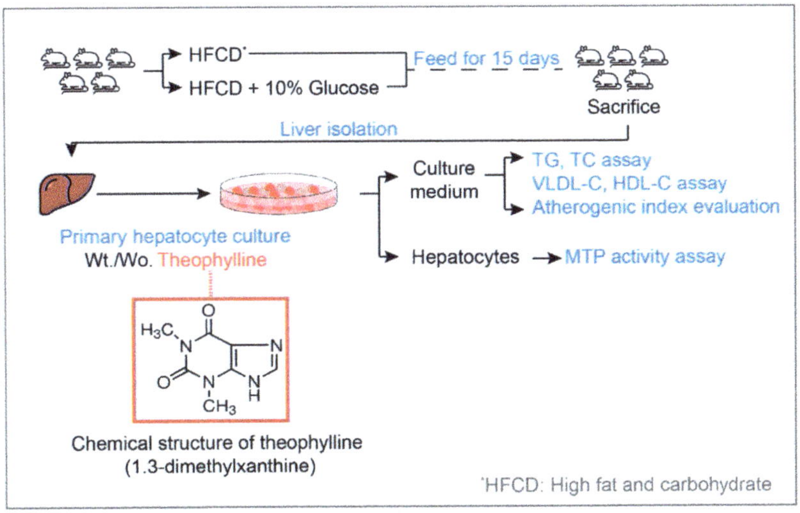


**Figure 1. Study design to investigate the effects of theophylline on CVD risk.** TG: triglyceride, TC: total cholesterol, HDL-C: high density lipoprotein-cholesterol, MTP: microsomal triglyceride transfer protein, Wt: with, Wo: without.

### *2.4. Determination of Theophylline Dose for Measuring Lipid Profiles and MTP Activity*

To evaluate the impact of theophylline treatment on CVD-risk lipids—specifically TG, TC, and LDL-C—in the blood of patients with asthma, a previous study administered 300 mg/day of theophylline over a 6-month period, approximately 180 days [16]. This regimen resulted in a cumulative concentration of 300 mmol of theophylline (Mw. 180) (approximately 1.67 mmol/day), which informed our experimental design. We utilized 0.3 μmol (100 μM) of theophylline to investigate the release of CVD risk lipids into the culture medium from rat hepatocytes, representing a theophylline dosage 10–6-fold lower than that (300 mmol) administered to asthmatic patients [16].

The primary metric for evaluating the effect of theophylline on the secretion of TG and TC in hepatocytes is MTP activity. Therefore, we employed 10 mM theophylline, which has previously been utilized to evaluate the secretory activity of insulin in the pancreas, to investigate the influence of theophylline on MTP activity [27].

### *2.5. Rational for Using Hepatocytes to Study Blood Lipid Profiles*

Although plasma or serum lipid measurements such as triglycerides, total cholesterol, VLDL, LDL, and HDL are widely used to assess cardiovascular disease (CVD) risk, it is difficult to attribute these changes to specific sources. Circulating lipids originate from multiple pathways, including dietary chylomicrons absorbed from the intestine, VLDL and

HDL secreted by the liver, and lipids released from adipose tissue. Because the liver is the predominant source of atherogenic lipids [2–7], we investigated the potential anti-CVD mechanism of theophylline by examining hepatocyte-derived lipid secretion. Specifically, we measured VLDL-triglycerides, total cholesterol, and HDL-cholesterol in the culture medium of primary rat hepatocytes isolated from animals fed a high-fat/cholesterol diet (HFCD) or 10% glucose as lipid precursors. These culture medium lipid profiles closely reflect plasma or serum lipid changes and therefore provide mechanistic insights into CVD risk.

### *2.6. Preparation of Rat Hepatocytes and Their Culture*

Freshly isolated hepatocytes were prepared from adult male Sprague-Dawley rats (200–250 g) administered with HFCD or HFCD plus 10% glucose for 15 d. Rats were not fasted until the post-mortem to protect any changes on TG level and MTP activity [26]. After 15 d of care, rats were anesthetized with 20% urethane before the post-mortem, and subsequently the hepatocytes were isolated by the collagenase-liver perfusion method of Berry and Friend [28]. Hepatocyte viability was assessed via trypan blue exclusion, yielding an average viability of 95%. Hepatocytes exhibiting 95% viability were washed three times with a washing buffer comprising Williams' Medium E, 23.8 mM $NaHCO_3$, 0.5 mM dexamethasone, and 15 mM 4-(2-hydroxyethyl)-1-piperazineethanesulfonic acid (pH 7.4), supplemented with 1000 U/L penicillin, 1 mg/L streptomycin, and 1 mM insulin. A total of $10^6$ hepatocytes per well, suspended in culture medium (washing buffer supplemented with 10% fetal bovine serum), were placed in culture dishes (Falcon®; Becton Dickinson and Company, Franklin Lakes, NJ, USA) and incubated at 37 °C under 5% $CO_2$ for 3 h. The monolayer was washed twice with culture medium and supplemented with 3 mL of fresh culture medium. The hepatocytes were subsequently incubated at 37 °C for 6 h, with or without theophylline, after which the incubation was terminated on ice. The culture medium from the hepatocytes in the HFCD group (Group 1) was utilized to measure TG, TC, and HDL-C levels. Hepatocytes prepared from both the HFCD (Group 1) and HFCD plus 10% glucose (Group 2) groups were used to assess MTP activity.

### *2.7. Determination of Lipid Levels in Hepatocyte Culture Medium*

Lipid levels in the hepatocyte culture medium were quantified using assay kits from Nissui Pharmaceutical (Tokyo, Japan).

TG levels were measured enzymatically using lipoprotein lipase, glycerol kinase, and glycerol peroxidase. TC was also determined enzymatically using cholesterol esterase, cholesterol oxidase, and peroxidase.

HDL-C levels were assessed enzymatically following the precipitation of VLDL with a complex of sodium phosphotungstic acid and $MgCl_2$. VLDL cholesterol (VLDL-C) levels were calculated as follows: VLDL-C = 1/5 TG, as derived from the Friedewald formula: LDL-C = TC − HDL-C − 1/5 TG [29]. The atherogenic index (AI) was calculated using the formula TC − HDL-C/HDL-C. In addition, TG/HDL-C and TC/HDL-C ratios were computed to evaluate the impact of theophylline on CVD risk.

### *2.8. Preparation of Hepatocellular Lysates and Measurement of MTP Activity*

To investigate the effects of theophylline on MTP activity, hepatocellular lysates were prepared. Hepatocytes ($10^6$ cells/well) were sonicated on ice for two bursts of 10 s each using an ultrasonic homogenizer (Bandelin Sonopuls HD2070; Bandelin Electronic GmbH & Co. KG, Berlin, Germany) set to 30% power.

To measure MTP activity, hepatocellular lysates (200 μg protein) were incubated with gentle agitation at 37 °C for 2 h in the presence of 10 μL of a donor molecule containing a fluorescent neutral lipid and 10 μL of an acceptor molecule, as outlined in the MTP assay kit manual (Calbiochem®, Merck KGaA, Darmstadt, Germany). MTP activity was assessed by measuring fluorescence at 535 nm, with excitation at 465 nm, and calculated by subtracting the blank fluorescence value from each sample using a standard curve.

### 2.9. Protein Assay

It is well established that the presence of lipids (e.g., phospholipids, VLDL, LDL) interferes with protein quantification when using the BCA method [30–32]. In contrast, chelating agents such as EDTA and EGTA, which are known to interfere with the Lowry method, are absent in hepatocyte culture medium and lysates. Therefore, to avoid lipid-derived interference, we measured protein concentrations using the Lowry method [33]. These measurements were subsequently used to calculate the absolute lipid concentration in the culture medium and the absolute activity of MTP in hepatocytes.

### 2.10. Statistical Analysis

All statistical analyses were performed using GraphPad Prism 8. Prior to data analysis, normality was first checked with the Shapiro–Wilk test and homogeneity of variance with the Brown-Forsythe test. Comparisons of means between two groups were performed using Student's *t*-test, and comparisons of three or more groups were subjected to one-way analysis of variance (ANOVA) followed by post hoc analysis with Dunnett's multiple comparison test. All data were expressed as mean ± standard deviation (mean ± SD) of at least three replicate experiments, and were considered statistically significant if $p$-value < 0.05.

## 3. Results

### 3.1. Intake of Feed and Lipids Producing Materials

As shown in Table 2, rats in the HFCD group consumed 20.4 ± 1.5 g of feed per day and gained 3.3 ± 0.1 g of body weight per day. In contrast, rats in the HFCD plus glucose group consumed significantly less feed (14.8 ± 1.7 g/day) but gained greater body weight (3.6 ± 0.9 g/day). Consequently, the feed efficiency ratio (FER) of the HFCD plus glucose group (0.24 ± 0.05) was significantly higher than that of the HFCD group (0.16 ± 0.01), indicating that additional glucose intake (10.6 ± 0.9 g/day) enhanced weight gain relative to feed intake (Table 2). Although rats in the HFCD plus glucose group consumed less dietary fat, oil, and carbohydrates than those in the HFCD group, the additional glucose intake likely served as a major substrate for lipid synthesis.

**Table 2.** Daily feed intake and lipid-producing substrate concentration in rats.

| Animal Group | Feed Intake (g/d) | Body Weight Gain (g/d) | FER | Intake of Lipid-Producing Materials of Feed (g/d) | | | |
|---|---|---|---|---|---|---|---|
| | | | | Fat and Oil (270 g/kg-HFCD) | Carbohydrates (243.8 g/kg-HFCD) | Glucose (g/d) | Total |
| HFCD | (1) 20.4 ± 1.5 | 3.3 ± 0.1 | 0.16 ± 0.1 | (3) 5.51 ± 0.41 | (5) 4.97 ± 0.41 | - | 10.48 ± 0.8 |
| HFCD + glucose | (2) 14.8 ± 1.7 *** | 3.6 ± 0.9 * | 0.24 ± 0.5 | (4) 4.00 ± 0.46 | (6) 3.60 ± 0.41 ** | 10.6 ± 0.9 | 18.2 ± 1.7 *** |

* $p < 0.05$, ** $p < 0.01$, *** $p < 0.001$. Data are given as means ± S.D. ($n = 5$). HFCD, high fat and carbohydrate. FER, feed efficiency ratio. FER = body weight gain (g/d)/feed intake (g/d). (3) = (1) × 270 g/kg-HFCD. (4) = (2) × 270 g/kg-HFCD. (5) = (1) × 243.8 g/kg-HFCD. (6) = (2) × 243.8 g/kg-HFCD. Fat and oil (270 g/kg-HFCD) and carbohydrates (243.8 g/kg-HFCD) were derived from Table 1.

### 3.2. Effects of Theophylline on TG Release from Hepatocytes into the Culture Medium

Hepatocytes ($10^6$ cells/well) derived from rats fed an HFCD released TG (67 ± 1.0 × $10^2$ ng/protein-mg) into the culture medium; nevertheless, theophylline treatment reduced this release to 63 ± 0.8 × $10^2$ ng/protein-mg (Figure 2A). This represents a decrease in culture medium TG levels of approximately 6% compared with that released by the untreated HFCD hepatocytes (Figure 2A). These findings suggest that theophylline inhibits TG release from hepatocytes obtained from rats fed an HFCD as a standard pellet diet.

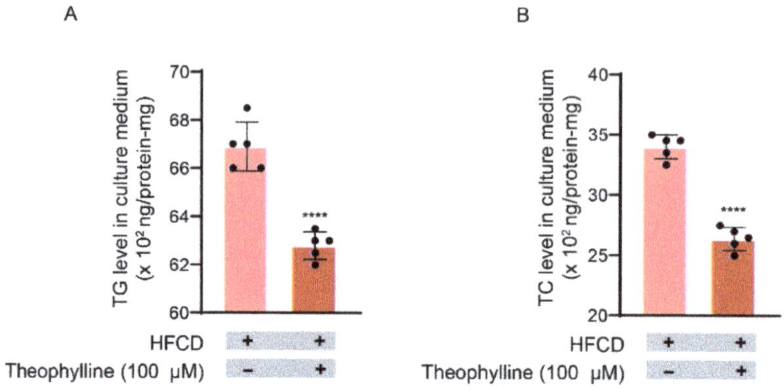


**Figure 2. Effect of theophylline on the release of triglyceride and total cholesterol (TC) into hepatocytes culture medium.** The levels of TG and TC were determined as described in Materials and Methods. (**A**) Level of TG. (**B**) Level of TC. TG, triglyceride; TC, total cholesterol. HFCD, high fat and carbohydrate diets. Data are given as means ± SD., $n = 5$. **** $p < 0.0001$.

### 3.3. *Effects of Theophylline on TC Release from Hepatocytes into the Culture Medium*

As illustrated in Figure 2B, hepatocytes ($10^6$ cells/well) from rats on an HFCD released TC ($34 \pm 1.0 \times 10^2$ ng/protein-mg) into the culture medium; nonetheless, theophylline treatment decreased this release to $26 \pm 1.0 \times 10^2$ ng/protein-mg (Figure 2B). Consequently, these results suggest that theophylline elicits a 24% reduction in TC release from hepatocytes derived from rats fed an HFCD as a standard pellet diet.

### 3.4. *Effect of Theophylline on MTP Activity in Hepatocytes*

MTP activity serves as an index for evaluating the secretion of TG and TC from hepatocytes [8,34]. Therefore, we examined whether the inhibition of TG and TC secretion by theophylline (Figure 2A,B) correlates with a reduction in MTP activity. As depicted in Figure 3, MTP activity in hepatocytes from rats fed an HFCD supplemented with 10% glucose significantly increased, reaching 27.8 ± 0.8 nM/protein-mg/min, an increase of 25% compared with that observed in hepatocytes on an HFCD alone (221 ± 0.2 nM/protein-mg/min). This implies that the addition of 10% glucose to the HFCD enhances MTP activity in hepatocytes, subsequently increasing the secretion of TG and TC into the culture medium.

To further elucidate the inhibitory effect of theophylline on TG and TC secretion, we investigated its impact on MTP activity in rat hepatocytes administered the HFCD plus 10% glucose, which is known to elevate MTP activity [10]. As illustrated in Figure 3, theophylline reduced the elevated MTP activity from 27.8 ± 0.8 nM/protein-mg/min to 26.0 ± 0.2 nM/protein-mg/min, indicating a 6.9% inhibition of MTP activity induced by the HFCD plus 10% glucose. These findings suggest that the inhibition of TG and TC secretion by theophylline (Figure 2A,B) is at least partially attributable to its inhibitory effect on MTP activity.

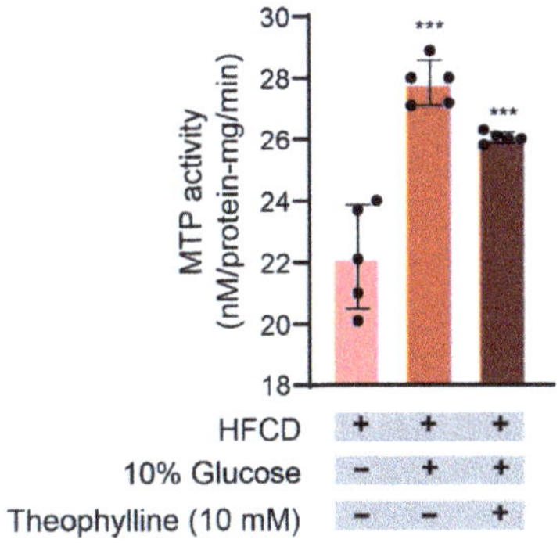


**Figure 3. Effects of theophylline on MTP activity in hepatocytes.** MTP activity was determined as described in Materials and Methods. HFCD, high fat and carbohydrate diets. MTP, microsomal triglyceride transfer protein. Data are given as means ± SD., $n = 5$. *** $p < 0.001$.

### *3.5. Effects of Theophylline on VLDL Release from Hepatocytes into the Culture Medium*

TC is secreted from hepatocytes into the bloodstream in the form of TC-poor VLDL and subsequently metabolized into TC-rich LDL. While it is pertinent to investigate theophylline's effect on CVD risk by measuring TC-rich LDL, it is important to note that the culture medium used in this study for rat primary hepatocytes did not contain the requisite enzymes (e.g., lipoprotein lipase, hepatic lipase, etc.) required to metabolize TC-poor VLDL into TC-rich LDL. Therefore, we assessed theophylline's potential inhibitory effect on CVD risk by measuring TC-containing VLDL, specifically VLDL-C, which represents the initial lipoprotein that transitions into TC-rich LDL and is exclusively secreted by hepatocytes.

As depicted in Figure 4A, hepatocytes ($10^6$ cells/well) derived from HFCD-fed rats released VLDL-C at a rate of $13.4 \pm 0.2 \times 10^2$ ng/protein-mg into the culture medium. However, theophylline reduced this release to $12.6 \pm 0.2 \times 10^2$ ng/protein-mg (Figure 4A). These findings indicate that theophylline inhibits VLDL-C release from HFCD-fed rat hepatocytes by 6.0% (Figure 4A), thereby reflecting the compound's inhibitory effect on TC secretion (Figure 2B).

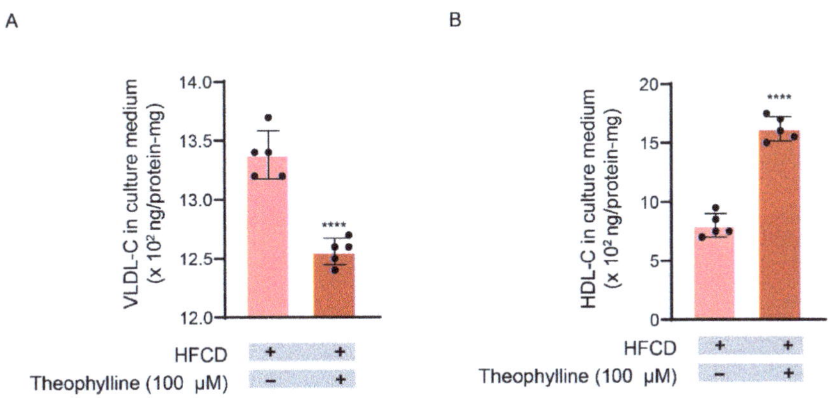


**Figure 4. Effects of theophylline on the release of VLDL-C and HDL-C into hepatocytes culture medium.** The level of VLDL-C was calculated and the level of HDL-C was determined as described in Materials and Methods. (**A**) Level of VLDL-C. (**B**) Level of HDL-C. HFCD, high fat and carbohydrate diets; VLDL-C, very low density lipoprotein; HDL-C, high density lipoprotein. Data are given as means ± SD., $n = 5$. **** $p < 0.0001$.

### 3.6. Theophylline's Effect on HDL Release from Hepatocytes

Hepatocytes also released HDL independent of MTP activity. In this study, we evaluated the impact of theophylline on HDL release by measuring the concentration of HDL-C in the hepatocyte culture medium. As illustrated in Figure 4B, HFCD-fed rat hepatocytes released HDL-C at a rate of $8 \pm 1.0 \times 10^2$ ng/protein-mg, which significantly increased to $16 \pm 1.0 \times 10^2$ ng/protein-mg following theophylline treatment. This suggests that theophylline enhances HDL release from HFCD-fed rat hepatocytes by up to 92.8% (Figure 4B).

### 3.7. Theophylline's Inhibitory Effect on CVD Risk Lipoproteins and Their Lipids

The impact of theophylline on CVD risk was evaluated using TG, TC, and HDL-C concentrations, which were utilized to calculate the AI, TC/HDL-C ratio, and TG/HDL-C ratio. The average AI derived from HFCD-fed rat hepatocytes was 0.89 (Figure 5A). However, theophylline reduced this value to an average of 0.63, suggesting an inhibitory effect on CVD risk by elevating HDL release compared with VLDL secretion from hepatocytes. This is particularly relevant as TC, used in the AI calculation (TC − HDL-C/HDL-C) is present in both VLDL and HDL. In addition, the TC/HDL-C ratio ($1.89 \pm 0.1$) derived from HFCD-fed rat hepatocytes significantly decreased to $1.63 \pm 0.1$ with theophylline treatment (Figure 5B). Similarly, the TG/HDL-C ratio decreased from $8.45 \pm 0.9$ to $3.89 \pm 0.3$ following theophylline treatment (Figure 5C). These results indicate that theophylline may mitigate CVD risk by promoting HDL-C secretion rather than TG, as evidenced by a 51.2% reduction in the TG/HDL-C ratio (Figure 5C).

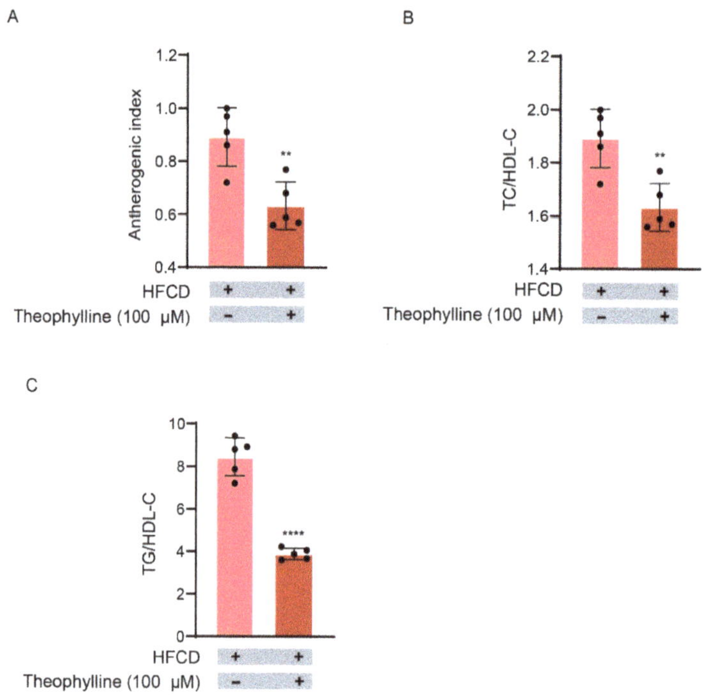


**Figure 5. Effects of theophylline on indexes of CVD risk.** Atherogenic index (AI), TC/HDL-C ratio and TG/HDL-C were calculated with the levels of TG, TC and HDL-C. (**A**) Atherogenic index. (**B**) TC/HDL-C ratio. (**C**) TG/HDL-C. HFCD, high fat carbohydrates diets. Data are given as means ± SD., $n = 5$. ** $p < 0.01$ and **** $p < 0.0001$.

## 4. Discussion

The pathological processes underlying the development of CVD are significantly influenced by elevated levels of both TG and TC (i.e., cholesterol esters and free cholesterol) released from hepatocytes. Consequently, numerous researchers have focused on the exploration and development of pharmacological drugs that can effectively lower both TG and TC levels simultaneously, as this reduction is crucial for the treatment and prevention of CVD. Despite several studies suggesting that theophylline exerts no preventive effect on CVD owing to its inability to lower blood levels of TG and TC concurrently [13,15,35], our research indicates otherwise. We found that theophylline inhibits the concurrent secretion of TG and TC from hepatocytes isolated from HFCD-fed rats, typically promoting the increased synthesis of these lipids.

These findings suggest that theophylline inhibits the secretion of TG and TC by suppressing MTP activity in hepatocytes. MTP is responsible for assembling TG into VLDL, along with TC (i.e., free cholesterol and cholesterol esters), phospholipids, and ApoB, for secretion into the blood [5,8]. This assertion is corroborated by our observation wherein theophylline was found to inhibit MTP activity stimulated by the combined intake of the HFCD and 10% glucose, further elevating hepatocellular TG and TC levels beyond those induced by the HFCD alone [10]. Moreover, our results reflect that the secretion of TG and TC increased proportionately with the endogenous production of these lipids in the liver or hepatocytes when an additional 10% glucose was provided alongside the HFCD. This observation aligns with previous studies demonstrating that high-carbohydrate diets can lead to hypertriglyceridemia [36].

Theophylline reduces the secretion of TG from hepatocytes into the culture medium primarily through two mechanisms. First, theophylline increases the intracellular levels of cAMP by inhibiting the activity of cAMP PDE in rat liver [37]. Elevated cAMP levels inhibit the secretory activity of MTP, which is responsible for the export of TG-rich VLDL from hepatocytes [11]. Second, although we did not determine whether theophylline increases cAMP production in primary hepatocytes, it is hypothesized that theophylline, as a PDE inhibitor, enhances the activities of cAMP-dependent adipose triglyceride lipase and hormone-sensitive lipase in hepatocytes, leading to increased TG degradation [38–40]. Supporting this, theophylline-based KMUP-1, also a PDE inhibitor, heightens cAMP production, promoting TG degradation in the liver [35]. Furthermore, hepatocellular cAMP is identified as a molecular target for drugs aimed at treating non-alcoholic fatty liver disease induced by HFCD intake or alcoholic fatty liver disease [40].

The mechanism by which theophylline reduces the secretion of TC from hepatocytes into their culture medium can be elucidated as follows: First, theophylline increases intracellular cAMP levels, which may inhibit the VLDL-C secretory activity of MTP. Second, the elevated cAMP levels may enhance the activity of cAMP-dependent cholesteryl ester hydrolase in liver cells, thereby promoting the hydrolysis of cholesterol esters [41–44]. The precise nature of the relationship between theophylline-induced cAMP elevation and the inhibition of TG-rich VLDL and VLDL-C secretion by MTP remains unclear. It is uncertain whether theophylline-elevated cAMP directly inhibits MTP secretory activity or if the reduction in TG and TC levels results from decreased secretion mediated by elevated cAMP. However, our previous study [11], demonstrated that when rat liver homogenates with lower cAMP and higher TG levels than normal were incubated with dibutyryl-cAMP (db-cAMP), a cAMP analogue, MTP activity decreased in a dose-dependent manner. This finding suggests that increased cAMP in hepatocytes correlates with decreased TG and TC secretion associated with VLDL [45]. Therefore, theophylline, as a PDE inhibitor, presumably suppresses MTP's secretory activity by promoting cAMP production in hepatocytes, thereby inhibiting TG and TC secretion.

TC comprises both cholesterol esters and free cholesterol; the increased cAMP levels induced by theophylline specifically promote the hydrolysis of cholesterol esters into free cholesterol and free fatty acids. Therefore, the decline in TC secretion observed with theophylline treatment reflects a reduction in the secretion of cholesterol esters. The free fatty acids and free cholesterol generated from the hydrolysis of cholesterol esters in the liver are used to produce ATP and bile acids, which are necessary for metabolic functions [41]. Furthermore, free cholesterol is incorporated into VLDL and HDL and can be secreted from the liver into the culture medium or bloodstream.

Accordingly, monitoring changes in TG and TC levels in the hepatocyte culture medium, as well as VLDL and HDL concentrations in the blood, is instrumental in assessing the effects of theophylline on hypertriglyceridemia and hypercholesterolemia. Among the VLDL and HDL secreted by the liver, VLDL is particularly rich in TG; thus, the reduction in TG levels observed in the culture medium following theophylline treatment indicates that theophylline inhibits the secretion of TG-rich VLDL from the liver.

The TG-rich VLDL secreted into the bloodstream is metabolized into intermediate-density lipoprotein (IDL) by lipoprotein lipase (LPL), which is released from muscle and adipose tissues, facilitating TG decomposition. Subsequently, the TG molecules in IDL are further decomposed into LDL by hepatic lipase secreted from the liver [46]. Notably, during the metabolism of TG-rich VLDL to LDL, TG levels decrease, whereas TC levels increase. This implies that elevated TG-rich VLDL levels in the blood correspond to an increase in LDL, which is characterized by high TC content, ultimately contributing to hypercholesterolemia pathological condition associated with CVD.

Therefore, the reduction in TG levels in the culture medium owing to theophylline suggests that it may inhibit the secretion of TG-rich VLDL, thereby decreasing LDL production and lowering TC levels. Notably, the culture medium used for primary hepatocytes in this experiment lacked LPL, which is necessary for the breakdown of TG in VLDL, IDL, and LDL, resulting in the absence of TC-increasing lipoproteins.

Consequently, the observed decrease in TC in the hepatocyte culture medium following theophylline treatment reflects a reduction in VLDL-C. In fact, theophylline has been revealed to lower VLDL cholesterol levels in the hepatocyte culture medium, suggesting a consequential decline in LDL cholesterol levels, thereby potentially alleviating CVD risk.

In contrast to VLDL and LDL, HDL is a blood lipoprotein associated with an attenuated risk of CVD. Therefore, numerous researchers are investigating substances or drugs that can lower both VLDL and LDL in the blood while simultaneously elevating HDL levels. Theophylline has also been explored for these therapeutic effects; however, it is known not to increase the concentration of HDL in the blood [13,15,16].

Notwithstanding, in this study, we observed that theophylline enhanced the secretion of HDL-C, an indicator of HDL, from primary hepatocytes that had been stimulated to secrete elevated levels of TG and TC following HFCD administration to rats. This finding suggests that theophylline may mitigate HFCD-induced CVD risk. HDL secretion is known to be stimulated by cAMP-phosphorylated ABCA1 in hepatocytes [12,47]. Therefore, theophylline, as a cAMP promoter, plausibly enhances HDL release by phosphorylating ABCA1 in hepatocytes. On evaluating the effects of theophylline on CVD risk indicators, we found it to significantly reduce both the TC/HDL-C and TG/HDL-C ratios in the culture medium of HFCD-fed rat primary hepatocytes.

These findings suggest that theophylline decreases the secretion of TG-rich VLDL containing TC while simultaneously increasing HDL-C secretion, indicating a potential reduction in CVD risk. This conclusion is further supported by the observation that theophylline lowers the AI, a recognized marker of CVD risk.

Importantly, several independent studies have already provided supportive histological evidence regarding the hepatic effects of theophylline. For instance, Dong et al. [48] demonstrated that theophylline reduced hepatic fat accumulation in NAFLD models, Liu et al. [49] reported its protective effects against liver fibrosis and fat deposition in obese mice, and Hussein et al. [50] showed that theophylline alleviated immunologically induced hepatic inflammation. These findings suggest that theophylline exerts consistent beneficial effects on liver pathology, which are in line with our observations.

In summary, our study demonstrates that theophylline, a PDE inhibitor, suppresses MTP activity in hepatocytes, thereby reducing the secretion of VLDL containing triglycerides and cholesterol while promoting HDL secretion. These findings were corroborated in rat hepatocytes, which readily develop hypertriglyceridemia and hypercholesterolemia under HFCD feeding. When considered together with prior reports on the hepatoprotective effects of theophylline, the overall evidence supports the possibility that theophylline may contribute to lowering CVD risk, at least in part, by improving liver function.

## 5. Conclusions

The present ex vivo investigation reveals that nutritionally attainable concentrations of tea-derived theophylline ($\leq$100 μM) directly inhibit MTP activity in rat hepatocytes primed with an HFCD. This inhibition curtails the secretion of atherogenic TG, TC, and VLDL-C while simultaneously doubling HDL-C output, thereby improving the TC/HDL-C and TG/HDL-C ratios as well as the AI. Collectively, these data position theophylline as a promising modulator of dyslipidemia and a potential adjunct for CVD risk reduction. Further confirmation through in vivo models and human dietary intervention trials is warranted.

The beneficial effects of theophylline observed in this study were achieved without evidence of cytotoxicity or alterations in cell viability, underscoring the specificity of its modulatory action on lipid metabolism. Considering that theophylline is naturally abundant in common dietary sources, such as green tea, cacao, and kola nuts, its regular consumption may provide a practical and accessible means of managing dyslipidemia and associated cardiovascular risks. Furthermore, the physiologically relevant concentrations employed in this study underscore the translational potential of these findings to everyday dietary practices and the development of nutraceutical products.

While these ex vivo results are promising, we recognize the limitations of directly translating findings from cell-based models to clinical practice. One limitation is the relatively small sample size of five rats per group. This was chosen in accordance with the 3Rs principle, and the study was sufficiently powered to detect highly significant effects (e.g., $p < 0.01$ and $p < 0.0001$). Additionally, this study focused on a model of diet-induced dyslipidemia and therefore did not include a normal diet control group. This limits the generalization of our findings to physiological conditions in a healthy state. Nevertheless, future studies with larger animal cohorts would strengthen the generalizability of our conclusions. Furthermore, we exclusively used male rats to minimize metabolic variability arising from the female estrous cycle, allowing for a clearer interpretation of theophylline's direct effects. As sex-based differences in lipid metabolism are well-documented, investigating these effects in female animals is a valuable direction for future research. Therefore, future research should extend these findings to animal models and human clinical trials to rigorously evaluate theophylline's effectiveness, optimal dosing regimens, and long-term safety across diverse populations. Confirmation through comprehensive in vivo experiments and well-controlled human dietary intervention studies will be critical to fully validate theophylline's role as an adjunctive therapeutic strategy for CVD risk reduction and its incorporation into dietary guidelines and public health recommendations.

**Author Contributions:** Conceptualization, H.-J.P. and J.-H.S.; methodology, H.J.; software, H.J.; validation, M.-K.P. and H.J.; formal analysis, H.J.; investigation, J.-S.B.; resources, H.-J.P.; data curation, M.-K.P.; writing—original draft preparation, M.-K.P.; writing—review and editing, M.-K.P.; visualization, M.-K.P.; supervision, H.-J.P.; project administration, H.-J.P.; funding acquisition, J.-H.S. All authors have read and agreed to the published version of the manuscript.

**Funding:** This work was supported by Korea Basic Science Institute (National Research Facilities and Equipment Center, 2021R1A6C101A416) funded by the Ministry of Education, the project to train professional personnel in biological materials by the Ministry of Environment, Republic of Korea, and the Regional Innovation System & Education (RISE) Glocal 30 program through the Daegu RISE Center, funded by the Ministry of Education (MOE) and the Daegu, Republic of Korea (2025-RISE-03-001).

**Institutional Review Board Statement:** The animal study protocol was approved by the Institutional Animal Care and Use Committee (IACUC) of Kyungpook National University (protocol code KNU 2025-0169 and date of approval 14 April 2025).

**Informed Consent Statement:** Not applicable.

**Data Availability Statement:** The raw data supporting the conclusions of this article will be made available by the authors on request.

**Acknowledgments:** We express our appreciation and especially thank the support and cooperation of the project to train professional personnel in biological materials by the Ministry of Environment, Republic of Korea. The authors have reviewed and edited the output and take full responsibility for the content of this publication.

**Conflicts of Interest:** Author Min-Kyu Park and Jae-Ho Shin were employed by the company Microbalance Inc. The remaining authors declare that the research was conducted in the absence of any commercial or financial relationships that could be construed as a potential conflict of interest.

## Abbreviations

The following abbreviations are used in this manuscript:

| | |
|---|---|
| CVD | Cardiovascular disease |
| VLDL | Very low density lipoprotein |
| MTP | Microsomal triglyceride transfer protein |
| cAMP | Cyclic adenosine monophosphate |
| HFCD | High-fat, high-carbohydrate diet |
| TG | Triglyceride |
| TC | Total cholesterol |
| VLDL-C | VLDL-cholesterol |
| HDL-C | HDL-cholesterol |
| AI | Atherogenic index |

## References

1. Tsao, C.W.; Aday, A.W.; Almarzooq, Z.I.; Anderson, C.A.; Arora, P.; Avery, C.L.; Baker-Smith, C.M.; Beaton, A.Z.; Boehme, A.K.; Buxton, A.E.; et al. Heart disease and stroke statistics—2023 update: A report from the American Heart Association. *Circulation* **2023**, *147*, e93. [CrossRef]
2. Higgins, J.A.; Hutson, J.L. The roles of Golgi and endoplasmic reticulum in the synthesis and assembly of lipoprotein lipids in rat hepatocytes. *J. Lipid Res.* **1984**, *25*, 1295–1305. [CrossRef] [PubMed]
3. Gibbons, G.F. Assembly and secretion of hepatic very-low-density lipoprotein. *Biochem. J.* **1990**, *268*, 1–13. [CrossRef]
4. Vance, J.E.; Vance, D.E. Lipoprotein assembly and secretion by hepatocytes. *Annu. Rev. Nutr.* **1990**, *10*, 337–356. [CrossRef]
5. Griffin, B.A.; Zampelas, A. Influence of dietary fatty acids on the atherogenic lipoprotein phenotype. *Nutr. Res. Rev.* **1995**, *8*, 1–26. [CrossRef] [PubMed]
6. Benoist, F.; Nicodeme, E.; Grand-Perret, T. Microsomal triacylglycerol transfer protein prevents presecretory degradation of apolipoprotein B-100: A dithiothreitol-sensitive protease is involved. *Eur. J. Biochem.* **1996**, *240*, 713–720. [CrossRef] [PubMed]

7. Gordon, D.A.; Jamil, H. Progress towards understanding the role of microsomal triglyceride transfer protein in apolipoprotein-B lipoprotein assembly. *Biochim. Biophys. Acta Mol. Cell Biol. Lipids* **2000**, *1486*, 72–83. [CrossRef]
8. Atzel, A.; Wetterau, J.R. Mechanism of microsomal triglyceride transfer protein catalyzed lipid transport. *Biochemistry* **1993**, *32*, 10444–10450. [CrossRef]
9. Cho, H.-J.; Kang, H.-C.; Choi, S.-A.; Ju, Y.-C.; Lee, H.-S.; Park, H.-J. The possible role of $Ca^{2+}$ on the activation of microsomal triglyceride transfer protein in rat hepatocytes. *Biol. Pharm. Bull.* **2005**, *28*, 1418–1423. [CrossRef]
10. Cho, H.-J.; Kang, H.-C.; Ju, Y.-C.; Lee, H.-S.; Kim, H.-S.; Park, H.-J. Augmentation of $Ca^{2+}$ induced microsomal triglyceride transfer protein activity by glucose supply enhances hypertriglyceridemia in vivo. *Biol. Pharm. Bull.* **2006**, *29*, 889–895. [CrossRef]
11. Cho, H.-J.; Kim, H.-S.; Yu, Y.-B.; Kang, H.-C.; Lee, D.-H.; Rhee, M.-H.; Cho, J.-Y.; Park, H.-J. The opposite correlation between calcium ion and cyclic-AMP regarding the activation of microsomal triglyceride transfer protein in rat liver. *BMB Rep.* **2009**, *42*, 642–647. [CrossRef]
12. Mulay, V.; Wood, P.; Rentero, C.; Enrich, C.; Grewal, T. Signal transduction pathways provide opportunities to enhance HDL and apoAI-dependent reverse cholesterol transport. *Curr. Pharm. Biotechnol.* **2012**, *13*, 352–364. [CrossRef]
13. Uzuner, N.; Karaman, Ö.; Saydam, N.; Güner, G. Lipoprotein profile in long term theophylline administration in children with asthma. *Allergol. Immunopathol.* **2002**, *30*, 79–84. [CrossRef] [PubMed]
14. Apgar, J.L.; Tarka, S.M. Methylxanthine composition and consumption patterns of cocoa and chocolate products. In *Caffeine*; CRC Press: Boca Raton, FL, USA, 2019; pp. 163–192. [CrossRef]
15. Inoue, H.; Kobayashi-Hattori, K.; Horiuchi, Y.; Oishi, Y.; Arai, S.; Takita, T. Regulation of the body fat percentage in developmental-stage rats by methylxanthine derivatives in a high-fat diet. *Biosci. Biotechnol. Biochem.* **2006**, *70*, 1134–1139. [CrossRef] [PubMed]
16. Allayee, H.; Hartiala, J.; Lee, W.; Mehrabian, M.; Irvin, C.G.; Conti, D.V.; Lima, J.J. The effect of montelukast and low-dose theophylline on cardiovascular disease risk factors in asthmatics. *Chest* **2007**, *132*, 868–874. [CrossRef]
17. Rustaeus, S.; Lindberg, K.; Stillemark, P.; Claesson, C.; Asp, L.; Larsson, T.; Borén, J.; Olofsson, S.-O. Assembly of very low density lipoprotein: A two-step process of apolipoprotein B core lipidation. *J. Nutr.* **1999**, *129*, 463S–466S. [CrossRef] [PubMed]
18. Shelness, G.S.; Ingram, M.F.; Huang, X.F.; Delozier, J.A. Apolipoprotein B in the rough endoplasmic reticulum: Translation, translocation and the initiation of lipoprotein assembly. *J. Nutr.* **1999**, *129*, 456S–462S. [CrossRef]
19. Song, Y.; Lee, S.-J.; Jang, S.-H.; Ha, J.H.; Song, Y.M.; Ko, Y.-G.; Kim, H.-D.; Min, W.; Kang, S.N.; Cho, J.-H. Sasa borealis stem extract attenuates hepatic steatosis in high-fat diet-induced obese rats. *Nutrients* **2014**, *6*, 2179–2195. [CrossRef]
20. Jung, M.-S.; Lee, S.-J.; Song, Y.; Jang, S.-H.; Min, W.; Won, C.-K.; Kim, H.-D.; Kim, T.H.; Cho, J.-H. Rubus crataegifolius Bunge regulates adipogenesis through Akt and inhibits high-fat diet-induced obesity in rats. *Nutr. Metab.* **2016**, *13*, 29. [CrossRef]
21. Ondee, T.; Pongpirul, K.; Udompornpitak, K.; Sukkummee, W.; Lertmongkolaksorn, T.; Senaprom, S.; Leelahavanichkul, A. High fructose causes more prominent liver steatohepatitis with leaky gut similar to high glucose administration in mice and attenuation by Lactiplantibacillus plantarum dfa1. *Nutrients* **2023**, *15*, 1462. [CrossRef]
22. Mokhtari, I.; Moumou, M.; Mokhtari, C.; Harnafi, M.; Milenkovic, D.; Amrani, S.; Harnafi, H. Nutritional composition and effect of loquat fruit (*Eriobotrya japonica* L. var. Navela) on lipid metabolism and liver steatosis in high-fat high-sucrose diet-fed mice. *Prev. Nutr. Food Sci.* **2024**, *29*, 256–265. [CrossRef]
23. Kim, H.-J.; Son, J.; Jin, E.; Lee, J.; Park, S. Effects of exercise and L-arginine inake on inflammation in fat diet induced obese rats. *J. Exerc. Nutr. B* **2016**, *20*, 36–40. [CrossRef]
24. Kim, S.Y.; Lee, J. Exercise training suppresses vascular fibrosis in aging obesity induced rats. *J. Exerc. Nutr. Biochem.* **2014**, *18*, 175–180. [CrossRef]
25. Yang, L.Y.; Kuksis, A.; Myher, J.J.; Steiner, G. Contribution of de novo fatty acid synthesis to very low density lipoprotein triacylglycerols: Evidence from mass isotopomer distribution analysis of fatty acids synthesized from $[^2H_6]$ ethanol. *J. Lipid Res.* **1996**, *37*, 262–274. [CrossRef]
26. Nossen, J.O.; Rustan, A.C.; Drevon, C.A. Calcium-antagonists inhibit secretion of very-low-density lipoprotein from cultured rat hepatocytes. *Biochem. J.* **1987**, *247*, 433–439. [CrossRef]
27. Burr, I.M.; Balant, L.; Stauffacher, W.; Renold, A.E. Perifusion of rat pancreatic tissue in vitro: Substrate modification of theophylline-induced biphasic insulin release. *J. Clin. Investig.* **1970**, *49*, 2097–2105. [CrossRef]
28. Berry, M.N.; Friend, D.S. High-yield preparation of isolated rat liver parenchymal cells: A biochemical and fine structural study. *J. Cell Biol.* **1969**, *43*, 506–520. [CrossRef]
29. Friedewald, W.T.; Levy, R.I.; Fredrickson, D.S. Estimation of the concentration of low-density lipoprotein cholesterol in plasma, without use of the preparative ultracentrifuge. *Clin. Chem.* **1972**, *18*, 499–502. [CrossRef]
30. Kessler, R.J.; Fanestil, D.D. Interference by lipids in the determination of protein using bicinchoninic acid. *Anal. Biochem.* **1986**, *159*, 138–142. [CrossRef]
31. Morton, R.E.; Evans, T.A. Modification of the bicinchoninic acid protein assay to eliminate lipid interference in deyermining lipoprotein protein content. *Anal. Biochem.* **1992**, *204*, 332–334. [CrossRef]
32. Thermo Scientific. *Introduction Pierce® BCA Protein Assay Kit*; Thermo Scientific: Waltham, MA, USA, 2013.

33. Lowry, O.H.; Rosebrough, N.J.; Farr, A.L.; Randall, R.J. Protein measurement with the Folin phenol reagent. *J. Biol. Chem.* **1951**, *193*, 265–275. [CrossRef]
34. Thierer, J.H.; Foresti, O.; Yadav, P.K.; Wilson, M.H.; Moll, T.O.; Shen, M.-C.; Busch-Nentwich, E.M.; Morash, M.; Mohlke, K.L.; Rawls, J.F. Pla2g12b drives expansion of triglyceride-rich lipoproteins. *Nat. Commun.* **2024**, *15*, 2095. [CrossRef]
35. Wu, B.-N.; Kuo, K.-K.; Chen, Y.-H.; Chang, C.-T.; Huang, H.-T.; Chai, C.-Y.; Dai, Z.-K.; Chen, I.-J. Theophylline-based KMUP-1 improves steatohepatitis via MMP-9/IL-10 and lipolysis via HSL/p-HSL in obese mice. *Int. J. Mol. Sci.* **2016**, *17*, 1345. [CrossRef]
36. Chong, M.F.; Fielding, B.A.; Frayn, K.N. Metabolic interaction of dietary sugars and plasma lipids with a focus on mechanisms and de novo lipogenesis. *Proc. Nutr. Soc.* **2007**, *66*, 52–59. [CrossRef]
37. Hickie, R.A.; Walker, C.M.; Datta, A. Increased activity of low-$K_m$ cyclic adenosine 3′:5′-monophosphate phosphodiesterase in plasma membranes of Morris hepatoma 5123tc (h). *Cancer Res.* **1975**, *35*, 601–605.
38. Quiroga, A.D.; Lehner, R. Liver triacylglycerol lipases. *Biochim. Biophys. Acta Mol. Cell Biol. Lipids* **2012**, *1821*, 762–769. [CrossRef]
39. Schott, M.B.; Rasineni, K.; Weller, S.G.; Schulze, R.J.; Sletten, A.C.; Casey, C.A.; McNiven, M.A. β-Adrenergic induction of lipolysis in hepatocytes is inhibited by ethanol exposure. *J. Biol. Chem.* **2017**, *292*, 11815–11828. [CrossRef]
40. Wahlang, B.; McClain, C.J.; Barve, S.; Gobejishvili, L. Role of cAMP and phosphodiesterase signaling in liver health and disease. *Cell. Signal.* **2018**, *49*, 105–115. [CrossRef]
41. Botham, K.M. Cyclic AMP and the regulation of cholesterol metabolism. *Biochem. Soc. Trans.* **1992**, *20*, 454–459. [CrossRef]
42. Hoang, V.-Q.; Suckling, K.E.; Cho-Chui, Y.; Botham, K.M. The effect of cyclic AMP analogues and glucagon on cholesteryl ester synthesis and hydrolysis in cultured hamster hepatocytes. *FEBS Lett.* **1993**, *329*, 17–20. [CrossRef]
43. Botham, K.M.; Hoang, V.-Q.; Jones, A.K.; Martinez, M.J.; Ochoa, B.; Suckling, K.E. Comparison of the effects of cyclic AMP analogues on cholesterol metabolism in cultured rat and hamster hepatocytes. *Comp. Biochem. Physiol. B Biochem. Mol. Biol.* **1996**, *113*, 185–191. [CrossRef]
44. Hernández, M.L.; Martínez, M.J.; Ruiz, J.I.; Ochoa, B. Stimulation of microsomal cholesterol ester hydrolase by glucagon, cyclic AMP analogues, and vasopressin in isolated rat hepatocytes. *Lipids* **1996**, *31*, 269–276. [CrossRef]
45. Björnsson, O.G.; Sparks, J.D.; Sparks, C.E.; Gibbons, G.F. Regulation of VLDL secretion in primary culture of rat hepatocytes: Involvement of cAMP and cAMP-dependent protein kinases. *Eur. J. Clin. Investig.* **1994**, *24*, 137–148. [CrossRef]
46. Taskinen, M.-R.; Kuusi, T. Enzymes involved in triglyceride hydrolysis. *Baillieres Clin. Endocrinol. Metab.* **1987**, *1*, 639–666. [CrossRef] [PubMed]
47. Tsujita, M.; Wu, C.-A.; Abe-Dohmae, S.; Usui, S.; Okazaki, M.; Yokoyama, S. On the hepatic mechanism of HDL assembly by the ABCA1/apoA-I pathway. *J. Lipid Res.* **2005**, *46*, 154–162. [CrossRef] [PubMed]
48. Dong, J.; Du, X.; Yang, R.; Shan, L.; Lu, X.; Shen, Y.; Li, Y.; Duan, S.; Du, Z.; Fueng, J.; et al. Tramnscriptomic analysis revels the mechanisms underlying the differential effects of caffeine, theophylline, and theobromine in regulating hepatic fat accumulation. *Food Funct.* **2025**, *16* (Suppl. 1), 2503–2514. [CrossRef]
49. Liu, T.-T.; Liu, X.-T.; Huang, G.-L.; Liu, L.; Chen, Q.-X.; Wang, Q. Theophylline extracyed from Fu Brick Tea affects the metabolism of preadipocytes and body fat in mice as a pancreatic lipase inhibitor. *Int. J. Mol. Sci.* **2022**, *23*, 2525. [CrossRef]
50. Hussein, R.M.; Elsirafy, M.; Wahba, Y.S.; Kawy, H.S.A.; Hasanin, A.H.; Galal, G. Theophylline, an old drug with multi-faceted effects: Its potential benefits in immunological liver injury in rats. *Life Sci.* **2015**, *136*, 100–107. [CrossRef]

MDPI AG
Grosspeteranlage 5
4052 Basel
Switzerland
Tel.: +41 61 683 77 34

*Biomedicines* Editorial Office
E-mail: biomedicines@mdpi.com
www.mdpi.com/journal/biomedicines

www.ingramcontent.com/pod-product-compliance
Lightning Source LLC
Chambersburg PA
CBHW061521210926
51850CB00008B/1993
* 9 7 8 3 7 2 5 8 6 7 7 4 5 *